Security for Multihop Wireless Networks

T0225527

OTHER COMMUNICATIONS BOOKS FROM AUERBACH

Advances in Biometrics for Secure Human Authentication and Recognition
Edited by Dakshina Ranjan Kisku, Phalguni Gupta, and Jamuna Kanta Sing
ISBN 978-1-4665-8242-2

Anonymous Communication Networks: Protecting Privacy on the Web
Kun Peng
ISBN 978-1-4398-8157-6

Case Studies in Enterprise Systems, Complex Systems, and System of Systems Engineering
Edited by Alex Gorod, Brian E. White, Vernon Ireland, S. Jimmy Gandhi, and Brian Sauser
ISBN 978-1-4665-0239-0

Cyber-Physical Systems: Integrated Computing and Engineering Design
Fei Hu
ISBN 978-1-4665-7700-8

Evolutionary Dynamics of Complex Communications Networks
Vasileios Karyotis, Eleni Stai, and Symeon Papavassiliou
ISBN 978-1-4665-1840-7

Extending the Internet of Things: Towards a Global and Seemless Integration
Antonio J. Jara-Valera
ISBN 9781466518483

Fading and Interference Mitigation in Wireless Communications
Stefan Panic
ISBN 978-1-4665-0841-5

The Future of Wireless Networks: Architectures, Protocols, and Services
Edited by Mohesen Guizani, Hsiao-Hwa Chen, and Chonggang Wang
ISBN 978-1-4822-2094-0

Green Networking and Communications: ICT for Sustainability
Edited by Shafiullah Khan and Jaime Lloret Mauri
ISBN 978-1-4665-6874-7

Image Encryption: A Communication Perspective
Fathi E. Abd El-Samie, Hossam Eldin H. Ahmed, Ibrahim F. Elashry, Mai H. Shahieen, Osama S. Faragallah, El-Sayed M. El-Rabaie, and Saleh A. Alshebeili
ISBN 978-1-4665-7698-8

Intrusion Detection in Wireless Ad-Hoc Networks
Nabendu Chaki and Rituparna Chaki
ISBN 978-1-4665-1565-9

Intrusion Detection Networks: A Key to Collaborative Security
Carol Fung and Raouf Boutaba
ISBN 978-1-4665-6412-1

MIMO Processing for 4G and Beyond: Fundamentals and Evolution
Edited by Mário Marques da Silva and Francisco A.T.B.N. Monteiro
ISBN 978-1-4665-9807-2

Modeling and Analysis of P2P Content Distribution Systems
Yipeng Zhou and Dah Ming Chiu
ISBN 978-1-4822-1920-3

Network Innovation through OpenFlow and SDN: Principles and Design
Edited by Fei Hu
ISBN 978-1-4665-7209-6

Pervasive Computing: Concepts, Technologies and Applications
Guo Minyi, Zhou Jingyu, Tang Feilong, and Shen Yao
ISBN 978-1-4665-9627-6

Physical Layer Security in Wireless Communications
Edited by Xiangyun Zhou. Lingyang Song, and Yan Zhang
ISBN 978-1-4665-6700-9

Security for Multihop Wireless Networks
Shafiullah Khan and Jaime Lloret Mauri
ISBN 978-1-4665-7803-6

Self-Healing Systems and Wireless Networks Management
Junaid Ahsenali Chaudhry
ISBN 978-1-4665-5648-5

Unit and Ubiquitous Internet of Things
Huansheng Ning
ISBN 978-1-4665-6166-3

AUERBACH PUBLICATIONS
www.auerbach-publications.com
To Order Call: 1-800-272-7737 • Fax: 1-800-374-3401
E-mail: orders@crcpress.com

Security for Multihop Wireless Networks

Shafiullah Khan and Jaime Lloret Mauri

CRC Press
Taylor & Francis Group
Boca Raton   London   New York

CRC Press is an imprint of the
Taylor & Francis Group, an **Informa** business

CRC Press
Taylor & Francis Group
6000 Broken Sound Parkway NW, Suite 300
Boca Raton, FL 33487-2742

First issued in paperback 2016

© 2014 by Taylor & Francis Group, LLC
CRC Press is an imprint of Taylor & Francis Group, an Informa business

No claim to original U.S. Government works

Version Date: 20140228

ISBN 13: 978-1-138-03393-1 (pbk)
ISBN 13: 978-1-4665-7803-6 (hbk)

Visit the Taylor & Francis Web site at
http://www.taylorandfrancis.com

and the CRC Press Web site at
http://www.crcpress.com

Contents

SECTION II SECURITY IN WIRELESS SENSOR NETWORKS

SECTION III SECURITY IN OTHER AD HOC NETWORKS

Preface

First of all, we would like to express our gratitude to the publisher Taylor & Francis for giving us full support in the timely completion of this book. We also thank all the authors for their chapter contributions to this book, including those whose manuscripts we were not able to include due to the rigorous review-based selection process. The authors of the accepted chapters come from many parts of the world, including Algeria, Australia, Canada, China, Colombia, France, Germany, India, Italy, Malaysia, New Zealand, Nigeria, Singapore, Spain, and the United States.

It is well understood nowadays that wireless networks have become a part of our daily technical life. Though the impact of wireless networking was more or less assessed since the advent of basic wireless technologies, today's vast and dynamic features of various wireless applications might not have been accurately envisaged. Today, the types of wireless networks range from cellular networks to ad hoc networks, infrastructure-based networks to infrastructure-less networks, short-range networks to long-range direct communication wireless networks, static wireless networks to mobile networks, and so on. Hence, while initiating this book project, choosing a plain title seemed to be a challenge but it also allowed different topics on wireless security to be compiled in a single volume.

This book is mainly targeted at researchers, postgraduate students in universities, academics, and industry practitioners or professionals. Elementary information about wireless security is not a priority of this book. Hence, some chapters include detailed research works and results on wireless network security. The book provides a broad coverage of wireless security issues.

The contributions identify various vulnerabilities in the physical layer, MAC layer, network layer, transport layer, and application layer, and focus on ways of strengthening security mechanisms and services throughout the layers. Instead of simply including chapters like a regular textbook, we have mainly focused on research-based outcomes. Hence, while addressing all the relevant issues and works in various layers, we basically lined up the chapters from *easy-to-read* survey-type articles to detailed investigation-related works.

Owing to the nature of research works, some of the concepts and future vision may not seem to be fully practical considering the state of the art. Still, to capture a snapshot of the current status, past gains, and future possibilities in the field of wireless network security, the book should be a good and timely collection. It could also be used as reference material as all the chapters include citations to the latest research trends and findings.

The book is divided into three sections:

Section I focuses on security issues and solutions in mobile ad hoc networks;
Section II presents a detailed overview of security challenges in wireless sensor networks; and
Section III is about security mechanisms in other networks.

Shafiullah Khan
Institute of Information Technology (IIT)
Kohat University of Science and Technology (KUST)
Kohat City, K.P.K, Pakistan
skhan@kust.edu.pk

Jaime Lloret Mauri
Integrated Management Coastal Research Institute
Polytechnic University of Valencia
Valencia, Spain
jlloret@dcom.upv.es

Contributors

Sasan Adibi
Business IT and Logistics
Royal Melbourne Institute of
Technology (RMIT)
Melbourne, Australia

Jasone Astorga
Department of Communications
Engineering
University of the Basque Country
UPV/EHU
Bilbao, Spain

Dauda Ayanda
Department of Computer Science and
Engineering
Obafemi Awolowo University
Ile-Ife, Nigeria

Z. Abu Bakar
UTM-MIMOS Center of Excellence
Universiti Teknologi Malaysia
Johor, Malaysia

Rajesh P. Barnwal
Information Technology Group
CSIR-Central Mechanical Engineering
Research Institute
West Bengal, India

Aymen Boudguiga
Institut Mines-Télécom
Télécom SudParis
Evry, France

Alejandro Buchmann
Technische Universität Darmstadt
Darmstadt, Germany

Carlos T. Calafate
Departamento de Informática de
Sistemas y Computadores
Universitat Politècnica
de València
Valencia, Spain

Juan-Carlos Cano
Departamento de Informática de
Sistemas y Computadores
Universitat Politècnica
de València
Valencia, Spain

Sandra Céspedes
ICESI University
Valle, Colombia

Gerard Chalhoub
LIMOS-CNRS
Clermont Université
Aubière, France

Shuo Chen
School of Electrical and Electronic
Engineering
Nanyang Technological University
Singapore

Long Cheng
College of Information Science and
    Engineering
Northeastern University
Shenyang, China

Roberto Doriguzzi Corin
CREATE-NET
Trento, Italy

J.E. Díaz-Verdejo
Department of Signal Theory
University of Granada – CITIC
Granada, Spain

M. Esa
UTM-MIMOS Center of Excellence
Universiti Teknologi Malaysia
Johor, Malaysia

Mohamed Amine Ferrag
Department of Computer Science
University of Badji Mokhtar
Annaba, Algeria

N. Fisal
UTM-MIMOS Center of Excellence
Universiti Teknologi Malaysia
Johor, Malaysia

Marc Fischlin
Technische Universität Darmstadt
Darmstadt, Germany

P. García-Teodoro
Department of Signal Theory
University of Granada – CITIC
Granada, Spain

Salim Ghanemi
Department of Computer Science
University of Badji Mokhtar
Annaba, Algeria

Soumya K. Ghosh
School of Information Technology
Indian Institute of Technology
    Kharagpur
West Bengal, India

Enrique Hernández-Orallo
Departamento de Informática de
    Sistemas y Computadores
Universitat Politècnica de València
Valencia, Spain

Marivi Higuero
University of the Basque Country
UPV/EHU
Bilbao, Spain

Eduardo Jacob
University of the Basque Country
UPV/EHU
Bilbao, Spain

Daniel Jacobi
Technische Universität Darmstadt
Darmstadt, Germany

S. Kamilah
UTM-MIMOS Center of Excellence
Universiti Teknologi Malaysia
Johor, Malaysia

A. S. Khan
UTM-MIMOS Center of Excellence
Universiti Teknologi Malaysia
Johor, Malaysia

Surraya Khanum
School of Information and
    Communication Technology
Griffith University
Gold Coast, Australia

M. Bala Krishna
University School of Information and
Communication Technology
GGS Indraprastha University
New Delhi, India

Maryline Laurent
Institut Mines-Télécom
Télécom SudParis
Evry, France

Shuai Li
Department of Electrical and
Computer Engineering
Stevens Institute of Technology
Hoboken, New Jersey

Maode Ma
School of Electrical and Electronic
Engineering
Nanyang Technological University
Singapore

G. Maciá-Fernández
Department of Signal Theory
University of Granada – CITIC
Granada, Spain

R. Magán-Carrión
Department of Signal Theory
University of Granada – CITIC
Granada, Spain

Ismail Mansour
LIMOS-CNRS
Clermont Université
Aubière, France

Pietro Manzoni
Departamento de Informática de
Sistemas y Computadores
Universitat Politècnica de València
Valencia, Spain

W. Maqbool
UTM-MIMOS Center of
Excellence
Universiti Teknologi Malaysia
Johor, Malaysia

Michel Misson
LIMOS-CNRS
Clermont Université
Aubière, France

Vallipuram Muthukkumarasamy
School of Information and
Communication Technology
Griffith University
Gold Coast, Australia

Mehdi Nafa
Department of Computer Science
University of Badji Mokhtar
Annaba, Algeria

Giovanni Russello
Department of Computer Science
University of Auckland
Auckland, New Zealand

Elio Salvadori
CREATE-NET
Trento, Italy

L. Sánchez-Casado
Department of Signal Theory
University of Granada – CITIC
Granada, Spain

Manuel D. Serrat-Olmos
Departamento de Informática de
Sistemas y Computadores
Universitat Politècnica de València
Valencia, Spain

Xuemin (Sherman) Shen
Department of Electrical and
    Computer Engineering
University of Waterloo
Waterloo, Ontario, Canada

Sanaa Taha
Kitchener
Ontario, Canada

Nerea Toledo
University of the Basque
    Country
UPV/EHU
Bilbao, Spain

Muhammad Usman
School of Information and
    Communication Technology
Griffith University
Gold Coast, Australia

Yunpeng Wang
Department of Electrical and
    Computer Engineering
University of Miami
Coral Gables, Florida

Xin-Wen Wu
School of Information and
    Communication Technology
Griffith University
Gold Coast, Australia

Suleiman Zubair
UTM-MIMOS Center of Excellence
Universiti Teknologi Malaysia
Johor, Malaysia

SECURITY IN MOBILE AD HOC NETWORKS

Chapter 1

Taxonomy and Holistic Detection of Security Attacks in MANETs

P. García-Teodoro, L. Sánchez-Casado, and
G. Maciá-Fernández

Contents

1.1 Introduction

Since the appearance of ad hoc networks in the 1980s, interest in this field has increased among the research community. Thus, the number of specialized publications on this topic has exponentially grown, as shown in Figure 1.1. The principal

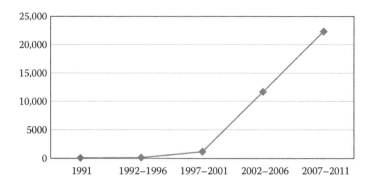

Figure 1.1   Evolution of the number of contributions directly or indirectly related to ad hoc networks. (From ACM, IEEE, ScienceDirect.)

features of these networks include the following, which simultaneously constitute their main weak spots:

- Use of wireless transmissions, which provide high accessibility but create many security risks because of the "open" nature of the channel.
- Lack of a fixed transport infrastructure (routers and access points), which generally implies more complex procedures to carry out communications.
- Establishment of multi-hop communications, that is, network nodes operate as retransmission elements to conform the source–destination routes.

Beyond the engineering challenges posed by this technology, for example, designing efficient mechanisms to access the medium or constructing source–destination routes, security is a principal concern in ad hoc networks. Their inherent open and distributed nature makes the networks sensitive and vulnerable to many security attacks. This vulnerability is noticeably more emphasized in MANETs (Mobile Ad Hoc Networks), that is, ad hoc networks where the locations of the nodes are not fixed [1,2], as mobility introduces another factor that makes it possible to conceal even simple attacks.

Despite a huge amount of contributions describing attacks and vulnerabilities, the number of works aimed at introducing a certain order in the field of security attacks in MANETs is however reduced. As discussed below, most works provide theoretical classifications that are not really useful for building practical security attack-related detection systems. In this context, this chapter introduces a novel taxonomy with two main goals: organize and classify security attacks in MANETs from a practical perspective, and provide guidelines to build detection approaches that are not attack-specific but that are holistic with better performance in the comprehensiveness, flexibility, and effectiveness of the global process.

With these two objectives in mind, the chapter is organized as follows. Section 1.2 briefly presents most known attacks currently reported for MANETs. Mainly motivated by the absence of a classification for grouping attacks from a practical perspective, Section 1.3 introduces a novel taxonomy. With this foundation, Section 1.4 subsequently describes guidelines for developing a flexible holistic detection system. We justify our proposal from the convenience of developing more comprehensive and robust detection schemes than those usually considered when facing individual attacks. A functional architecture is also proposed in Section 1.4 for the holistic detector. Section 1.5 provides some preliminary results of our developments by presenting packet-dropping attacks as a case study. Finally, Section 1.6 summarizes the main conclusions of the work and highlights some future research lines.

1.2 Security Attacks in MANETs

Several works are centered on studying security attacks in MANETs. Table 1.1 collates the most commonly reported attacks in the form of specialized literature [3–7], some of which are specific for such environments and others not. The table shows various attacks and briefly describes their behavior and effects.

In many existent attacks, several types, sub-types, and variants coexist, with subtle differences among the attacks. *Sybil, impersonation,* and *man-in-the-middle* attacks are similar in some aspects, as are *flooding* (i.e., *SYN flooding, HELLO flooding*) and *denial-of-service* attacks. And what about *fabrication* or *modification* versus *link spoofing*?

Such a diversity of reported attacks, artificial in some sense from our perspective, can be justified by several causes. The difference between some attacks is often a mere matter of degree or *impact*, as in *blackhole* versus *grayhole*. In other cases, the difference is due to a minor connotation, as in *routing cache poisoning* versus *link spoofing* and *link-broken error*. In summary, several attacks are, after all, the same attack.

Differentiation among attacks is sometimes extreme. Though *blackhole* and *selfish* attacks exhibit similar behavior (i.e., packets are dropped and not forwarded), they are separate attacks because of their (assumed) *motivation*: "maliciousness" in the first case and "egoism" in the second. The final damage to the environment hardly differed in each case.

The *purpose* of an attack is sometimes "mistaken" for the *procedure* that leads its execution. We think that the attack called *tampering* should not be categorized as an attack *per se*, because it only represents accessing a node to compromise it in some sense. Something similar occurs with the *wormhole* attack. In this case, the existence of a fast link between two nodes is exploited to direct network traffic through them. We can thus conclude that *wormhole* is not an attack, as the link

Table 1.1 Main Attacks in MANETs Reported in the Literature

Attack (in alphabetical order)	Description
Blackhole	Sending fake routing information that claims an optimum route to make other nodes relay data packets through the malicious node. In a second step, this node could drop or discard traffic.
Collision	Generating selective interferences to disrupt MAC mechanisms, which affects the capture of a channel for legitimate transmissions.
Delay	Adding a time delay to packet retransmissions.
Denial of service (DoS)	Exhausting network resources, which degrades network performance.
Eavesdropping	Listening to private communications, i.e., intercepting data.
Exhaustion	Repeated collisions and/or continuous retransmissions to occupy the channel.
Fabrication	Creating packets, normally aimed to fool authentication mechanisms.
Flooding	Consuming significant network resources, e.g., by injecting many junk packets.
Grayhole, a.k.a. Selective forwarding	Blackhole attack where the node drops packets selectively, e.g., with a certain probability, one packet every certain time, or only packets corresponding to specific flows.
HELLO flooding	Massive sending of HELLO packets to overwhelm neighbors.
Impersonation	Adopting the legitimate identity of a third node or application, which results in network misrepresentation.
Jamming (random, periodic, ...)	Generating signal interferences, which provokes communication disruption.
Jellyfish	Introducing time delays to TCP retransmissions, which decreases end-to-end performance.
Link spoofing	Advertising fake links with non-neighbors, thus disrupting routing operations.

Table 1.1 (continued) Main Attacks in MANETs Reported in the Literature

Attack (in alphabetical order)	Description
Link withholding	Ignoring a link advertisement, which can result in node isolation.
Link-broken error	Sending fake control messages, which gives rise to connectivity loss.
Man-in-the middle	Actuation between the sender and the receiver by impersonating one or both.
Modification	Modifying packets, thus disrupting the integrity of the exchanged messages.
Replay	Recording and resending third nodes' valid messages.
Routing cache poisoning	Faking routing table information, thus disrupting the routing function.
Routing table overflow	Advertising an excessive number of routes to non-existing nodes, which prevents neighbors from creating new legitimate routes.
Rushing	Artificial quick retransmission of routing packets, which can result in building fake routes.
Selfish	Bypassing certain protocols rules to save resources (e.g., battery), which decreases network performance.
Sinkhole	Sending fake routing information that claims an optimum route to make other nodes route data packets through the malicious node to inspect the traffic.
Sleep deprivation, *a.k.a.* Resource consumption	Repeated collisions that induce the node to continuous retransmissions, thus causing its death.
Sybil	Adopting multiple identities, e.g., becoming a legitimate part of the network.
SYN flooding	Creating many half-opened TCP connections, which provokes target resource exhaustion.
Tampering	Physically manipulating a node to affect some functionality or compromise it.
Wormhole	Two colluding attackers record packets at one location and replay them at another using a private high-speed link.

between the nodes really exists. A different issue is the attackers' subsequent action over the traffic to which they have access (e.g., dropping, modification).

Another diversification step in studying attack types in MANETs concerns those research works that analyze attacks against certain **specific protocols**, as is the case of attacks targeting certain routing protocols, such as AODV and OLSR [8,9]. In this situation, analyzing general features that characterizes the attacks are completely avoided; their specific causes and effects over the network services are instead investigated (e.g., faking/flooding/creating RREP (route reply), RREQ (route request), HELLO, and TC (topology control) packets).

Considering the above points, we can determine the necessity for establishing different classes of MANET attacks to improve and clarify this research field. This classification would facilitate, on the one hand, a better understanding of the vulnerabilities present in such environments. On the other hand, establishing different classes would leverage the definition and deployment of more flexible and effective detection approaches, which is the main goal of this contribution.

From this perspective, some classifications have already been proposed in the literature. They are briefly described below.

Attacks in MANETs are often classified as **active** or **passive** [10]. Because almost all currently known attacks are active, except *eavesdropping*, such a classification is not useful.

Attacks can be also classified as **external** or **internal** [11]. A node "inside" the system must perform an attack that sends authenticated data, while an external attacker can perform a *jamming* attack. However, as this classification is more theoretical than real (e.g., an internal attacker can also perform a *jamming* attack, and an external attacker can execute an impersonation attack), we again conclude that the external versus internal classification does not provide considerably discriminative information about attacks.

Another criterion used to organize MANET attacks is the network **layer** in which they occur [5]. Though this classification is widely accepted, it does not provide a usable classification for deploying practical holistic detection systems, as is our goal. Eavesdropping on MAC frames is different from eavesdropping on route discovering packets in AODV, because the knowledge required to perform the attack is different. However, the philosophy of the attack is exactly the same in both cases, and accordingly, the associated solutions should present similar characteristics. Similarly, a *SYN flooding* attack (in TCP) is, beyond its specific implementation, completely similar to a *HELLO flooding* attack (against OLSR). Therefore, as seen in *rushing* at the network layer versus *jellyfish* at the transport layer, to create a reasonable organization, the granularity level introduced by classifying attacks according to the target layer does not make sense for an organization aiming to practically implement detection systems.

Because of the low practical utility of the current classifications, further proposals have appeared in the specialized literature. Jawandhiya [12] classifies attacks according to the target security service (i.e., confidentiality, integrity, availability, access control), which seeks to clarify the consequences of the attacks on the network. Cardenas et al. [13] and Jain et al. [14] classify MANET DoS (denial of service) attacks on the basis of other criteria, including the goal of the attack. Conversely, Xiaojuan et al. [15] propose a game theory-based classification, while Yu et al. [16] deploy a classification related to trust mechanisms.

In the above context, we introduce a novel taxonomy for MANET attacks with two main features. First, adhering to the definition of "taxonomy," several successive criteria are considered to classify the existing attack types, from disposing of a common *root* to deriving the existent *species*.

Conversely, defining the *species* allows us to derive a set of detection features that, beyond the particularities of each attack, allows simplifying the design of a holistic detection environment.

1.3 Attack Taxonomy

Figure 1.2 shows the attack taxonomy in MANETs proposed here. From a common *root*, security attacks in MANETs, successive groups are obtained for the known attacks until each specific variant or *species* is derived. The different subsequent criteria used are *action of the attacker, effect of the attack, procedure of the attack,* and *function/service attacked.*

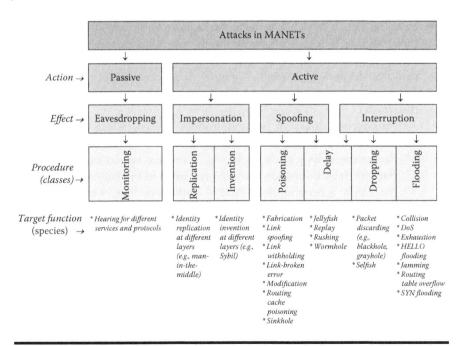

Figure 1.2 Taxonomy of security attacks in MANETs.

First, we classify attacks as active or passive, depending on the **action carried out by the attacker** to execute the attack. If the attack affects the system in some way, it is active; otherwise, it is passive.

A second level in the classification considers the **effect of the attack** on the system. Passive attacks imply *hearing information*, which can directly affect data confidentiality. Conversely, active attacks can produce three principal effects, which are referred either to the entities taking part in the communication, or to the transmitted information between them, or to the quality of the service provided. These effects are as follows:

- *Impersonation*, which can affect the authenticity of the participating entities in some ways
- *Spoofing*, which is a risk to the integrity of the information
- *Interruption*, that is, degradation (total or partial) of the quality of the service. This attack affects the availability service

A third level to differentiate MANET attacks is based on the **procedure carried out to execute the attack**. Passive attacks are possible by simply sensing or monitoring the channel. Instead, active attacks can be executed in several ways:

- Impersonation, that is, using an identity already in the system (*replication*) or a different one (*invention*).
- Spoofing, by: (a) directly introducing fake information into communications (*poisoning*), or (b) capturing packets and forwarding them with an artificial time delay, either positive or negative. Both cases allow disposing of "wrong" data at nodes and, consequently, "wrong" decision making.
- Three principal procedures can lead to the (total or partial) interruption of the service:
  - Introducing delays in communications (delay), which increases the response times involved.
  - Packet discarding (dropping), which implies wasted resources when transmitting information that does not reach its final destination.
  - Artificially generating some kind of traffic (flooding), which will block the normal access to the resources of the system.

At this point, the *species* of attacks in our taxonomy are obtained by applying a fourth classification criterion: the **target function** attacked, that is, the service and/or protocol that is the objective of the attack. There are many attack sub-types/variants, so different functions (and associated protocols) are implemented in the network. Implementing an *eavesdropping* attack is different at the physical and MAC layers because of the specific knowledge required in the second case to interpret the MAC frames and the associated timing. Similarly, an *impersonation* attack (*replication* or *invention*) can be performed at different levels, including MAC identification, TCP port, IP address, among others. Therefore, it is necessary to know the specific syntax of the corresponding function/service to detect the attack.

In summary, we argue that every reported MANET attack can be classified as belonging to one of the seven categories obtained from the multi-dimensional classification *action → effect → procedure*, as shown in Figure 1.2. The existence of these classes allows us to transcend particular differences due to the consideration of specific services and/or protocols.

Some attacks can be denoted as complex because they involve more than one type (class) of attack. Most authors define the *blackhole* attack as the combination of two processes: first, routing tables are poisoned to make the malicious node part of the routes; second, the malicious node drops packets in some way (e.g., selectively). Conversely, the *tampering* attack reported in Table 1.1 does not fit in the classification proposed in Figure 1.2, because we are not considering this type of attack as such, but only the action of manipulating a node. The interesting question from an attack detection perspective in this case is as follows: what is the goal of manipulating a node? Is it getting access to private credentials, modifying the radio frequency features, altering its identity, or something else entirely?

Different attacker nodes can deploy attacks individually or in *collusion*. This fact is not introduced in our classification, as the only objective here is to make detecting the malicious activity more difficult by "scattering" the action among several attackers. A collusion attack paradigm is that of the distributed DoS (DDoS) versus a traditional DoS.

1.4 Holistic Detection

As discussed above, there are many types, sub-types, variants, and sub-variants reported in the literature about security attacks in MANETs. The efforts carried out by the research community to understand and solve the different attacks are remarkable.

However, the absence of valid classifications from a practical perspective can be seen in the inexistence of studies that try to extract common attack features. On the contrary, such works and proposals in the literature are quite scanty for each attack analyzed. This fact greatly disperses the efforts made by the community and wastes global detection capabilities that can face different attacks, which is more important to us.

To alleviate this limitation, we propose and argue here the convenience of developing a holistic detection approach. This approach would allow the definition of a more comprehensive and complete detection system by generalizing specific attack detection capabilities. To do this, we suggest extracting a set of features common to all attacks of a given class such that, independent of the particularities of the network layer involved (i.e., independent of the target function attacked, or *specie*), the system would be able to detect the associated type of occurring attack. For clarity, consider the *jellyfish* and *rushing* attacks. Though they correspond to malicious events at different layers (transport and network, respectively), it seems logical that the possible evidences derived from their occurrences are similar, because they are both *delay*-typed, from the point of view of the procedure that supports their execution; for example, the packets might be forwarded (by the corresponding layer and protocol analyzed) with a given time delay.

In summary, by avoiding specific details about the layer, protocol, or service, system detection capabilities would increase, because a system would be able to determine the occurrence of more attacks by simply considering the same general procedures. Determining a specific attack (*specie*), if desired, would require a more in-depth analysis of the gathered information.

Although we do not try to be exhaustive in the subsequent analysis, we certainly consider it is necessary to perform a preliminary study about defining some general features to characterize each of the seven previously established attack classes, that is, a set of "parameters" that, independent of the service and protocol analyzed, allows us to conclude the occurrence of a certain action type against the normal system operation. The analysis of such a characterization is tentatively as follows:

■ *Monitoring*: Because the attacker does not actively participate in this case, we can conclude that this kind of attacks will not be detected from observable events, which is generally accepted. Of course, this circumstance is applicable to monitoring any service and protocol.

■ *Replication*: The existence of duplicated identities (i.e., MAC, IP) can be estimated using the suspicious location of the entities involved in the communications. Their location can be determined, among other possibilities, by studying the corresponding GPS (global positioning system) positions, identifying the existence of inadequate neighbor relationships (i.e., entries in ARP—address resolution protocol—and routing tables), and analyzing the trip times involved in the affected transmissions.

■ *Invention*: Although the inexistence of duplicated identities in this case can make detecting the attacker more difficult, a single device/user that uses several identities can be determined, as in the *replication* case, by studying the location of the communication entities.

Conversely, the potential observation of repeated/duplicated transmissions, as in the man-in-the-middle attack, can also conclude in the existence of impersonation-related behaviors.

■ *Poisoning*: *Poisoning* attacks often seek to alter the routing tables of the intermediate nodes, adapting the source–destination routes according to the attacker's interest. Some possible measures for detecting such anomalous behaviors can be based on monitoring certain relevant individual entries and/or table sizes.

Other detection options could be based on the correlation of the locations of the nodes to determine the existence of possible incoherencies in the tables.

■ *Delay*: Detecting *delay*-related attacks can be performed using a main circumstance: verifying the existence of duplicated packets in the network with a substantial temporal lag between them. However, how does one determine this temporal lag? For that, we can monitor and trace every packet around the network to detect its subsequent retransmissions.

A more plausible option, however, is to use timestamps in communications to facilitate verifying the accomplishment of the implemented protocols and services.

■ *Dropping*: Malicious packet discarding can be detected using some facts, including a deficient rate in packet retransmissions by a node (i.e., forwarded packets versus received packets), the unjustified (e.g., because of the inexistence of collisions) increase of retransmissions for a service, or the increase of retransmissions associated with upper layers.

■ *Flooding*: This kind of attack seeks to saturate resources by usually sending a high volume of traffic. Hence, the detection process involved in this case should be mainly based on analyzing the transmission rates generated by the entities over time. Because resource exhaustion can be performed selectively, effective detection procedures must be based on protocol or flow analysis.

As indicated above, our goal now is to suggest some possibilities for developing holistic capacities. A more detailed and specific proposal must be necessarily obtained from studying different detection solutions in the literature.

1.4.1 Detection Architecture

At this point, we tentatively define an architecture for the aforementioned holistic detection framework. As Figure 1.3 shows, the system has three main modules:

1. *Sensor unit* (su): The different protocols and services susceptible to be monitored require deploying specific sensing elements around the target environment. The sensors will report individual and independent events.

2. *Pre-processing unit* (pu): The information obtained by each sensor must be analyzed and corrclated to extract a set of *meta-features* that, transcending protocol and service particularities, would be able to partially represent the network system behavior.

3. *Detection unit* (du): The *meta-features* estimated for the several services and/ or layers must be subsequently analyzed and correlated to estimate the global system behavior and, from it, the likely occurrence of malicious actions. An efficient global event correlation would potentially allow detecting collusion attacks in the network. Many authors have pointed out this idea and have advocated using cross-layer solutions to model the environment as a whole instead of a disjointed set of events.

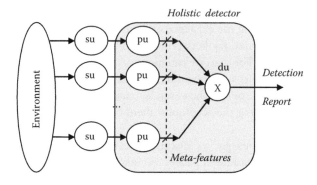

Figure 1.3 Conceptual architecture of a holistic detection system.

The most relevant issues among the above-mentioned ones are the *meta-features* that represent the "conceptual" events in the system. Moreover, and no less important, correlations between different meta-features should be estimated, which, as in the cross-layer solutions, try to model the environment as a whole instead of a disjointed set of events.

Other aspects that are being further studied refer to the type of architecture to be used, for example, centralized, distributed, or isolated, and the necessary communication among different architectural elements. In all these cases, well-known solutions appeared in the literature can be tentatively adopted.

1.5 Preliminary Holistic Detection Results: Packet Dropping as a Case Study

A key problem in MANETs is that of malicious elimination of packets in networks (*dropping* specie in Figure 1.2). For that a malicious node tries to introduce itself in the multi-hop origin–destination routes to control communications and, most of the cases, altering transmissions by discarding (selectively or not) packets instead of forwarding them. This type of attack constitutes a major security concern, because of its easy implementation and high impact in communications and services. In fact, these attacks are widely studied in the specialized literature and several detection solutions proposed for them [17]. Each solution is usually specific for one of the mentioned variants of dropping attacks, different parameters being considered in every particular detection process: RREP and RREQ packets and number of sequences for blackholes in refs. [18] and [19]; RREQ packets, number of sequences and destination, and source divergence for sinkholes in refs. [20] and [21]; the number of data packets sent and received, ratio of RTS and CTS packets and probe packets for grayhole in ref. [22]; and ratio of test packets sent and received for selfish in ref. [23].

We have developed a more general detection approach for the dropping attack specie than those usually available in the literature. The proposed scheme deals with a holistic detection in the sense that, as established in Section 1.4, the final decision is performed over a set of meta-features representing certain behavioral states of the monitored system instead of over some protocol- or service-specific features. Thus, the different steps involved in the detection are as follows:

1. Deployment of sensor units to calculate some specific and independent parameters for the nodes in the network:
   a. Number of RTS packets sent, $\#RTS_{sent}$,
   b. Number of CTS packets received, $\#CTS_{recv}$,
   c. Number of data packets received for retransmission, $\#DATA_{recv}$,
   d. Number of data packets forwarded, $\#DATA_{forw}$, and
   e. Number of AODV RREQ packets sent, $\#RREQ_{sent}$.
   That is, particular parameters regarding link, network, and application layers of the MANET operation are considered here.
2. Obtaining three meta-features, which constitute the core of the holistic detector:
   a. *Availability* (Av), which is related with the fact that the node is able or not able to access the medium due to situations such as collisions and channel errors that affect RTS and CTS packets. Availability is obtained as a probabilistic term from $\#RTS_{sent}$ and $\#CTS_{recv}$.
   b. *Connectivity* (Cn), which is mainly related with situations of mobility, congestion, and node failures with regard to the existence or nonexistence of a path to the destination. Connectivity is estimated from $\#RTS_{sent}$, $\#CTS_{recv}$, and $\#RREQ_{sent}$ also as a probability.
   c. *Forwarding* (Fw), which is the ratio of data packets retransmitted obtained from $\#DATA_{recv}$ and $\#DATA_{forw}$.
   It is important to note that *Av*, *Cn*, and *Fw* are defined after analyzing the "dropping problem" from a conceptual perspective. That is, to conclude that a given node is acting as a malicious dropper, we must first determine that the medium is available for the node; second, it has an effective route to the destination; and third, the node does not retransmit the packets it must.
3. Decision making, which, in consequence with the last paragraph, is logically performed from the meta-features as follows:

$$\text{Behavior (node)} = \begin{cases} \text{dropper} & \text{if } Av \text{ and } Cn \text{ and } \overline{Fw} \\ \text{normal} & \text{otherwise} \end{cases}$$

After studying our proposal in several simulated scenarios with NS2 (*http://www.isi.edu/nsnam/ns*), where the packet-dropping attack variants are implemented and different conditions regarding mobility and number of malicious

Table 1.2 Comparison of Detection Capabilities for Different Approaches against Dropping Attacks, Where the Last One Corresponds to Our Proposal

Detection Approach	True-Positive Rate (%)	False-Positive Rate (%)
Blackhole-specific	≤82	≥12
Sinkhole-specific	≥92	≤3
Grayhole-specific	≥88	≤7
Selfish-specific	≤75	≤2.2
Holistic	≥97.3	≤3.5

nodes in the network are analyzed, the obtained true-positive rate exceeds in all the cases, 97.3%, meanwhile the false-positive rate always remains below 3.5%. These results confirm the detection capabilities of our model, both in terms of performance figures in comparison with other proposals (see Table 1.2) and, from the perspective of this chapter, important in terms of the proposed holistic detection capacities.

Although it is not presented in terms of a holistic approach, interested readers can find in ref. [24] more technical details about the specific processing and detection results regarding our methodology.

1.6 Conclusions and Future Work

This chapter explains a brief survey of principal attacks in MANETs. Taking into account the existing confusion in this field, a taxonomy for grouping the attacks into classes is introduced afterwards. Besides categorizing the attacks in one or more such classes, the taxonomy is set out with the main aim of developing holistic detection schemes, which would allow the definition of more comprehensive and complete detection systems by generalizing specific attack detection capabilities. To do this, we suggest extracting a set of features common to all attacks of a given class such that, independent of the particularities of the network layer involved (i.e., independent of the target function attacked, or *specie*), the system would be able to detect the associated type of occurring attack.

As a proof of concept, the so-called dropping attacks have been studied from a holistic detection perspective. The detection rates obtained in comparison with other attack-specific approaches available in the specialized literature is evidence of the promising nature of our approach.

We are now planning to extend our proposal to other kinds of attacks, such as poisoning-, delay-, and flooding-related attacks. Another relevant research study to be conducted in the near future is about the architecture to be implemented in the holistic approach: centralized, distributed, or isolated. Each of them, together with the meta-features to be considered for each class of attacks, will be carefully studied and evaluated.

Acknowledgment

This work has been partially supported by the Spanish MICINN through project TEC2011-22579.

References

1. K.I. Lakhtaria: Technological advancements and applications in mobile ad-hoc networks: Research trends. *IGI Global*, 2012.
2. R.R. Roy: Mobile ad hoc networks. Chapter 1 in *Handbook of Mobile Ad Hoc Networks for Mobility Models*, Springer, Boston, 3–22, 2011.
3. D. Martins, H. Guyenne: Wireless sensor network attacks and security mechanisms: A short survey. *International Conference on Network-Based Information Systems*, pp. 313–320, Takayama, Japan, 2010.
4. B. Kannhavong, H. Nakayama, Y. Nemoto, N. Kato: A survey of routing attacks in mobile ad hoc networks. *IEEE Wireless Communications*, 14(5), 85–91, 2007.
5. B. Wu, J. Chen, J. Wu, M. Cardei: "A survey on attacks and countermeasures in mobile ad hoc networks. Chapter 12 in *Wireless/Mobile Network Security*, Eds. Y. Xiao, X. Shen, and D.Z. Du, Springer, New York, 2006.
6. R. Priya Ankala, D. Kavitha, D. Haritha: Mobile agent based routing in MANETs—Attacks & defences. *Network Protocols and Algorithms*, 3(4), 108–121, 2011.
7. K. Sahadevaiah, P. Reddy: Impact of security attacks on a new security protocol for mobile ad hoc networks. *Network Protocols and Algorithms*, 3(4), 122–140, 2011.
8. M. Saeed Alkatheiri, L. Jianwei, A.R. Sangi: AODV routing protocol under several routing attacks in MANETs. *International Conference on Communication Technology*, pp. 614–618, Jinan, China, 2011.
9. M. Salehi, H. Samavati, M. Dehghan: Performance assessment of OLSR protocol under routing attacks. *International Conference for Internet Technology and Secured Transactions*, pp. 791–796, Abu Dhabi, 2011.
10. A. El-Mousa, A. Suyyagh: AD HOC networks security challenges. *International Multi-Conference on Systems, Signals and Devices*, pp. 1–6, Amman, Jordan, 2010.
11. R. Mishra, S. Sharma: Vulnerabilities and security for Ad-hoc networks. *International Conference on Networking and Information Technology*, pp. 192–196, 2010.
12. P.M. Jawandhiya: A survey of mobile ad hoc network attacks. *International Journal of Engineering Science and Technology*, 2(9), 4063–4071, 2010.
13. A. Cardenas, T. Roosta, S. Sastry: Rethinking security properties, threat models, and the design space in sensor networks: A case study in SCADA systems. *Ad Hoc Networks*, 7, 1434–1447, 2009.
14. A.K. Jain, V. Tokekar: Classification of denial of service attacks in mobile ad hoc networks. *International Conference on Computational Intelligence and Communication Systems*, pp. 256–260, Bali, Indonesia, 2011.

15. L. Xiaojuan, H. Dao, K. Sarauki: Classification of attacks in wireless ad hoc networks: A game theoretic view. *International Conference on Networked Computing and Advanced Information Management*, pp. 144–149, Gyeongju, Korea, 2011.
16. Y. Yu, K. Li, W. Zhou, P. Li: Trust mechanisms in wireless sensor networks: Attack analysis and countermeasures. *Journal of Network and Computer Applications*, 35, 867–880, 2012.
17. S. Djahel, F. Nait-abdesselam, Z. Zhang: Mitigating packet dropping problem in mobile ad hoc networks: Proposals and challenges. *IEEE Communications Surveys & Tutorials*, 13, 658–672, 2011.
18. S. Kurosawa, H. Nakayama, N. Kato, A. Jamalipour, Y. Nemoto: Detecting black-hole attack on AODV-based mobile ad hoc networks by dynamic learning method. *International Journal of Network Security*, 5(3), 338–346, 2007.
19. M.Y. Su, K.L. Chiang, W.C. Liao: Mitigation of black-hole nodes in mobile ad hoc networks. *International Symposium on Parallel and Distributed Processing with Applications (ISPA)*, pp.162–167, Taipei, China, September 2010.
20. W. Shim, G. Kim, S. Kim: A distributed sinkhole detection method using cluster analysis. *Expert Systems with Applications*, 37(12), 8486–8491, 2010.
21. G. Kim, Y. Han, S. Kim: A cooperative-sinkhole detection method for mobile ad hoc networks. *AEU—International Journal of Electronics and Communications*, 64(5), 390–397, 2010.
22. J. Sen, M.G. Chandra, S.G. Harihara, H. Reddy, P. Balamuralidhar: A mechanism for detection of gray hole attack in mobile ad hoc networks. *6th International Conference on Information, Communications & Signal Processing*, pp. 1–5, Singapore, December 2007.
23. L.W. Wu, R.F. Yu: A threshold-based method for selfish nodes detection in MANET. *International Computer Symposium (ICS)*, pp. 875–882, Tainan, China, December 2010.
24. L. Sánchez-Casado, G. Maciá-Fernández, P. García-Teodoro: An efficient cross-layer approach for malicious packet dropping detection in MANETs. *11th IEEE International Conference on Trust, Security and Privacy in Computing and Communications (TrustCom)*, pp. 231–238, Liverpool, UK, 2012.

Chapter 2

Security and Privacy for Routing Protocols in Mobile Ad Hoc Networks

Mohamed Amine Ferrag, Mehdi Nafa,
and Salim Ghanemi

Contents

2.1 Introduction

Since their introduction, the wireless local area networks (LANs) have attracted the interest of professionals, faced with the needs of mobility and network connectivity to their organization. 802.11 networks, standardized by the IEEE in 1997 [1] has rapidly become until, in some cases, replace traditional wired networks like Ethernet. Since their arrival on the market, the steady evolution of their performance and lower their cost of acquisition helped accelerate their dissemination.

The IEEE 802.11 provides two modes: infrastructure mode and ad hoc mode. Infrastructure mode, also called cellular mode, uses a topology built around fixed access points. The latter is responsible for managing exchanges between mobile nodes located in their area transceiver. Multiple access points can be interconnected by a backbone network, called the distribution system to provide connections to a larger number of nodes or increase the space of node mobility. Ad hoc mode, it establishes an exchange point to point between two mobile nodes. If two nodes do not share the same areas transceiver, a direct connection is impossible. In this case, the intermediate nodes are used to establish a path between the source and destination nodes. These networks, whose architecture evolves according to the movement and appearance of nodes, are called Mobile Ad Hoc Network (MANET) or spontaneous networks.

In some contexts, users can benefit from the features of MANET to exchange information. A frequently cited example, in civil and military, is an ad hoc network formed by the interconnections between moving vehicles. In the industrial networks catch (sensor networks) can form a MANET to adapt to different environments. But many other situations of everyday life are adapted to the use of MANET. Consider, for example, a network created for the purposes and duration of a meeting of participants from different organizations or a network created between students and their teacher in a classroom for the duration of a course.

There are primarily two types of routing protocol in ad hoc networks. The first type is reactive routing, for example, dynamic source routing (DSR) [38,39], and ad hoc on demand distance vector (AODV) [6] that initiate the search for a route when trying to reach a destination that is not contained in the routing table. The second type is proactive routing, for example, optimized link state routing (OLSR) [7], and destination sequenced distance vector (DSDV) [40] that regularly update the information in the routing table with route discovery requests. Note that the hybrid algorithms exist. Distance vector algorithms and link state algorithms are the two types of algorithms that are used to maintain accurate routing tables and make use of both protocols under different conditions. In general, the route discovery is as follows. When a node wants to transmit a message to another node, it broadcasts a route request in different denominations according to the protocols. The route is built when this route to the recipient can return a message to the sender. The road is then set and the actual data exchange can take place in the absence of authentication, confidentiality, integrity, and so on. Ensuring smooth running of these protocols can greatly compromise network stability.

The issue of privacy is almost solved, but major gaps remain in terms of routing protocols. For this reason, the subject of this chapter focuses on the security of routing protocols in MANET. We chose AODV routing protocol because they are most widely used in the ad hoc community. MANET is subject to a number of attacks. For example, an attacker in the MANET may not be willing to route packets to other nodes. On the other hand, more sophisticated attacks against MANET routing can disrupt the route discovery. In addition, they may interfere with maintaining ride disobedient routing protocols. Blackhole, Byzantine, wormhole, and spoofing are illustrations of various attack threats for MANET. For the purposes of group communication security, cryptography has been integrated in MANETs. Among them, the most popular techniques are symmetric and public key infrastructure (PKI).

This chapter is devoted mainly to the state-of-the-art security and privacy for routing protocols in MANETs. First, we present the MANETs, their characteristics, application domains, and the 802.11 MAC used in these networks. Then, we present the needs of network security. In addition, we present the basic concept of ad hoc routing protocols including examples of protocols and the security threats they face. Second, we choose AODV routing protocol and we describe the different classification of vulnerabilities and attacks against AODV in MANETs. We mention the main lines of security AODV in mobile ad hoc. Then, we study different approaches to security environment dedicated to AODV in MANETs.

2.2 Mobile Ad Hoc Networks

MANETs continue to evolve with the development of mobile technology. Mobile devices become increasingly smaller and more powerful in terms of processing power and data storage. This allows nodes to ensure applications and more advanced services. This allows nodes to ensure applications and more advanced services, that is, connection requests, routing, and security. In this section, we present the MANETs, their characteristics, and the standard "IEEE 802.11b" in ad hoc mode characteristics of MANET.

Ad hoc mobile networks not only possess the same characteristics as mobile networks, but also a number of characteristics that are unique and differentiate them from others. We can cite some key features:

- *Lack of infrastructure*: No base station or access point, all nodes in the network move into a distributed environment without an access point or a point of attachment to the entire network. Nodes play the role of both an active player in the network transmitter and receiver and also as a router to relay communications from other network nodes.
- *Dynamic network topology*: The network nodes are autonomous and able to move arbitrarily. Mobility means that the network topology is dynamic because it can change at any time quickly and randomly. This topology change has an impact on the connections or links unidirectional and bidirectional nodes.
- *Limited resources*: Energy sources such as batteries are needed for the communication of mobile nodes. Unfortunately, these sources have a limited lifetime and depend on the depletion treatment performed at the node operations such as transmission, reception, and complex calculations. Therefore, the energy constitutes a real problem. Management mechanisms of energy are required for nodes in order to conserve energy and increase their lifetime. So any solution for MANETs must take into account the energy constraint.
- *Physical links*: On the basis of the wireless communication technologies essential to the establishment of an ad hoc network.
- Modeling of MANET.

A MANET can be modeled by a graph (Figure 2.1) $Gt = (Vt, Et)$ where:

- Vt is the set of nodes.
- Et models all the connections between these nodes.
- If $e = (u, v) \in Et$, this means that the nodes u and v are able to communicate directly at time t.

2.2.1 Applications in MANET

The ad hoc mobile network has managed to establish itself as a promising technology. Their characteristics, and in particular mobility and lack of infrastructure, expand their areas of application. We can cite the strengths of MANETs as follows:

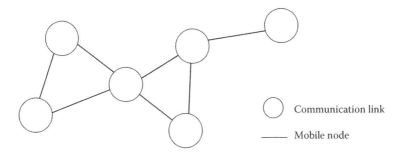

Figure 2.1   Modeling of an ad hoc network.

- *Network at low cost*: Wired networks are expensive from the economic view-point because they require wiring and infrastructure deployment. However, the MANETs can be deployed anywhere, especially in areas where wired networks cannot be made for reasons of geographical difficulties. Thus, ad hoc networks are becoming an alternative to reduce financial costs.
- *Network hostile place*: An example of a hostile place is on the battlefield. In environments with difficult access such as mountainous regions, ad hoc mobile networks are very convenient.
- *Wireless infrastructure*: In the case of a local situation that does not require wired network resources; for example, conferences or meetings. It is not necessary to go through the wired network for internal members.

2.2.1.1 Military Applications

Ad hoc mobile networks are designed based on applications for operational rations and military nature. These networks are suitable for hostile environments, because they are dynamic and rapidly deployable. Nodes in this type of communication are soldiers and armored vehicles.

2.2.1.2 Operations of Rescue

Ad hoc wireless networks are also used in rescue operations, especially during earthquakes or other disasters. These networks can be rapidly deployed on land operations and ensure the communication link between rescuers.

2.2.1.3 Corporate Network

The ease of deployment of these networks and their reduced cost interests more companies. This ensures a high mobility of agents, data sharing, and conferences. For example, at a meeting or conference, the speaker can communicate with all participants and create an interactive debate.

LLC 802.2			
MAC 802.11			
FHSS	DSSS	IR	....

Figure 2.2   Standardization of LANs by IEEE.

2.2.2 Standard "IEEE 802.11b" in Ad Hoc Mode

The standard "IEEE 802.11" [1] defines two operational modes: (i) infrastructure method, which requires the presence of specialized equipment called an access point, and control point-to-point communications between hosts on a network. Infrastructure method is the most common configuration where is used, for example, to allow mobile workstations to access corporate networks or public networks (hotspots); and ad hoc method, presented in ref. [2], show that applications still belong to the field of research.

- LLC: Logical link control (data link layer)
- MAC: Medium access control (data link layer)
- FHSS: Frequency hopping spread spectrum (physical layer)
- IR: InfraRouge (physical layer)

Standardization of LANs by IEEE in Figure 2.2 covers the physical layer and data link reference model OSI. The data link layer used by the IEEE is composed of two sublayers: the layer under LLC and the MAC. Layer "LLC" or "IEEE 802.2" is common to all 802 standards of IEEE. Layer IEEE 802.11 is a common layer to several physical layers.

2.3 Needs of Security

Basic needs for safe mobile ad hoc are more or less the same as for wired or wireless infrastructure. The main objective of network security is to ensure that the applications and protected information used as input and generated as output will not be affected by unintentional bad breaks

Functional elements needed to build a security system network are

- Confidentiality
- Authentication

- Authorization
- Message integrity
- Nonrepudiation

2.3.1 Confidentiality

Confidentiality is to keep information away from those who are not allowed to know (content invisible). This is usually what comes to mind when talking about network security. Cryptography is typically used to complete this. At the network layer, we can cite the ESP protocol (encapsulating security payload [3]), which ensures confidentiality to IP datagram by encrypting. ESP is a protocol belonging to IPsec (Internet protocol security [4]). Encryption algorithms, whether symmetric or asymmetric, require an encryption key to encrypt the message before sending it to the destination.

2.3.2 Authentication

Authentication verifies the identity of an entity or a node in the network. This is an essential step to control access to network resources. Without authentication, a node attacker can easily impersonate another node in order to enjoy the privileges assigned to that node or to carry out attacks in the identity of the node and harm the reputation of victim node. In the context of wired networks or wireless networks with infrastructure, the authentication process is based on a trusted third party where all network entities trust. The trusted third party is only a certification authority that distributes certificates to nodes that have the right of access to a network service. This authentication scheme is centralized, and is known as public key infrastructure (PKI) [5]. Applying directly to the PKI model, MANET is not possible for reasons of dynamic change and frequent network topology because the authentication service availability is limited due to the capacity of the nodes.

2.3.3 Authorization

Authorization is the access control to the network or system resources, so only specific authenticated nodes are allowed to access the specific resources. This type of control provides selective access to resources.

2.3.4 Message Integrity

Message integrity refers to the condition that the received message is not altered en route compared to the message originally issued. This is protection against threats that can cause unauthorized modification of system configuration or data. Integrity services are designed to ensure the proper functioning of resources and transmission. This service provides protection against deliberate or accidental alteration and unauthorized system functions (system integrity) and information (data integrity).

In the wireless network, the message can be changed for reasons not malicious, such as corruption of the packet at the radio propagation.

2.3.5 Nonrepudiation

Nonrepudiation guarantees that the message sender is the legitimate sender of the received message and the sender cannot deny sending the message. In other words, the nonrepudiation of origin proves that the data has been sent, and nonrepudiation of the arrival proves that they were received.

These five basics of network security are implemented as hardware and software in the network devices. Based on these five elements of safety function, it is possible to examine the various emerging technologies to implement specific security needs. In the next section, we present the main principles and characteristics of routing protocols in MANET.

2.4 Classification of Routing Protocols in MANET

A routing protocol determines the path between two mobile nodes based on a predefined strategy. In MANET, the routing protocol, distributed across all mobile nodes is more to minimize the time on the road, also called latency, and the use of resources for this operation. We limit our discussion to routing protocols point-to-point, also called unicast routing protocols.

When two mobile nodes in MANET directly exchange data packets without using intermediate mobile nodes, the connection is direct. If the path between source and destination nodes requires the presence of several intermediate nodes, the connection is called a multihop.

In a network with fixed architecture, routes to the various networks are predefined and maintained by fixed interconnection equipment called routers. The dynamic architecture of MANET, resulting from the movement, the emergence of mobile nodes, or the state of the physical connection, requires regular updating of routing tables located in each mobile node. To establish a route between two nodes mobile in MANET, the basic mechanism is known as flooding. Flooding is performed by successive broadcasts to all neighbors of each node. It forward the packet to all nodes in the network.

Complementary mechanisms of control can be used to avoid closures or duplication of packets. This flooding mechanism, very expensive in network resources, can be applied to very small networks. Routing protocols aim to limit the spread of packet flooding. Depending on the role played by the mobile nodes in the dissemination of messages, we can achieve a first differentiation of routing protocols:

- When all nodes have the same functionality, the protocol is referred to as uniform

■ If some nodes have special features in the dissemination of messages, the protocol is referred to as nonuniform

We can also distinguish the standard protocols by their mode of operation. Come in fixed architecture, both technical link-state and distance vector have a routing table at each node combines the address of the next node, the vector, the number of intermediate nodes, and the distance. Routing protocols use link-state, a database that allows them to build the network topology and knows the path to all mobile nodes in MANET. Usually a metric based on several parameters relating to connections is used to select the best route.

To reduce the network load due to updated packages, some routing protocols trigger the search for a route only when it is requested. The time to obtain a route is longer. Protocols that use this operating mode are called reactive. To reduce the number of control messages required to discover routes, routing protocols that are nonuniform select some mobile nodes to create dynamic and hierarchical architectures. Thus, protocols for the selection of neighbors are that each mobile node discharges the routing function to a subset of direct neighbors. While for partitioning protocols, the network is cutting areas in which routing is performed by a single master node. Some of these protocols, called hybrid, jointly use routing link state and distance vector routing.

The main routing protocols are presented according to their mode of operation in Figure 2.3.

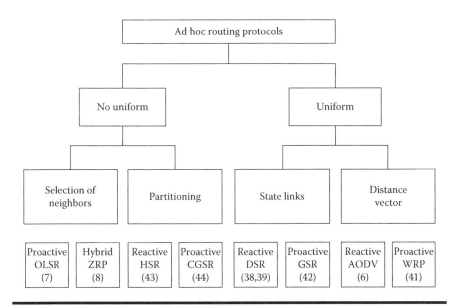

Figure 2.3   Classification of routing protocols in MANET.

Since July 2003, only two routing protocols, AODV [6] and OLSR [7] are subject to an RFC. Each protocol belongs to a family of protocols and each of them has a strategy. We take a routing protocol in each routing type. For reactive protocols, proactive protocols, and hybrid protocols, we take AODV, OLSR, and zone-routing protocol (ZRP), respectively.

2.4.1 Ad Hoc On Demand Distance Vector

AODV [6] is a reactive protocol, uniform, and oriented destination. The route chosen is bidirectional and is the shortest path (in number of nodes) between the source and destination. Each node maintains a routing table that stores the entries for a destination:

- The identifier of the destination
- The identifier of the next node to the destination
- The number of nodes to the destination

When a source node "S" has data to send to the destination node "D," but does not have the entire path to the destination, it should initiate a Route "REQuest" (RREQ). This packet is flooded through the network. Each node receiving an "RREQ," is then forwarded to its neighbors, if it has not already done so at least once, or if it is not the destination node of the message, and under the condition that counter time to live (TTL) is not zero.

The package "RREQ" includes the identifier of the source (SrcID), the identifier of the destination (DESTID), sequence number of the source (SrcSeqNum), sequence number of the destination (DesSeqNum), identification of broadcast (BcastID), and TTL. The "DesSeqNum" indicates the freshness of the road accepted by the source. When a node receives an RREQ packet, and if the recipient is or knows a valid path to the destination, it sends the message source REPly Route response (RREP). Otherwise, it rebroadcasts the RREQ to its neighbors.

The intermediate node is able to determine if a path is valid or not by comparing its own sequence number with the DesSeqNum in the RREQ. An intermediate node is able to know if the RREQ is already covered or not by comparing the BcastID to SrcID. During the broadcast of RREQ, each intermediate node injects the node address and the preceding BcastID. A timer is used to delete the entry in the case where an RREP packet is not received before the expiration date. When receiving a packet RREP, an intermediate node also stores information about the node from which the packet is received. This node is the next hop in the data transmission to the destination node that initiated the RREP.

2.4.2 OLSR Protocol

OLSR protocol [7] is a proactive protocol based on the regular exchange of information on the network topology. The algorithm is optimized by reducing the size and number of messages exchanged: only particular nodes, the MPR broadcast control messages to the entire network.

All nodes periodically send HELLO messages to their neighbors (HELLO_INTERVAL timer) on each node. OLSR uses a single message format, transported by user datagram protocol (UDP). The header specifies whether the message should be sent only in the immediate vicinity or to the entire network. Each node stores the description of its neighborhood: two-hop neighbors interface, MPR and MS. This description is updated each receiving a HELLO message, and old information is erased.

OLSR routing is based on the routing of messages by the nodes that have a symmetric neighborhood. A link can participate in a way that it is symmetrical. Routing to remote stations over a hop (1 + N hops) is due to the MPR, which periodically broadcast messages topology control (TC). A sequence number can eliminate duplicates. These messages are used to maintain a table in each station topology. The routing table is built and updated from the information contained in the interface table and the table adjacent to the topology uses a shortest path algorithm. The metric considered is the number of hops.

2.4.3 Zone Routing Protocol

Routing protocols hybrids are the result of a combination of the advantages of the two previous families, which offers a good compromise between road and persistence of network overload. In general, hybrid proposals exploit the network hierarchy using reactive and proactive protocols and are then integrated at various levels of the hierarchy.

The ZRP routing protocol [8] uses two approaches proactive, that is, intrazone routing protocol (IARP) for routing in the same area and reactive inter-zone routing protocol (IERP) between areas. The IARP maintains information on the status of link nodes at a distance d. In the case where the source node and destination are in the same area, the road is then immediately available. Most existing proactive protocols could be used as IARP in ZRP. In the next section, we present vulnerabilities and protection tools of MANETs. We study in particular the vulnerability of the AODV routing protocol.

2.5 Vulnerabilities and Protection Tools of MANETs

MANETs are vulnerable to the same attacks as other types of networks and they are easier to implement because in an ad hoc network, we cannot control access to the transmission medium, or define the limits of network.

An ad hoc routing protocol designed with a lack of security controls, which increases the risk of attacks that can be orchestrated by external or internal nodes, take into account the position of the attacking node to the MANET.

■ The external attackers are nodes that are not part of the network. In this case, the implementation of cryptographic mechanisms can solve the problem: only nodes with the necessary permissions can access the network or decrypt the content.

■ The internal attackers are nodes that are a legitimate part of the network. These nodes have the authorizations and cryptographic material necessary to belong to the network and other nodes make them a priori confidence.

For any position (internal or external), the node attacker uses several techniques to disrupt the functioning of the protocol. The combination of these techniques may result in a more sophisticated attack. These techniques known as elementals are presented below:

■ *Replay of messages*: Node attacker records a sequence of traffic it injects later in the MANET.

■ *Changing messages*: Node attacker modifies one or more fields of the message before forwarding.

■ *Deleting messages*: Node attacker deletes messages.

■ *Manufacture of messages*: Node attacker produces a message and is injected into the MANET.

As we have seen, routing protocols operate in two distinct phases: a phase of discovered network topology in which control information about network topology knowledge are exchanged, then a phase retransmission of data messages in which the data are sent from a source to a destination. Unlike wired networks, where routing operations are usually performed by physical equipment interconnection and managed by a dedicated administration legitimate, in MANET, these operations are entirely the responsibility of the nodes that compose them. This operating characteristic raises many security concerns. In the remainder of this section, we present the vulnerability of the routing protocol AODV.

2.5.1 Vulnerability of AODV

In [9,10], the authors propose two approaches to analyze attacks against the AODV routing protocol, called respectively *anomalous basic events* and *atomic misuse*.

The first approach classifies attacks against AODV into two categories: (1) atomic, resulting from manipulation of a single message routing and (2) composed, defined as a collection of atomic actions. The second approach classifies attacks into a series of elementary events. Each set can contain one or more operations performed in order as follows: (1) receiving a packet, (2) changing packet fields, and (3) the retransmission of the packet is one set of three operations. Table 2.1 provides a list of attacks against the AODV routing protocol.

Table 2.1 Attacks against the AODV Routing Protocol and Their Influence on the Security Property

Attacks	Influence on the Security Property				Target	Result
	Confidentiality	Authentication	Integrity	Availability		
Detour attack [11]			X		Nodes in the direct vicinity of the opponent	Not participate in routing
Blackhole attack [13]			X		Nodes specific	Creating a tunnel and disrupt routing
Wormhole attack [20]			X		Subset of nodes close to the hole	Creating a tunnel and disrupt routing
Routing table poisoning [12]			X		Subset of node	Division of routing
Sybil attack [14]	X	X	X		Nodes specific	Create of multiple identities
Man-in-the-middle attack [15]	X	X	X		Nodes specific	Use impersonation
Rushing attack [16]			X		Nodes in the direct vicinity of the opponent	Attract traffic
Resource consumption [17]	X	X		X	All nodes	Weakening battery
Routing table overflow [18]				X	Nodes in the direct vicinity of the opponent	Overflow of routing table
Location disclosure [19]	X	X			Nodes forming the path to a destination	Discover the location of mobile nodes

Source: Adapted from M.A. Ferrag, M. Nafa and S. Ghanmi. A new security mechanism for ad-hoc on-demand distance vector in mobile ad hoc social networks, *7th Workshop on Wireless and Mobile Ad-Hoc Networks (WMAN 2013) in conjunction with the Conference on Networked Systems NetSys/KIVS*, Stuttgart, Germany, Mar. 11–15, 2013.

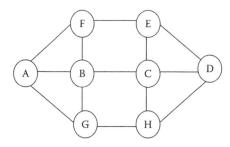

Figure 2.4 An example of a MANET containing mobile nodes attackers using AODV.

We present in Figure 2.4, a MANET containing nodes attackers using AODV as routing protocol, and we create different attacks as follows:

2.5.1.1 Detour Attack

- "A" = source, "D" = destination
- "A" sends the packet "RREQ" to request the road
- "B" makes the roads no longer pass through it

2.5.1.2 Routing Table Poisoning

- "A" = source, "D" = destination
- "F" and "E" send packets back to "A" for say that the optimal path is "A, F, E, D"
- The injection of false routing messages causes no-optimal network congestion, or a division of the network

2.5.1.3 Sybil Attack

- "A" = source, "E" = destination
- "B" takes the identity of "C"
- The attacker has many identities and behaves as if it were a set of nodes

2.5.1.4 Man-in-the-Middle Attack

- "A" = source, E" = destination
- "B" takes the identity of the "E"
- "B" takes the identity of "A"
- The attacker impersonates the destination toward the source and the source toward the destination without any of them realizing that he is being attacked

2.5.1.5 Rushing Attack

- "A" = source, "E" = destination
- "F" forwards messages faster
- The attacker relays messages faster for the road that passes by it is taken

2.5.1.6 Resource Consumption

- "A" = source, "E" = destination
- "B" modifies packets for messages passing through it

2.5.1.7 Routing Table Overflow

- "A" = source, "E" = destination
- "F" beyond the routing table of the target

2.5.1.8 Location Disclosure

- "A" = source, "E" = destination
- "B" tend to obtain the identity of the nodes forming the path to "E"

2.5.1.9 Worm/BlackHole Attack

- "A" = source, "E" = destination
- "F" create a false link between "A" and "E"
- Transmission of erroneous information, this leads to disruption of routing and loss of connectivity

After seeing the creation of the attacks against the AODV routing protocol, we present different security mechanisms for routing protocol AODV.

2.5.2 Security Mechanisms for Routing Protocol AODV

Many solutions have been proposed to secure ad hoc routing protocols. We classify these solutions into two categories (Figure 2.5):

1. Proactive security systems, in the sense that mechanisms are established in advance to ensure the safety enhancing system resilience to attacks with solutions based on cryptography and cons-measurement at the physical layer.
2. Reactive security systems that react according to the behavior of the neighborhood and is divided into reputation management solutions and trust and directional antennas.

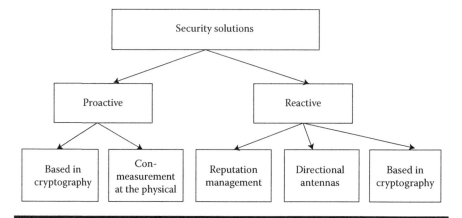

Figure 2.5 The mechanism for secure ad hoc routing protocols.

2.5.2.1 Solutions Based on Cryptography

Solutions based on cryptography are often used against external attackers. Some approaches are based on cryptographic hashes strings for authentication. This is the case of "SEAD" [21], a proactive routing protocol based on DSDV that protects against attacks by changing the sequence number and the number of hops in the update messages. The same authors also propose an extension to DSR called "ARIADNE" [22]. This solution provides an authentication point-to-point routing messages using the hash function secret key (hash-based message authentication code [HMAC]). However, to ensure secure authentication, ARIADNE is based on TESLA [23], a protocol to ensure secure authentication during broadcasts.

In [24] proposes a message-based secure OLSR (M-SOLSR), which uses the signature to ensure that the message comes from a trusted node and, thus, ensures the authentication of the source. In [25], the authors propose a secure routing scheme based on the method ARAN where not only authenticate nodes from end to end; they offer a nonrepudiation service using pre-established certificates distributed by a trusted server. Similar to this, another security method proposed in [26] where the authors combine the use of chain hashes and signatures to ensure source authentication and integrity of messages.

These security mechanisms based on cryptographic methods are used to secure the transfer of data so that only the relevant nodes can recover original data. A message is transformed to be incomprehensible to unauthorized persons and to prevent any changes. The proper recipient can retrieve the data clearly and is assured that the message has not been altered. The changes in question are based on mathematical functions known as cryptographic algorithms. We distinguish two types of cryptographic algorithms: reversible ones using the same key (symmetric) and those using two different keys (asymmetric).

2.5.2.1.1 Symmetric Cryptography

A mechanism of the symmetric encryption secret key allows entities (generally two) to perform encryption/decryption using a single key. In this sense, anyone can encrypt messages and so whoever is aware of the secret key can decipher. This encryption mode behaves like a closed box with a lock where entities wishing to communicate must have each a key allowing them to open and close the box. In Figure 2.6a, the node A wants to send a message to B. So it encrypts the message with the key k. Upon receiving the message, B decrypts the message using the same key k. It is important to note that there is an initial phase where two nodes agree on the secret key that will be used. Among the symmetric encryption, algorithms are the best known "Data Encryption Standard (DES)" [27].

2.5.2.1.2 Asymmetric Cryptography

A mechanism of asymmetric encryption is a function of two encryption keys: a public key that is provided to all those wishing to send a confidential message to

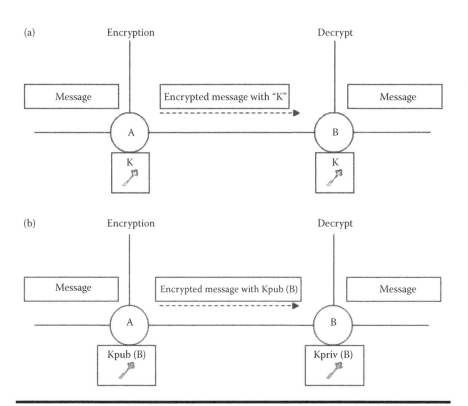

Figure 2.6 **(a) Symmetric encryption and (b) asymmetric encryption.**

one entity and a private key used to decrypt messages intended for it. These two keys are related in such a way that only the private key can decrypt a message encrypted by the corresponding public key. In Figure 2.6b, the node A wants to send the message to B. It then retrieves the public key of the destination Kpub (B) and encrypts the message. The encrypted message is sent over the communication channel. Destination B decrypts the received message using his secret key Kpriv (B) to obtain the plaintext message sent by A. A public key algorithm is the most popular "RSA" [28] named after its developers, Rivest, Shamir, and Adleman.

2.5.2.1.3 Cryptographic Hash Functions

To ensure the integrity of messages, a cryptographic hash function can be used. This is a nonreversible function with the following properties:

■ Unique and fixed size.
■ It is impossible to recover the original message from the digest.
■ Knowing a given message and imprint using a hash function, it is very difficult to generate another message that gives the same impression.
■ It is impossible to find two messages producing the same hash.

2.5.2.1.4 Message Authentication Code

The algorithm message authentication code (MAC) is a special case of hash function. As a hash function, the MAC algorithm produces a digest of fixed size (called message authentication code–MAC). But unlike the latter, it uses a secret key in addition.

MAC can be used to simultaneously verify the data integrity like any other hash function and the authenticity of the source data. In practice, a calculation method-based MAC hash function that is more sophisticated and safer is used: it is the HMAC [29] has been the subject of the RFC 2104.

2.5.2.2 Systems Reputation Management

The proposed solutions are based on reputation systems. In [30], the authors define the reputation as the perception that another node has about its intentions. Thus, a node's behavior cannot decide on its reputation, but can influence its own reputation. A numerical value is associated with the reputation of a node whose calculation is different from a solution to another, but that is mainly based on neighborhood watch and the first-hand information and/or second hand information obtained.

One of the first works [31] proposed to use a watchdog and a miss path to mitigate the consequences of dishonest behavior. While the former is used to monitor the neighborhood and detect nontransmission of a packet from a neighbor, the second allows the reputation management and routing evaluating each path used which avoids those with nodes attackers. Trusted AODV (TAODV) [32] is AODV extended by adding a model of reputation management, a specific routing protocol for reputation information management system and key.

Confidant [33] (Cooperation of Nodes Fairness in Dynamic Ad Hoc Networks) proposes to combine the use of positive and negative assessments. Thus, each node monitors the behavior of its neighbors while keeping track of estimate (positive or negative) reputation as well as the views of neighbors. These estimates are then combined to give a level of confidence in node directions through which the packet will be routed.

2.5.2.3 Directional Antennas

Nodes equipped with directional antennas using sectors (8 in numbers, namely N, S, E, W, NE, NW, SE, and SW) to communicate between them. A node that receives a message from one of its neighbors gets a rough (N, S, E, W) on its position. It knows the relative orientation with respect to its neighbor. These additional bits of information (angle of arrival of the signal) are exploited in some way to facilitate the detection/discovery of wormhole.

In [34], the authors propose a method for checking the neighborhood using directional antennas. Neighboring nodes examine the direction of the received signal for each of the other nodes and share a witness. The neighbor relation is only confirmed when the directions of all the pairs match.

2.5.2.4 Counter Measurement at the Physical Layer

The first work deals with the wormhole attack based on a hardware and signal processing technique. It is suggested that it may be a secret method of modulating radio signal bits. Only authorized nodes can demodulate the signal. Vulnerability of this method is that it is not kept in a safe space. This can lead to unauthorized adversaries to compromise legitimate nodes in the network to obtain the necessary access or opponents authorized to disclose their knowledge of the method. (It may be considered complementary mechanisms security code modulation/demodulation such as obfuscation, or the starch environment resistant to weathering). In terms of safety, this method only allows a defense against wormhole attacks led by outside node attackers (unauthorized) to the network, that is to say, nodes that do not have key cryptography. There is also the question of establishing/negotiating the secret method between nodes in the network legitimately [26].

2.6 Future Research Directions

The routing function is paramount in ad hoc networks. It is therefore a first requirement to protect against tampering, deliberate or not. However, this chapter considers leaving open a number of additional studies in the following directions:

- *Test the detection system in a real environment.* It would be interesting to see the behavior of the proposed system in a real environment.
- *Propose monitoring mechanisms.* It would be interesting to set up mechanisms to monitor the implementation of routing protocols to compare the messages exchanged between the nodes in ad hoc networks.
- *Add control messages.* It seems interesting to use other types of control messages to detect attacks in ad hoc network.
- *Studying unknown attacks.* It would be interesting to propose methods for detecting unknown attacks that will appear in the future based on machine learning methods of nodes.
- *The integration of routing mechanisms in the physical layer and application.* In ad hoc networks, routing mechanisms in the network layer, we believe it is worth sharing mechanisms in the physical layer and application.
- *The integration of social networks in ad hoc.* With the growing trend of social networking, we think of inventing a special routing protocol for use in the ad hoc network.

2.7 Conclusions

In this chapter, we presented the problem of security for routing protocols in MANET, especially AODV. In the first part, we presented the MANETs, their characteristics, application domains, and the 802.11 MAC used in these networks. We addressed the security needs. In Section 2.4, we presented the basic concept of ad hoc routing protocols including examples of protocols and security threats they face. We chose AODV routing protocol and described the different classifications of vulnerabilities and attacks against AODV in MANETs. We mentioned the main lines of security AODV in mobile ad hoc. Then, we studied different approaches to security environment dedicated to AODV in MANETs.

References

1. IEEE Institute of Electrical and Electronics Engineers. http://standards.ieee.org/get-ieee802/802.11.html (Last accessed: April 2, 2013).
2. Matthew S.GAST. 802.11 Wireless Networks—The definitive Guide, O'REILLY, 2002.

3. IP Encapsulating Security Payload (ESP), RFC2406, 1998.

4. Security Architecture for the Internet Protocol, RFC2401, 1998.

5. Chokhani, W. Ford, Internet X.509 public key infrastructure certificate policy and certification practices framework, RFC3647, 2003.

6. C. Perkins, E. Belding-Royer, and S. Das. Ad hoc on-demand distance vector (AODV) routing. http://tools.ietf.org/html/rfc356, July 2003. (Last accessed: April 2, 2013).

7. T. Clausen, P Jacquet. Optimized link state routing protocol. http://tools.ietf. org/html/draft-ietf-manet-olsr-11, IETF RFC 3626 2003. (Last accessed: April 2, 2013).

8. Z.J Haas, and M.R Pearlman. *The Zone Routing Protocol: A Hybrid Framework for Routing in Ad Hoc Networks, in Ad Hoc Networks*, edited by C.E. Perkins, Addison Wesley, 2000.

9. P. Ning and K. Sun. How to misuse AODV: a case study of insider attacks against mobile ad-hoc routing protocols. In *Proc. IEEE Systems, Man and Cybernetics Society, Information Assurance Workshop (IAW'03)*, pages 60–67. IEEE, June 2003.

10. Y. Huang and W. Lee. Attack analysis and detection for Ad Hoc routing protocols. In *Proc. of 7th International Symposium on Recent Advances in Intrusion Detection RAID*, Sophia Antipolis, France, volume 3224, pages 125–145. Springer, September 2004.

11. L. Guang, C. Assi, and A. Benslimane. Interlayer attacks in mobile Ad hoc networks. In *Proc. of the Second International Conference on Mobile ad-hoc and sensor networks (MSN 2006)*, Hong Kong, China, pages 436–448. Springer, December 2006.

12. T. Condie, V. Kacholia, S. Sankararaman, J. Hellerstein, and P. Maniatis. Induced churn as shelter from routing-table poisoning. In *Proc. of the 13th Annual Network and Distributed System Security Symposium* (NDSS'06), 2006.

13. H. Deng, W. Li, and D.P. Agrawal. Routing security in wireless ad hoc networks. *IEEE Communications Magazine*, 40(10): 70–75, 2002.

14. J. Douceur. The Sybil attack. *Peer-to-Peer Systems*, 1: 251–260, 2002.

15. C.Y. Tseng, P. Balasubramanyam, C. Ko, R. Limprasittiporn, J. Rowe, and K. Levitt. A specification-based intrusion detection system for AODV. In *Proc. 1st ACM workshop on Security of ad hoc and sensor networks* (SASN'03), pages 125–134. ACM, October 2003.

16. Y.C. Hu, A. Perrig, and D.B. Johnson. Rushing attacks and defense in wireless ad hoc network routing protocols. In *Proc. of the 2nd ACM workshop on Wireless security*, page 40. ACM, 2003.

17. I. Stamouli, P.G. Argyroudis, and H. Tewari. Real-time intrusion detection for ad hoc networks. In *Proc. Of the Sixth IEEE International Symposium on a World of Wireless Mobile and Multimedia Networks (WoWMoM'05)*, pages 374–380. IEEE Computer Society, 2005.

18. S. Gupte and M. Singhal. Secure routing in mobile wireless ad hoc networks. *Ad Hoc Networks*, 1(1): 151–174, 2003.

19. G. Vigna, S. Gwalani, K. Srinivasan, E.M. Belding-Royer, and R.A. Kemmerer. An intrusion detection tool for AODV-based ad hoc wireless networks. In *Proc. of the 20th Annual Computer Security Applications Conference (ACSAC'04)*, Tucson, Arizona, pages 16–27. IEEE Computer Society, 2004.

20. Y.C. Hu, A. Perrig, and DB Johnson. Wormhole attacks in wireless networks. *IEEE Journal on Selected Areas in Communications*, 24(2): 370–380, 2006.

21. Y.C. Hu, D.B. Johnson, and A. Perrig. SEAD : Secure efficient distance vector routing for mobile wireless ad hoc networks. *Ad Hoc Networks*, 1(1): 175–192, 2003.

22. Y.C. Hu, A. Perrig, and D.B. Johnson. Ariadne: A secure on-demand routing protocol for ad hoc networks. *Wireless Networks (WiNet)*, 11: 21–38, 2005.

23. A. Perrig, R. Canetti, JD Tygar, and D. Song. The TESLA broadcast authentication protocol. *RSA CryptoBytes*, 5(2): 2–13, 2002.

24. C. Adjih, T. Clausen, P. Jacquet, A. Laouiti, P. Muhlethaler, and D. Raffo. Securing the OLSR protocol. In *Proc. of Med-Hoc-Net*, pages 25–27, 2003.

25. K. Sanzgiri, B. Dahill, B.N. Levine, C. Shields, and E.M. Belding-Royer. A secure routing protocol for ad hoc networks. In *Proc. 10th IEEE International Conference on Network Protocols (ICNP'02)*, pages 78–89, November 2002.

26. M.G. Zapata and N. Asokan. Securing ad hoc routing protocols. In *Proc. 1st ACM workshop on Wireless security (WiSE'02)*, pages 1–10. ACM, September 2002.

27. A. Kahate. *Cryptography and Network Security*. Tata McGraw-Hill, India, 2 edition, 2003.

28. R.L. Rivest, A. Shamir, and L. Adleman. A method for obtaining digital signatures and public-key cryptosystems. *Communications of the ACM*, 21(2): 126, 1978.

29. H. Krawczyk, M. Bellare, and R. Canetti. HMAC: Keyed-Hashing for Message Authentication. http://www.ietf.org/rfc/rfc2104.txt, February 1997. Internet Draft.

30. L. Mui, M. Mohtashemi, and A. Halberstadt. A computational model of trust and reputation. In *Proc. of the 35th Annual Hawaii International Conference on System Sciences (HICSS)*, Hawaii, pages 2431–2439, 2002.

31. S. Marti, T.J. Giuli, K. Lai, and M. Baker. Mitigating routing misbehavior in mobile ad hoc networks. In *Proc. 6th Annual International Conference on Mobile Computing and Networking (MobiCom'00)*, Boston, MA, pages 255–265. ACM, August 2000.

32. X. Li, MR Lyu, and J. Liu. A trust model based routing protocol for secure ad-hoc networks. In *Proc. Aerospace Conference*, (AC'04), volume 2, pages 1286–1295. IEEE, December 2004.

33. S. Buchegger and J.Y. Le Boudee. Self-policing mobile ad hoc networks by reputation systems. *IEEE Communications Magazine*, 43(7): 101–107, 2005.

34. L. Hu and D. Evans, Using directional antennas to prevent wormhole attacks, *Proc. Network and Distrib. Sys. Sec. Symp.*, San Diego, CA, Feb. 2004.

35. Nabil Ali Alrajeh et al, Secure routing protocol using cross-layer design and energy harvesting in wireless sensor networks. *International Journal of Distributed Sensor Networks* 2013, http://dx.doi.org/10.1155/2013/374796.

36. M.A. Ferrag, M. Nafa and S. Ghanmi. A New Security Mechanism for Ad-hoc On-demand Distance Vector in Mobile Ad Hoc Social Networks, 7th Workshop on Wireless and Mobile Ad-Hoc Networks (WMAN 2013) in conjunction with the Conference on Networked Systems NetSys/KIVS, Stuttgart, Germany, Mar. 11–15, 2013.

37. M.A. Ferrag. *Study of Attacks in Ad Hoc Networks*. LAP LAMBERT Academic Publishing, Germany, November 3, 2012.

38. D.B. Johnson and D.A. Maltz. Dynamic source routing in ad hoc wireless networks. In Thomasz Imielinski and Hank Korth, editors, *Mobile Computing*, volume 353, Chapter 5, 153–181. Kluwer Academic Publishers, USA, 1996.

39. D. Johnson, Y. Hu, and D. Maltz. The Dynamic Source Routing Protocol (DSR) for Mobile Ad Hoc Networks for IPv4.,2007. RFC4728.

40. C.E. Perkins and P. Bhagwat. Highly dynamic Destination-Sequenced Distance-Vector Routing (DSDV) for mobile computers. In *Proc. of the Conference on Communications Architectures, Protocols and Applications (SIGCOMM'94)*, 244–254. ACM, 1994.

41. S. Murthy and J.J. Garcia-Luna-Aceves, An efficient routing protocol for wireless networks. *ACM Mobile Networks and App. J.*, Special Issue on Routing in Mobile Communication Networks, 1(2), 183–97, 1996.
42. Tsu-Wei Chen and Mario Gerla. Global state routing: A new routing scheme for Ad-hoc wireless networks, *Proc. IEEE ICC'98*, pp.1369–79, 1998.
43. A. Iwata, C.-C. Chiang, G. Pei, M. Gerla, and T.-W. Chen. Scalable routing strategies for ad hoc wireless networks. *IEEE Journal on Selected Areas in Communications, Special Issue on Ad-Hoc Networks*, 17(8), 1369–79, 1999.
44. C.-C. Chiang. Routing in clustered multihop, mobile wireless networks with fading channel. *Proc. IEEE SICON'97*, Apr.1997, pp.197–211.

Chapter 3

Privacy Key Management Protocols and Their Analysis in Mobile Multihop Relay WiMAX Networks

A. S. Khan, N. Fisal, M. Esa, S. Kamilah,
Suleiman Zubair, W. Maqbool, and Z. Abu Bakar

Contents

3.1 Introduction

The IEEE 802.16 standard intends to provide broadband wireless access (BWA) for metropolitan area networks (MAN), and the last mile of the BWA is designed as a substitute for the cable and digital subscriber line (DSL). After adopting nomadic mobility, that is, IEEE 802.16e, the standard moves toward the mobile multihop relay (MMR) functions in IEEE 802.16j for coverage extension and throughput enhancement that is done simply by adding relays in-between multihop relay base stations (MR-BS) and the subscriber station (SS). However, insertion of new relays demands strict authentication and key management schemes between MR-BS, relay station (RS), and SS to secure the traffic from different adversaries especially when the multihop RS initially joins the network [1]. Thus, security is essential in wireless technologies to allow rapid adoption and enhance their maturity, while designing the standard, the IEEE 802.16 working group has taken into consideration the security aspects to avoid mistakes of IEEE 802.11. Standard security specifications can mainly be found within the medium access control (MAC) layer called the security sublayer. The separate security sublayer provides authentication, secure key exchange, and encryption [2,3].

In MMR networks, authentication protocols provide authenticity and secrecy using authentication messages and key management schemes in centralized as well as distributed security schemes [4]. Normally authentication is carried out by request and response messages within the nontransparent relay stations (N-RS) and the MR-BS in a client–server model [2]. In the current authentication protocols, these authentication messages are vulnerable to denial of service (DoS) attacks, replay attacks, man-in-the–middle (MitM) attacks, and interleaving attacks [5]. For instance, an authentication request message is sent in plaintext to facilitate the authentication procedures, thus eavesdropping can happen. MR-BS may face replay attack from an adversary who intercepts and saves the authentication message. Although the adversary cannot derive secret information as it does not have the corresponding secret key, it can replay this message multiple times and thus either exhaust the MR-BS capabilities or force the MR-BS to deny the N-RS who owns the certificate thus causing DoS [6]. To avoid replay attack in authentication request message, timestamp was introduced. The timestamp received from an authentication request is also included in the authentication response message to guarantee the SS that this message is the actual response. Timestamps from the MR-BS ensure its freshness. The key drawback with the addition of timestamps is it requires both entities to maintain time synchronization. As clocks are never exactly synchronized over a network, there has to be some interval during which a timestamp is valid. An adversary can replay the message during this interval and can easily gain success [7]. The authentication response message is also exposed to replay attack. The N-RS also faces fraudulence from an adversary who intercepts the message earlier. The adversary can make its own authentication response message using a self-generated secret, thus gaining control of the communication which is a typical MitM attack. To avoid MitM attack, mutual authentication is proposed. In mutual authentication, one additional message is sent from N-RS to the MR-BS to authenticate itself with the MR-BS. However, if the authentication response message is compromised during the mutual authentication schemes, it would lead toward the interleaving attacks [8].

Initial proposed works on the OD-2009 showed that the draft is vulnerable to DoS, replay attacks, MitM attacks, and the interleaving attacks [1]. Several works discussed the security issues with their suggested solutions on mobile worldwide interoperability for microwave access (WiMAX) [9–13]. For MMR networks, secure mutual authentication protocols against rogue RSs were proposed. The proposed solutions worked well for the centralized security mechanism; however, scalability becomes a main issue as the number of hops increase so does the message authentication overhead [14]. To avoid the issues of scalability, distributed security architecture for MMR networks has been proposed. In this proposed architecture tunnel mode, encrypting the traffic with MR-BS public key is used. A sort of dedicated tunnel is used through which encrypted data is transferred. No doubt this is a very powerful attempt to countermeasure the MAC layer attacks; however, message authentication overhead is a critical issue which limits the popularity of the algorithm [15]. Hybrid authentication and key distribution for MMR networks are proposed to countermeasure the issues of centralized and distributed security mechanisms. In this work at the initial phase when an RS is joining MMR network, a centralized authentication is used for mutual authentication, after gaining the authentication key for hop-by-hop authentication between RSs distributed authentication is used. No doubt the proposed scheme lessens the overall authentication overhead; however, still two problems exist. The first one is the authentication overhead at centralized authentication scenarios and second is the MAC layer attacks; the proposed scheme does not tackle the MAC layer attacks [16]. To mitigate such issues, a scheme is needed that can countermeasure the MAC layer attacks with minimum authentication overhead.

3.1.1 MMR WiMAX Networks

As discussed in Section 3.1, in IEEE 802.16j-2009, multihop relays are an elective deployment to support performance and coverage area in WiMAX networks. In multihop relay networks, BS can be modified to MR-BS. Communication within SS and MR-BS are relayed through RS, thus, enhancing the coverage area and efficiency of the network. Multihop relays are partially or fully under the supervision of MR-BS. This leads toward two different modes, namely, centralized and distributed scheduling modes. Relays with full MR-BS supervision are functioned under a centralized scheduling mode where MR-BS is fully responsible for all the decisions. Relays with partial MR-BS supervision function under distributed scheduling mode where all the decisions are taken by RS with the collaboration of MR-BS [17].

There are two relay categories: transparent and nontransparent. Nontransparent relays function in both centralized as well as distributed scheduling mode. But for transparent relays, it only can function in centralized scheduling mode. These relays can operate in three separate schemes depending on the processing of received signals. These schemes include amplify and forward, decode and forward, and estimate and forward. Decode and forward and amplify and forward relays are also termed as nontransparent relays and transparent relays, respectively [4,18]. These relays may be fixed in locations like mounted on the top of a building or mobile as utilized on vehicles [19]. As far as security matters are concerned, these relays work in two different security modes, that is, centralized security mode and distributed security mode [20] that are discussed later in this chapter.

In MMR WiMAX networks, two different RS operational mode are defined; transparent mode RS (T-RS) and nontransparent mode RS (N-RS). The key difference between these two relay modes of operation is the ability to generate and send control information to its subordinate stations, which are included in the frame header. The RS operating in transparent mode do not generate and transmit control information, but in nontransparent mode, the relays do generate and transmit its own control information to its subordinate stations. The frame header contains essential scheduling information, which the nodes use to determine when they are allowed to transmit and receive information. T-RS only forward the control information generated by the MR-BS, and hence does not extend the coverage area of the BS. However, T-RS can be used to enhance the system capacity in terms of throughput within the BS coverage area. T-RS has less complexity and is cheaper as compared with the N-RS. So, the T-RS only can operate in a centralized scheduling mode and for a topology of up to two hops only [21].

In transparent relaying, the SSs communicating with relays can receive and decode the control information from the MR-BS. So, the RSs serving those SSs are not required to transmit control information themselves. These SSs are in range of the MR-BS, but by using multiple hops with the aid of RSs, it can achieve higher throughput [22]. So, the goal of this type of relaying is to enhance network capacity in terms of throughput. This scheme of relaying is called *transparent relaying*;

because the SS is not aware of the existence of the RS. In the transparent relaying, all control information originates from the MR-BS. The N-RSs can operate on both centralized and distributed scheduling. When NT-RSs operate in distributed scheduling, they generate their own control information. However, when they are operating in a centralized scheduling, they only forward those provided by the BS. N-RSs can be used to provide cell coverage extension as well as capacity enhancement. N-RSs can operate in topologies larger than two hops in either a centralized or distributed scheduling mode. On the other hand, the improved functions of the N-RSs lead to increased complexity and, hence, a higher cost than T-RS. In addition, the transmission of the framing information can result in interference between neighboring RSs [23].

Furthermore, the N-RS can operate on amplify and forward scheme or decode and forward scheme. In amplify and forward, the N-RS acts as an analog repeater that only amplify the received signal and forward it to the next node. In nontransparent relaying, the SSs served with RSs cannot receive or decode the control information from the MR-BS. So, the relays serving these SSs must generate and transmit its own frame containing control information at the beginning of it. The end SS considers the serving RS as its BS and cannot deal with control information sent by MR-BS. These relays are called *nontransparent* because the SS synchronizes and receives control information from it. The SSs out of range of the MR-BS cannot even receive the control information sent by it, so the nontransparent RS can send to them its own frame and hence extend the MR-BS coverage [23].

3.1.2 Security Requirements and Issues of MMR WiMAX Networks

The security sublayer lies above the physical layer and below the medium access control common part sublayer (MAC CPS), which is encrypted, authenticated, and validated. However, header and control information added by the physical layer are not encrypted or authenticated. Thus, physical layer information attached to the higher layer packets is vulnerable to threats. The MAC management messages are sent in the clear to facilitate network operations. Thus, MAC header, MAC management messages such as DCD, DL-MAP, UCD, UL-MAP, RNG-REQ, RNG-RSP, PKM-REQ, PKM-RSP, SBC-REQ, and SBC-RSP which are sent unencrypted, give wide field for the attacker to play [16].

DoS attack on the BS may possibly ensue during the PKMv2 authentication because of the intense public key computational load, an attacker might simply flood the BS with messages and the BS could use up its computational resources, evaluating signatures and decrypting messages [24]. BS authentication process in PKMv2 is vulnerable to an interleaving attack. In this attack, the attacker impersonates a valid RS, exchange the first two messages of PKMv2 sequences with a valid BS, and then it replays these to the original, valid RS to gain the final PKMv2 messages. The attacker then uses the final message from the original RS to complete the original PKMv2 sequence with the BS. This results in unauthorized access to the network. As the number of hops increases in the distributed and nontransparent environment, unreliability increases and, thus, more powerful and complex attacks can be attempted [25]. In the case where the attack involves the BS, it is a little bit tricky for the adversary to get successful as BS is a much more intelligent device. However, in cases where RS is involved, as RS is not more complex and intelligent than the BS, the chances of different attacks for RS is higher than BS [4].

MMR WiMAX networks may need the following security function, which have not been widely studied by others until now. Localized and hop-by-hop authentication is required. In MMR WiMAX network, RS is introduced for coverage extension and throughput enhancement, for this purpose, localized and hop-by-hop authentication between RS, MS, and MR-BS should be supported. All the participating devices must be validated and authenticated by authentication, authorization, and accounting (AAA) servers through MR-BS, because digital certificates of participating devices are only registered in AAA server database. However, on the other hand, N-RS should authenticate other N-RS/MS on behalf of MR-BS [23]. Conventional MS should be used in MMR WiMAX network without any functional modification.

3.1.3 Security Scheme for MMR WiMAX Networks

In multihop relay WiMAX networks, two different security schemes have been proposed, namely, centralized security scheme and distributed security scheme. Both schemes are discussed briefly in the following section. The centralized security scheme normally resides in MR-BS in the multihop relay system where security association (SA) is established within RS and MR-BS without the participation of intermediate RS. The intermediate RS does not decrypt the user data payload or do any kind of authentication to the SS or other RS; it just relays what MR-BS transmits to it. MR-BS is responsible for managing all the keys related to SS or RS. Intermediate RS does not have any key information related to SS. In distributed security scheme, the authentication keys established within SS and MR-BS is transferred to intermediate RS, during the registration to network, intermediate RS based on its capability may be configured to work in distributed security mode. An intermediate RS operating in this scheme initiate the RSA-PKM authentication protocol within MR-BS and itself, once authorization key (AK) is established within these two entities, MR-BS securely transfer the relevant AKs of the other requesting RS/SS to this intermediate RS. This intermediate RS derives all necessary keys and starts RSA-PKM authentication protocol with other subordinate RS/SS. After receiving the relevant keys from MR-BS, intermediate RS will reencrypt the relayed MAC packet data unit (PDU) [26,27].

3.1.4 Security Vulnerabilities of WiMAX Network

As a promising broadband wireless technology, WiMAX has many silent advantages such as high data rates, quality of service, scalability, security, and mobility. Many sophisticated authentication and encryption techniques have been enhanced with WiMAX, but it is still exposed to various attacks. This section intends to discuss vulnerabilities and threats related to WiMAX networks.

3.1.4.1 Masquerading Threat

In this attack, one entity assumed the identity of another entity. WiMAX supports RSA/X.509 certificate-based security measures and these certificates are programmed by the vendor to a device. Thus, sniffing or spoofing can help this attack to happen. Specifically, with the help of two techniques, these attacks can be successful, that is, identity theft and the rogue BS. The identity theft adversary can reprogram a device with the legal device hardware address and this address can easily be stolen by interfering the MAC management messages. If identity theft attack is successful, the adversary may act as a rogue BS, thus they can easily intercept the entire information which leads toward the MitM attack or degraded service, or some times even DoS can occur [28,29].

3.1.4.2 Replay Attacks

In a replay attack, the adversary normally intercepts the Auth-Req message transmitted by the legitimate SS and saves that message. The adversary will use this message by resending it after a specific period of time to perform a replay attack against BS. However, the adversary may not decrypt the Auth-Rsp keying parameters but it will replay this message multiple times to exhaust the capabilities of the BS. This may cause DoS to the legitimate SS. On the other hand, SS also faced this type of attack even worse than the BS. The adversary may develop its own Auth-Rsp message by generating an AK and sending to the SS impersonating as a BS. Thus, it can gain the control over the complete communication; this is typical called a MitM attack [6].

3.1.4.3 Interleaving Attacks

Interleaving attack can easily be successful, if the attacker can modify the authentication response message sent by BS to SS just by replacing Cert (BS) and SIG (BS) with Cert (ATTACKER) and SIG (ATTACKER). This attack is typically a PKMv2 authentication protocol attack and can happen even with the presence of mutual authentication.

In session 1, SS (ATTACKER) represents an intruder impersonating the legitimate SS and sends an Auth-Req message to BS, which is a replayed message originally sent by the SS primary. BS will respond to the Auth-Req message in session 1 with message 2. To get authenticated, SS (ATTACKER) needs to respond to message 1 of session 3. However, at this stage, SS (ATTACKER) cannot respond to this message as it does not have the corresponding private key to decrypt the message to attain the AK to encrypt the Nonce challenge. SS (ATTACKER) will utilize a legitimate SS to respond to this challenge and force it to initiate another session 2 by giving a response to a legitimate SS with the same Nonce given by BS. The legitimate SS will transmit message 1 of session 3 to SS (ATTACKER), thus SS (ATTACKER) can use this message to finish the session 1 and get authenticated. However, for SS (ATTACKER) to be successful in this attack, only one problem is left, that is, the AK is derived from PRE-PAK with the MAC address of SS and BS. If AKs in all sessions are the same, SS (ATTACKER) needs to impersonate BSAddr which is quite simple in wireless networks [22,23].

3.1.4.4 MitM Attack

MitM attack is another technical and classical attack that is generally successful where mutual authentication is absent. Normally, if a replay attack is successful in Auth-Rsp message, that is, the adversary either successfully can decrypt the authentication response message or generate its own authentication message, then the adversary can simply get complete hold of the entire communication, which causes a typical MitM attack. Lack of mutual authentication leads toward the message modification and masquerading issues. Once a message is modified at any level of communication, an MitM attack can occur [8].

3.1.4.5 DoS Attack

Maximum vulnerabilities faced by IEEE 802.16 standard are due to DoS attacks, especially during network entry or MAC management communication related to authentication. There are several DoS attacks in the standard; especially DoS related to ranging request and response messages where an adversary may forge this message to minimize the power level of SS. This makes the SS face trouble in transmitting to the BS that generates repeated triggering of initial ranging procedures. Second, DoS based on Auth-Req and Auth-Rsp message are very common and discussed in replay attacks. Third, a DoS attack is based on Auth-invalid MAC management message. This message is sent by BS when SS and BS have already shared the AK and SS cannot successfully refresh or update the AK as this message is also not protected by HMAC, thus this attack can be used to invalidate the legal SS [30,31].

3.1.5 PKMv1 Authentication Protocols

An SS utilizes PKM protocol to attain the AK and other traffic keying parameters from the BS. PKM is also utilized to support re-authentication and key management. The PKM protocol uses X.509 digital certificates, RSA public key algorithms, and powerful algorithms for encryption especially 3DES and AES to facilitate key exchange between the SS and the BS with a client–server model. In PKM protocol, SS first authenticates itself with the BS obtained and establishes the AK using public key cryptography. Once these entities get authenticated, it registers to the network. The AK attained is used to secure the exchange of TEK1 [2,31].

A brief summary of PKM protocols is as follows: the SS starts the protocol and authenticates with a BS; however, SS also authenticates BS using PKMv2. PKMv2 helps in mutual authentication to avoid MitM attacks. Once the shared key (AK) is established, BS sends a secure association identifier list (SAID) to SS. These are the lists of services that are explicitly authorized to SS. In PKM protocol, SS utilizes Auth-Req and Auth-Rsp MAC management messages to obtain AK and SAID list, and utilizes key-Req and key-Rsp MAC management messages to attain the keying parameters corresponding to a specified SAID. PKM protocol is categorized into two PKMv1 and PKMv2. For unilateral and mutual authentication, PKMv1 and PKMv2 protocols are used respectively [1]. In the following section, a brief description and security analysis of OD-2009 authentication protocols for PKMv1 is discussed.

3.1.5.1 OD-2009 PKMv1

Using OD-2009 PKMv1 authentication protocols, any SS can initiate the authentication process by sending Auth-info and Auth-Req messages to BS. The Auth-info message contains the information of X.509 certificates of SS manufacture. As this message only contains information, BS may ignore this message. After Auth-info, SS sends the Auth-Req message that contains its certificates, capabilities, and basic connection identifier. In response to this message, BS validates the SS X.509 certificates, generates AK, encrypt the AK with public key of SS, and sends back to the requesting entity with a sequence number and lifetime of AK in Auth-Rsp message [1]. The OD-2009 PKMv1 authentication protocol is shown in Figure 3.1.

In this figure, CERT (SS manufacturer) and CERT (SS) are the X.509 certificates of SS manufacturer and SS itself, respectively. The basic fields of the X.509 certificate include the certificate version, serial number, signature, issuer, validity, subject, subject public-key info, issuer unique ID, subject unique ID, and extensions. The capabilities supports the lists of data encryption algorithms of SS. The basic connection identifier (BCID) of SS is normally equal to the primary SAID. KU_{SS} (AK) is an AK encrypted by the public key of SS. The sequence number is of 4 bit used for AK; lifetime is the expiry time of the AK which is 32 bits. Finally, a SAID list is the identifier for which SS is authorized to attain the keying parameters.

To start with the security analysis of OD-2009, it is analyzed that the Auth-info message is optional and only contains information, thus security analysis begins from the actual Auth-Req message. The Auth-Req message is sent in plaintext because as the public key of BS is not yet shared and to facilitate authentication,

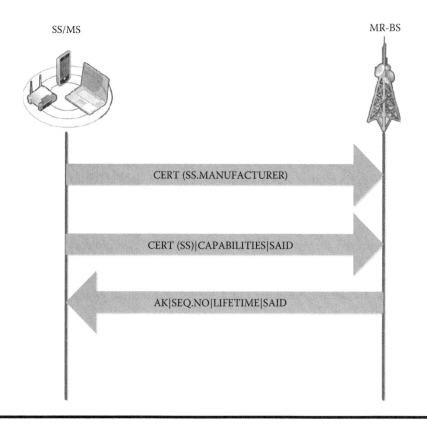

Figure 3.1 PKMv1 authentication protocol.

plaintext is preferable. Thus, to deal with the plaintext, eavesdropping is not a big issue. Thus, BS has to face a reply attack from the adversary who intercepts and save the Auth-Req messages sent by SS. However, the adversary eavesdropping the message cannot derive AK from the Auth-Rsp message, as it does not have the corresponding private key to decrypt the message. But, it will replay Auth-Req message multiple times to exhaust the BS's efficiency. This forces the BS to deny the SS, who holds that X.509 digital certificate. As BS normally sets a timeout value to reject Auth-Req from the same SS in a specific period of time, thus the legitimate Auth-Req message will also be disregarded. In this case, typical DoS attacks to the legitimate SS can occur [22,23].

Similarly, an Auth-Rsp message is also exposed to replay attacks which are worse than the Auth-Req message. In this case, the adversary can create its own Auth-Rsp message by generating an AK by itself, and send to the legitimate SS. Thus, it obtains the complete control over the communication of the victim SS. This is termed as an MitM attack. However, to deal with this type of attack, mutual authentication is needed, where SS need to authenticate BS as well [22,23].

3.1.6 PKMv2 Authentication Protocols

To avoid an MitM attack and to introduce mutual authentication, IEEE 802.16e proposed PKMv2, in which one additional message is added at the end of the original authentication protocol of PKM in IEEE 802.16 (also referred to as PKMv2). In fact, there are three optional protocols for an X.509 certificate: one-way, two-way, and three-way authentication. Although the original IEEE 802.16 (PKMvl) authentication protocol involves two messages, it is still a one-way authentication, because it only provides the SS certificate to the BS. SEN XU version can be regarded as two-way authentication, which provides mutual authentication between communication parties. The PKMv2 belongs to the three-way authentication, with a confirmation message from the SS to the BS. The following section discusses a brief description of OD-2009 and SEN XU authentication protocols with their security analysis with respect to PKMv2 [1,2,32,33].

3.1.6.1 OD-2009 PKMv2

Owing to lack of mutual authentication and to avoid an MitM attack, IEEE 802.16 standard proposed the PKMv2 which is shown in Figure 3.2.

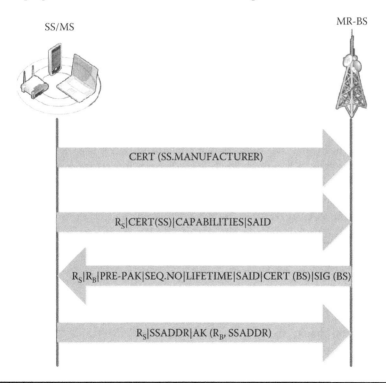

Figure 3.2 PKMv2 authentication protocol.

R_S and R_B are the 64-bit random number, respectively. As discussed earlier, standard utilized PKMv2-RSA authentication protocol generates the PRE-PAK to derive the AK. In the acknowledgment message from SS to BS, SS sends its MAC address to complete the three-way authentication with BS. For the security analysis, Auth-Info message is ignored due to its highly informative nature. The analysis began with the Auth-Req message. In this protocol, Auth-Req message is sent without the signature, which helps the adversary to easily impersonate or modify the message as discussed in the security analysis of PKMv1. This is simply referred to as a replay attack that leads toward the DoS or interleaving attack.

3.2 Proposed Security Mechanism

For the security mechanism of multihop WiMAX mesh networks, self-organized and efficient authentication and key management scheme (SEAKS) is proposed. SEAKS consists of two main functional modules that include authentication management and key management. The authentication management is incorporated with SEAKS–PKMv1 and v2 authentication mechanisms for single as well as for multihop and the re-authentication mechanisms. The key management consists of AK management and traffic encryption key (TEK) management. The authentication mechanism for single and multihop is termed as SEAKS protocol. Distributed authentication features of SEAKS protocol are illustrated in single as well as multihop authentication schemes. State machines for AK and TEK highlight the mechanism of localized key re-authentication and key management.

3.2.1 SEAKS Concept Design

SEAKS is based on a self-organized model using nontransparent, decode, and forward relays. SEAKS provides hybrid authentication schemes with distributed authentication and localized re-authentication and key maintenance. However, this technique not only helps in minimizing the overall authentication overhead on MR-BS and AAA server but also provides an efficient way to mitigate the vulnerabilities. The functional components of SEAKS are shown in Figure 3.3. The detailed and exhaustive discussion of SEAKS concept design will be discussed in the following section.

3.2.1.1 Authentication Management

Authentication management allows both SEAKS–PKMv1 and SEAKS–PKMv2 authentication protocols to authenticate and to perform key exchanges between the SS/RS/N-RS and the MR-BS using client–server mode. SEAKS authentication management provides a self-organized and cost-efficient mechanism for multiple N-RS to authenticate itself in distributed and hop-by-hop security control and also

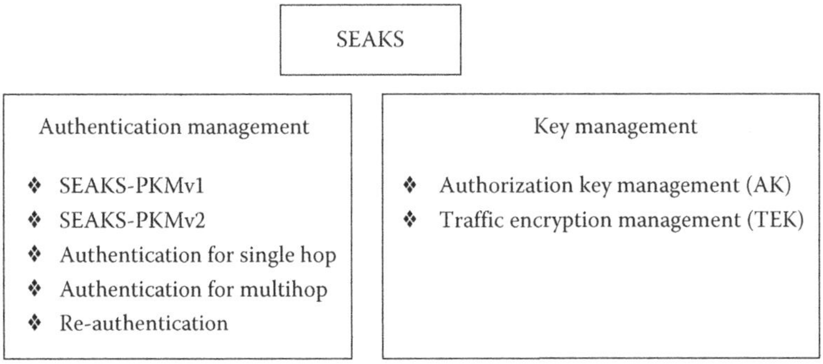

```
                        ┌─────────────────┐
                        │      SEAKS       │
                        └─────────────────┘

┌─────────────────────────────┐   ┌─────────────────────────────────┐
│  Authentication management  │   │         Key management          │
│                             │   │                                 │
│  ❖  SEAKS-PKMv1             │   │  ❖  Authorization key management (AK) │
│  ❖  SEAKS-PKMv2             │   │  ❖  Traffic encryption management (TEK) │
│  ❖  Authentication for single hop │ │                               │
│  ❖  Authentication for multihop │ │                                 │
│  ❖  Re-authentication       │   │                                 │
└─────────────────────────────┘   └─────────────────────────────────┘
```

Figure 3.3   Functional components of SEAKS.

allows re-authentication in localized security controls. This section will elaborate both versions of authentication protocol with their security analysis in detail. The security mechanisms and authentication protocols for OD-2009 and the SEN XU can be found in [1,2,4].

3.2.1.1.1 SEAKS Authentication Protocols and Their Security Analysis

In any security matter, two distinct functions must be carefully considered, that is, authentication and secrecy. Often, authentication is needed but not secrecy and vice versa. PKM protocols utilize three messages to get N-RS authenticated with MR-BS. The first two messages are Auth-Info and Auth-Req, while the third message is Auth-Reply as shown in Figure 3.4. Since the first message is highly informative and optional, this analysis will be on message 2. Message 2 is always sent in a plaintext as capabilities. SAID are already shared between MR-BS and N-RS during subscriber basic capabilities (SBC) and ranging process. Second, certificates must be sent in a plaintext as public key cannot be accessed by MR-BS. Message 2 is also highly vulnerable to all sorts of attacks. In this case, only authenticity of a message is required not the secrecy. The key goal is to transmit the message in such a way that *"attacker cannot alter or modify the message,"* thus, helping to avoid replay, DoS, and MitM attacks. Similarly, message 3 also exposes the SS to replay attacks even worse. To avoid the replay attacks in message 3, both authenticity and secrecy is required, which includes, *"message should not be modified and should come from the legitimate MR-BS."*

3.2.1.1.2 SEAKS–PKMv1 Authentication Protocols

The SEAKS–PKMv1 protocol is basically a unilateral authentication where SS/NRS does not authenticate the MR-BS. In this section, we initially elaborate the authentication steps of our proposed authentications protocol. Later, security

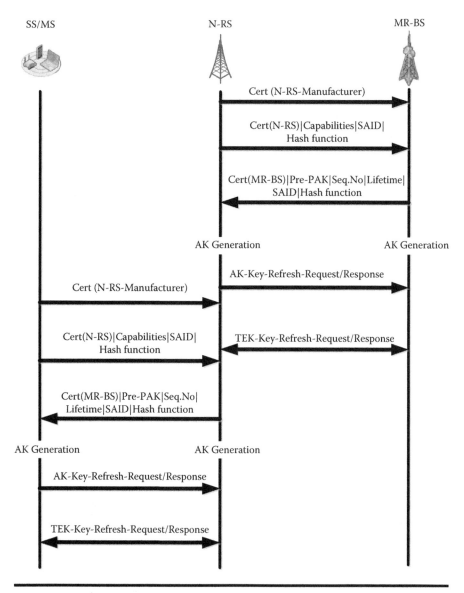

Figure 3.4 SEAKS–PKMv1 authentication protocols.

analysis will be carried out for each message. An N-RS begins authentication by sending an authentication information message that contains the N-RS manufacturer's X.509 certificates. As discussed earlier, this message is informative and is ignored by MR-BS.

Then the SS sends an authorization request message (Auth-Req) to its MR-BS that includes certificates of N-RS, capabilities, SAID, and hash function. In response to Auth-Req, the MR-BS validates the requesting N-RS's identity, determines the encryption algorithms and protocols to be shared with the N-RS, and then generates a Pre-PAK. Since, AK is the root of all other keys generations, the exchange of AK is not advisable. So Pre-PAK is generated by MR-BS and lets both parties generate their own AK. MR-BS then sends certificates of MR-BS, Pre-PAK, sequence number, life time, SAID, and hash function back to N-RS in an Auth-Rsp message. In this protocol, to resolve the issues of timestamp, signature, and the initial handling of public keys as discussed in [23] concept of hashing is used. The SEAKS–PKMv1 authentication protocol is well explained in Figure 3.4.

For security analysis of the protocol, N-RS sends its Auth-Req message in plaintext. Eavesdropping is not a problem since all the information is public and preferred to be sent in plaintext (P) to facilitate authentication as N-RS does not have MR-BS public key. To send the message 2 safely and unchanged, a hash function is preferred where N-RS will hash the plaintext P into 160 bits by using a secure hash algorithm (SHA) function and encrypt it with its own private key. The message containing both the plaintext P and the signed hashed function is sent to MR-BS that can be represented as follows:

$$P \mid D_N(H)$$

where P represents the plaintext, D_N represents the private key of N-RS, and H represents the hash function. N-RS sends the message within its allocated time period to MR-BS. MR-BS receives the message and computes the SHA hashed of the plaintext, it also decrypts the hashed value of plaintext by applying its public key to obtain the original hash value. MR-BS will check these hash values. If both agree with each other, the message is considered as valid and the MR-BS generates a key response message to SS. However, if an attacker intercepts this message and induces a replay attack to MR-BS by modifying this message, even if a single bit modification is carried out, this leads to the mismatch of both the hash values. In this case, MR-BS will silently discard the message.

Similarly, message 3 also exposes the SS to replay attacks, even worse. The SS also faces fraudulence from an adversary who intercepts its Auth-Req message. The adversary can make its own Auth-Reply message with the AK generated by him, thus gaining control of the communication of the victimized SS. This is a typical MitM attack, which demands the need of mutual authentication, that is, the SS needs to authenticate the BS as well.

To avoid the replay attacks in message 3, both authenticity and secrecy is required, which includes "*message should not be modified and should come from the legitimate MR-BS.*" To achieve these security goals, all the critical information will be encrypted by N-RS's public key and combined with the certificates of MR-BS to form one complete plaintext P, given by

$$D_N(\text{Critical information}) \mid \text{Cert MR} - \text{BS} = P$$

$$P \mid D_b(H)$$

After computing the SHA for the 'P' to obtain hash value, and encrypt it with the private key of MR-BS, both original 'P' and D_b (H) are sent to the N-RS. N-RS will receive the message, decrypt it using the MR-BS public key to obtain the original hash value of the 'P' computed by N-RS. N-RS will compute the hash value of 'P' using SHA and compare both hash values. When both values match, it shows authenticity and nonrepudiation of this message. To obtain the critical information that is encrypted by the public key of MR-BS, only N-RS has the corresponding private key to decrypt this information, which ensures the secrecy of the message. This SEAKS-PKMv1 protocol successfully provides Auth-Req and Auth-Res message with authenticity, nonrepudiation, and the secrecy.

3.2.1.1.3 SEAKS-PKMv2 Authentication Protocols

Owing to lack of mutual authentication in PKMv1, the IEEE 802.16 standard has proposed PKMv2 in which one additional message is added at the end of the original authentication protocol of PKMv1. However, PKMv2 belongs to the three-way authentication [1] with a confirmation message from the SS to the BS. Since the first message is optional and only informative, the security analysis begins from the next message. Message 2 is sent without the signature. Without the signature of the SS, the request message is easily modified or impersonated. This is similar to what was discussed in PKMv1 and again this is referred to as simply replay attack that can also result in DoS. Due to the lack of a signature in message 2, impersonation is not a problem, which leads to the interleaving attack. Interleaving attacks arise if an attacker can modify message 2, sent by the MR-BS to the legitimate N-RS by replacing the Cert MR-BS and SIG MR-BS with Cert (Attacker) and SIG (Attacker), respectively. Even with a signature from N-RS serving as message authentication, interleaving attack can still occur. SEAKS-PKMv2 authentication protocols helps to resolve the above-mentioned threats in an efficient manner. SEAKS-PKMv2 is basically forward and backward compatible and works with both IEEE 802.16e in distributed and nontransparent relay-based IEEE 802.16 networks. The protocol is explained in Figure 3.5.

In an authentication request message, instead of using signature or public key cryptography, SEAKS-PKMv2 protocol uses the hash function that not only helps in avoiding a replay attack, but also helps to counter interleaving attacks. Adding a hash function in message 3 also helps in avoiding impersonation. Modification of a message can be easily identified and the whole message will be silently discarded by the MR-BS. As far as an acknowledgment message for the MR-BS response message is concerned, only an AK encrypted by the public key of MR-BS with a random number is transmitted to MR-BS. This is to ensure the authenticity, nonrepudiation, and secrecy of this message.

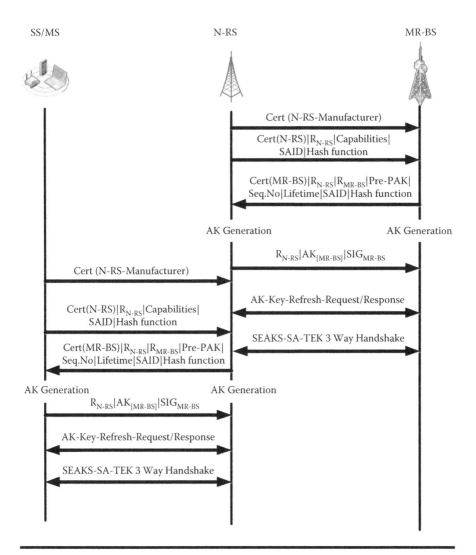

Figure 3.5   SEAKS–PKMv2 authentication protocol.

3.2.1.2 SEAKS Protocols and Their Security Analysis

The authentication mechanism residing at N-RS is responsible for getting AKs and a valid list of security association identifiers (SAIDS). N-RS is also responsible for authenticating itself with MR-BS and the neighboring N-RS. The state machine diagram for SEAKS authentication management is shown in Figure 3.6. The SEAKS authentication state machine also gives birth to not only AK and Re-Auth, but also TEK refreshment. The state machine for SEAKS consists of nine states and nine distinct events. The nine states includes Start, Auth-Wait, Authorized,

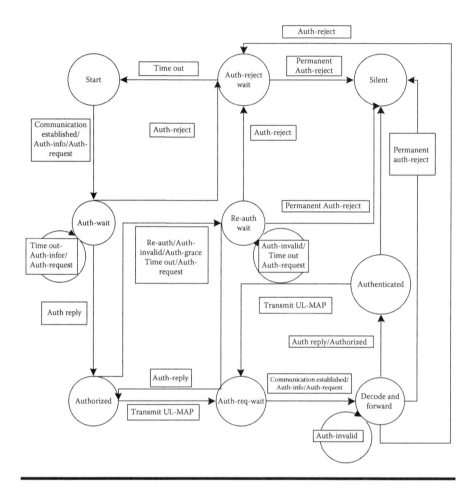

Figure 3.6 SEAKS authentication state machine.

Auth-Reject-Wait, Re-Auth-Wait, Re-Req-Wait, Silent, Decode and Forward, and Authenticated. The nine events include communication established, Time out, Transmit UL-MAP, Auth-Grace-Time out, Auth-Key-Authorized, Perm Auth-Reject, Auth-Reject, Re-Auth, and Auth-Invalid.

The state diagram illustrates the protocol messages transmitted and internal events generated for each of the models state transitions; however, the diagram does not indicate additional internal actions, such as clearing or the starting of timers that accompany the specific state transitions. SEAKS begins in the "Start" state; an initial state where no resources are allocated or used. A communication is established upon entering the start state, if the MAC has completed the basic capabilities negotiation. Once the communication is established, N-RS is now eligible to send Auth-info and Auth-Req messages to MR-BS to obtain an AK and the list of authorized SAIDs. The second state is Auth-Wait, where after sending authentication information and authentication request message to MR-BS, N-RS waits for the response.

If N-RS received an Auth-Reply message that contains the lists of valid SAIDs and AK, it moves to the authorized state. Otherwise, it will stay at the Auth-Wait state and wait for the Auth-replay. At Auth-wait state, if the time out occurs and Auth-replay is not received, it moves to Auth-reject phase. However, at Auth-reject wait, if time out occurs, authentication procedures will start from scratch and moves to the start state. If MR-BS sent a permanent Auth-reject at Auth-reject wait state, it moves to the silent state. Once N-RS is authorized, it starts transmitting UL-MAP and moves to Auth-Req-wait. If it receives Auth-info or Auth-request message, it moves to the decode and forward state. At this state, if the authentication request is invalid, it will remain at this state. Otherwise, it authenticates the requesting N-RS. At the decode and forward state, if Auth-rejection occurs, it moves to Auth-reject wait or if it receives permanent rejection from serving N-RS, it moves to a silent state. Once authenticated, the newly joined N-RS starts transmitting UL-MAP and waits for the authentication request message from any other N-RS.

3.2.1.2.1 Authentication Procedures for Single Hop

To understand the authentication procedures for single hop in MMR WiMAX networks, consider an N-RS$_1$, who wants to join the WiMAX networks. N-RS$_1$ sends its Auth-REQ message to the serving MR-BS. In response to an authorization request message, MR-BS validates the requesting N-RS's identity, determines the encryption algorithm and protocol support, activates an AK for N-RS$_1$, encrypts it with the N-RS$_1$'s public key, and sends it back to the N-RS$_1$ in an authentication response message. It also includes a 4-bit sequence number used to distinguish between successive generations of AKs, a lifetime, and the security identities for which N-RS$_1$ are authorized to obtain keying parameters. Once authenticated and the AK is obtained, N-RS$_1$ must periodically refresh its AK by reissuing an authentication request message to the MR-BS. During reauthorization cycle, to avoid service interruption, AKs have overlapping lifetimes. Both N-RS and MR-BS support up to two simultaneously active AKs during these transition periods. Authentication of N-RS$_1$ with MR-BS is shown in Figure 3.7. Once N-RS$_1$ achieves authorization, its starts a separate TEK for each SAID defined in the authentication response message.

3.2.1.2.2 Authentication Procedure for Multihop

Figure 3.8 illustrates the authentication procedure for multihop networks. Consider a second nontransparent relay station (N-RS$_2$) that wants to join the network. Due to its nontransparent nature, it is not in the coverage of MR-BS and only N-RS$_1$

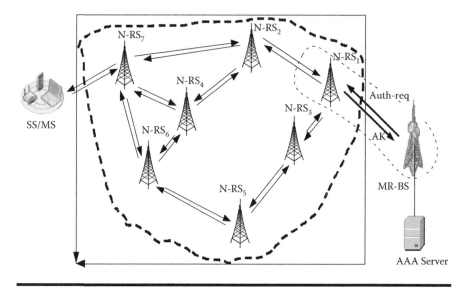

Figure 3.7   Authentication of N-RS$_1$ with MR-BS.

can listen to it. In this case, N-RS$_2$ listens to the UL-MAP from N-RS$_1$ and sends the Auth-Req message to N-RS$_1$.

However, any nontransparent node that wants to join the network must have to authenticate itself with MR-BS as MR-BS is directly attached to the AAA server. Meanwhile, N-RS$_1$ cannot authenticate N-RS$_2$ on behalf of MR-BS [1]. According to SEAKS, N-RS$_1$ received the authentication request message from N-RS$_2$ and sends it to MR-BS during the refreshing of the AK message. N-RS$_1$ receives MAC PDU of N-RS$_2$ and encapsulates it into its own PKM-REQ message of type 9 and code 4 [1,3]. According to Figure 3.8, MR-BS receives MAC PDU of N-RS$_1$, which is basically sent for refreshing AK. MR-BS will check the MAC header of N-RS$_1$. If relay auth request (RAR) is equal to 1, it means that there is one relay request inside MAC PDU. RAR is basically the reserve bit utilized for RAR indications. Once MR-BS obtains Auth-Req of N-RS$_2$, it validates its authenticity and activates AK$_2$ and other parameters, encrypts it with N-RS$_1$ public key and responds to N-RS$_1$ in its Auth-Rsp message. N-RS$_1$ receives N-RS$_2$'s security information, saves one copy of all the information into its table, generates AK$_{21}$, encrypts it with N-RS$_2$ public key, and sends its Auth-Rsp message to N-RS$_2$. Once N-RS$_2$ is authenticated, it will initiate a separate authorization and TEK with N-RS$_1$.

All N-RSs maintain a knowledge-shared table of recently exchanged AKs with its neighbors. If N-RS$_2$ fails to re-authenticate before the expiration of its current AK, N-RS$_1$ will wait until it sends an authentication request message. If N-RS$_2$ sends the authentication request message again, rather than sending this request to MR-BS, N-RS$_1$ will check its own table. If N-RS$_2$'s certificate is found within its table, it will validate N-RS$_2$ authenticity locally. Thus, enhancing the

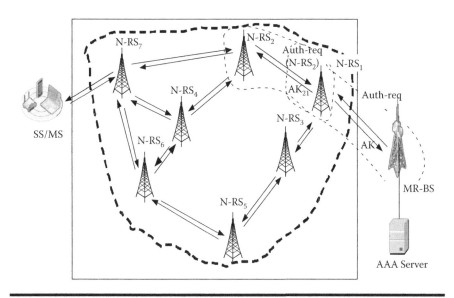

Figure 3.8   Authentication of N-RS$_2$ with N-RS/MR-BS.

communication cost efficiency in terms of authentication overhead, which lessens the overall complexity of the protocol. Figure 3.9 shows the authentication mechanism of more than two N-RS with MR-BS. In this case, if N-RS$_3$ wants to join the network, it will send the authentication request message to N-RS$_2$, as it is working in nontransparent mode. While sending the message, N-RS$_3$ will

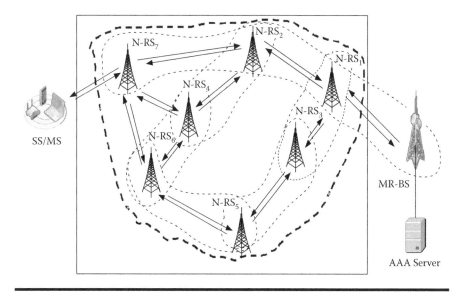

Figure 3.9   Authentication of N-RS$_n$ with N-RS$_1$/MR-BS.

set RAR = 1 inside the MAC header so that N-RS$_2$ can recognize that there is one authentication request message inside the MAC payload, and set the TYPE value = 8 and code = 4, which means it is a PKM–Auth-Req message. Once N-RS$_2$ receives this message, it will check RAR values. If the value is one, it will save the message to its table, and forward it to N-RS$_1$. Before sending, it will again set the RAR = 1.

Hence, there are two MAC messages present inside the MAC payload of N-RS$_2$, one is Auth-Req (code 4) and the other is Key-Req (code 5). N-RS$_1$ will receive this message and check RAR value; if it is one then it will copy the Auth-Req message to its table, otherwise it will ignore and forward it to MR-BS, which will receive the message and validate it. MR-BS will send back the authentication response message with type 9. Again here, there are two MAC messages inside the MAC payload, one is with Key-Reply (code 8) and the other is Auth-Reply (code 5) to N-RS$_1$. N-RS$_1$ checks the code values, if it is 5, it will send to N-RS$_2$. If 8, then it will use for its refreshing of keys. N-RS$_2$ again receives two MAC messages inside the payload, one is with code 5 and the other is with code 8. It will retain code 8 with itself and send the code 5 message to N-RS$_3$. Thus, N-RS$_3$ is authenticated with MR-BS in a distributed manner and maintains its keys locally as mentioned previously. Likewise, if any other N-RS such as N-RS$_4$ and N-RS$_5$ want to join the network, they will follow the same procedures. After a specific interval of time, all the N-RSs shared their knowledge tables, thus creating a self-organized environment. This self-organized environment is responsible for distributed authentication and localized re-authentication and key management.

3.3 Mathematical Analysis

In this mathematical analysis, comparison of the communication cost for both OD-2009 and SEAKS authentication protocol is derived. The communication cost of a single hop in MMR WiMAX network involves four types of communication processes within N-RS and MR-BS. The processes include distributed privacy and key management authentication messages, localized refreshing of the AK, refreshing of TEK, and re-authentication.

The total cost associated with the overall SEAKS protocols from source to destination, that is, from any N-RS to MR-BS is the number of messages transmitted to perform complete authentication procedures including multihop, uplink, and downlink communications. In particular, total communication cost per authentication can be defined by [34]

$$C = h \sum_{A=1}^{K} S_k \qquad (3.1)$$

where

C = Total communication cost

h = Hop count

K = Total number of messages

S_k = Size of bytes, where k is the integer from $1 - k$

In Equations 3.2 and 3.3, the total communication cost for single hop OD-2009 is calculated. PKM represents the authentication messages that include authentication request and response to and from N-RS and MR-BS. However, in this authentication message, the authentication information message is not included. Thus, for authentication, only two messages are transmitted, one for request and the other for the response. The AK represents the refreshing of AK, which is normally a key request and its corresponding response that also includes two messages. TEK represents the refreshing of TEK for each SAID and again includes two messages. RA stands for re-authentication and P stands for the probability of how many times this re-authentication occurs.

3.3.1 Cost Analysis of OD-2009 Authentication Protocol

The total communication cost for single hop OD-2009 can be written as

$$C = h\sum_{A=1}^{K=2} S_{PKM} + h\sum_{A=1}^{K=2} S_{AK} + h\sum_{A=1}^{K=2} S_{TEK} + \{(h\sum_{A=1}^{K=2} S_{RA})*P\} \tag{3.2}$$

$$C = h\sum_{A=1}^{K=2} (S_{PKM} + S_{AK} + S_{TEK} + \{(RA)*P\}) \tag{3.3}$$

The total cost for two hops communication is given by Equations 3.4 and 3.5.

$$C = 2h\sum_{A=1}^{K=4} S_{PKM} + 2h\sum_{A=1}^{K=4} S_{AK} + 2h\sum_{A=1}^{K=4} S_{TEK} + \{(2h\sum_{A=1}^{K=4} S_{RA})*P\} \tag{3.4}$$

$$C = 2h\sum_{A=1}^{K=4} (S_{PKM} + S_{AK} + S_{TEK} + \{(RA)*P\}) \tag{3.5}$$

The total cost for three hops communication is given by Equations 3.6 and 3.7.

$$C = 3h\sum_{A=1}^{K=6} S_{PKM} + 3h\sum_{A=1}^{K=6} S_{AK} + 3h\sum_{A=1}^{K=6} S_{TEK} + \{(3h\sum_{A=1}^{K=6} S_{RA})*P\} \tag{3.6}$$

$$C = 3h \sum_{A=1}^{K=6} (S_{PKM} + S_{AK} + S_{TEK} + \{(RA)*P\}) \qquad (3.7)$$

For *n* number of hops, Equation 3.8 is rewritten as

$$C = nh \sum_{A=1}^{K=n} S_{PKM} + nh \sum_{A=1}^{K=n} S_{AK} + nh \sum_{A=1}^{K=n} S_{TEK} + \{(nh \sum_{A=1}^{K=n} S_{RA})*P\} \qquad (3.8)$$

From the above Equations 3.2 through 3.8 for a different number of hops, it can be seen that as the number of hops increased, the number of messages transmitted also increased. Thus, it can be said that the total number of messages transmitted is equal to the total number of hops and total number of messages transmitted per single hop.

Total Number of messages transmitted = Number of Hops
$$\times \textit{Messages per single Hop}$$

3.3.2 Cost Analysis of SEAKS Authentication Protocol

As discussed in Section 3.3 SEAKS security protocols support distributed and hop-by-hop authentication, localized key management, and re-authentication. The communication costs for the SEAKS authentication protocol can be derived from the following Equations 3.9 and 3.10.

$$C = h \sum_{A=1}^{K=2} S_{PKM} + h \sum_{A=1}^{K=2} S_{AK} + h \sum_{A=1}^{K=2} S_{TEK} + \{(h \sum_{A=1}^{K=2} S_{RA})*P\} \qquad (3.9)$$

$$C = h \sum_{A=1}^{K=2} (S_{PKM} + S_{AK} + S_{TEK} + \{(RA)*P\}) \qquad (3.10)$$

The value of K is 2, which means the total messages transferred is two in a single hop. However, if the same equations are used for two hops, it can be seen from Equations 3.11 and 3.12 that the number of hops increase, but the total number of messages transferred is the same which is 2. For three hops, the numbers of hops increase to three, but the total number of messages transferred is still the same which is two, as shown in Equations 3.13 and 3.14. Thus, for *n* number hops, Equation 3.15 is obtained.

For two hops,

$$C = h \sum_{A=1}^{K=2} S_{PKM} + h \sum_{A=1}^{K=2} S_{AK} + h \sum_{A=1}^{K=2} S_{TEK} + \{(h \sum_{A=1}^{K=2} S_{RA})*P\} \qquad (3.11)$$

$$C = 2h\sum_{A=1}^{K=2}(S_{PKM} + S_{AK} + S_{TEK} + \{(RA)^*P\}) \tag{3.12}$$

For three hops,

$$C = 3h\sum_{A=1}^{K=2}S_{PKM} + 3h\sum_{A=1}^{K=2}S_{AK} + 3h\sum_{A=1}^{K=2}S_{TEK} + \{(3h\sum_{A=1}^{K=2}S_{RA})^*P\} \tag{3.13}$$

$$C = 3h\sum_{A=1}^{K=2}(S_{PKM} + S_{AK} + S_{TEK} + \{(RA)^*P\}) \tag{3.14}$$

For *n* number of hops,

$$C = h\sum_{A=1}^{K=2}S_{PKM} + h\sum_{A=1}^{K=2}S_{AK} + h\sum_{A=1}^{K=2}S_{TEK} + \{(h\sum_{A=1}^{K=2}S_{RA})^*P\} \tag{3.15}$$

Equations 3.9 through 3.15 show that as the number of hops increases, the number of messages remains constant. Thus, it can be said that the total number of messages transmitted is equal to the total number of messages transmitted per single hop.

Total Number of messages transmitted = Messages per single Hop

Figure 3.10 shows that for OD-2009, when the number of hops increases, the messages transmitted also increase tremendously. However, when compared with the proposed scheme, it can be seen that there is no change when the number of hops increase.

3.4 Future Research Direction

MMR WiMAX security protocols provide a platform for further enhancement toward future applications such as IEEE 802.16j-m, LTE-advanced, higher reliability networks, and machine-to-machine applications. However, future works can be carried out to enhance the performance of the proposed SEAKS protocols. The suggestions for future works are

■ To design a mechanism of pre-authentication for secure handover, N-RS may seek to use pre-authentication to facilitate an accelerated reentry at a particular target MR-BS that results in the establishment of an authorization key in N-RS and target MR-BS.

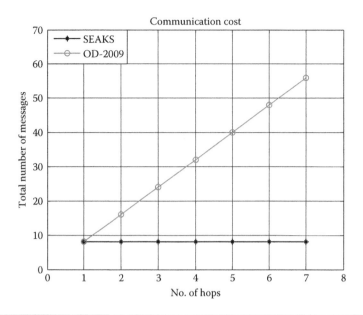

Figure 3.10   Comparison of communication cost in authentication protocols.

■ To develop a cross-layer design with the network and MAC layer to investigate the performance of IPSec in SEAKS protocols to enhance the security at routing level.
■ To investigate the performance of proposed protocols in host identity protocol (HIP) security architecture in machine-to-machine application where host identity also supports public key cryptography.
■ To develop a cross-layer design with the application layer and MAC layer to support RADIUS client/server protocol and provides centralized AAA management.

3.5 Conclusion

This paper addressed SEAKS for hop-by-hop authentication and key management schemes in nontransparent relay-based WiMAX networks. This scheme is suitable for both fixed as well as mobile nontransparent relays. SEAKS provides hybrid authentication schemes with distributed authentication and localized re-authentication and key maintenance. However, this technique not only helps in minimizing the overall authentication overhead on MR-BS and AAA servers, but also provides efficient ways to countermeasure the vulnerabilities. SEAKS provides authentication and key management. Two modified PKM authentication protocols have been developed; one for unilateral authentication which is SEAKS-PKMv1 and the other for mutual authentication which is SEAKS-PKMv2. These two authentication protocols are responsible for MR-BS and N-RSs to successfully authenticate each other and securely transfer the AK in a distributed manner. Once an AK is shared, the AK will be refreshed periodically. However, SEAKS protocols provide localized key and re-authentication management. Once authentication is completed successfully and the N-RS are registered to the network, N-RS starts a separate TEK for each (SAIDs) which is identified in the authorization reply messages from MR-BS. TEK management is responsible for the maintaining and refreshing of keys mechanism within N-RS and MR-BS. N-RS usually sends the key refresh request to MR-BS periodically and to avoid service interruption and unwanted re-authentication, MR-BS maintains two sets of keying materials per SAID. SEAKS protocol enhances the previous works by [1,32,33] in order to achieve high delivery ratio, minimum packet overhead and processing time, and a high delivery ratio when the number of rogue RSs increases. In general, the finding concludes that SEAKS authentication protocol provides high packet delivery ratio and spends less numbers of packet overhead as compared to SEN XU and OD-2009. SEAKS protocol improves processing time and show high packet delivery ratio when rogue RS increases. This chapter also concludes its efficiency with reference to other attacks such as replay attacks, DoS, MitM, and interleaving attacks in general.

References

1. IEEE Standard for Local and Metropolitan Area Networks Part 16: Air Interface for Broadband Wireless Access Systems Amendment 1: Multihop Relay Specification, *IEEE Std 802.16j-2009 (Amendment to IEEE Std 802.16-2009)*, pp. 1–290, 2009.
2. J. Chee and T. Ming, Improving security in the IEEE 802.16 standards, *Eighth International Conference Information on Technology: New Generations (ITNG 2011)*, pp. 408–412, 11–13 April 2011, Las Vegas, NV.
3. H. Chin-Tser and J. M. Chang, Responding to security issues in WiMAX networks, *IT Professional*, 10, 15–21, 2008.
4. A. S. Khan, N. Fisal, N. N. M. I. Ma'arof, F. E. I. Khalifa, and M. Abbas, Security issues and modified version of PKM protocol in non-transparent multihop relay in IEEE 802.16j networks, *International Review on Computers and Software*, 6, 104–109, 2011.
5. R. M. Hashmi, A. M. Siddiqui, M. Jabeen, and K. S. Alimgeer, Towards secure wireless MAN: Revisiting and evaluating authentication in WiMAX, in *Computer Networks and Information Technology (ICCNIT)*, 2011, pp. 165–173.
6. A. Altaf, R. Sirhindi, and A. Ahmed, A novel approach against DoS attacks in WiMAX authentication using visual cryptography, in *Emerging Security Information, Systems and Technologies, 2008. SECURWARE'08*. 2008, pp. 238–242.
7. C. Kolias, G. Kambourakis, and S. Gritzalis, Attacks and countermeasures on 802.16: Analysis and assessment, *Communications Surveys & Tutorials, IEEE*, 15, 487–514, 2012.
8. N. Xiong, F. Yang, H. Y. Li, J. H. Park, Y. Dai, and Y. Pan, Security analysis and improvements of IEEE standard 802.16 in next generation wireless metropolitan access network, *Wireless Communications and Mobile Computing*, 11, 163–175, 2011.

9. N. Chauhan and R. K. Yadav, Security analysis of identity based cryptography and certificate based in Wimax network using Omnet ++ simulator, in *Advanced Computing & Communication Technologies (ACCT)*, 2012, pp. 509–512.

10. M. Deva Priya, J. Sengathir, and M. L. Valarmathi, Root node attack in a WiMAX 802.16e network, in *Proceedings of the 2nd International Conference on Trendz in Information Sciences and Computing, TISC-2010*, pp. 186–191.

11. V. K. Gondi and N. Agoulmine, Security and mobility architecture for isolated wireless networks using WIMAX as an infrastructure, in *Integrated Network Management, 2009. IM'09. IFIP/IEEE*, 2009, pp. 279–282.

12. T. Iwata, M. Nishigaki, M. Shojaei, N. Movahhedinia, and B. Tork Ladani, An entropy based approach for DDoS attack detection in IEEE 802.16 based networks, in *Advances in Information and Computer Security* 7038: Springer, Berlin, 129–143, 2011.

13. B. N. Komu, M. Mzyece, and K. Djouani, SPIN-based verification of authentication protocols in WiMAX networks, Published in *IEEE Vehicular Technology Conference* (VTC Fall), 2012.

14. Y. Lee, G. Lee, H. Kim, and C. Jeong, Performance analysis of authentication and key distribution scheme for mobile multi-hop relay in IEEE 802.16j, *Personal and Ubiquitous Computing*, 16, 697–706, 2012.

15. P. Rengaraju, C. H. Lung, and A. Srinivasan, Design of distributed security architecture for multihop WiMAX networks, *2010 8th International Conference on Privacy, Security and Trust (PST 2010)*, pp. 54–61, 17–19 Aug. 2010, Ottawa, ON.

16. H. Jie and H. Chin-Tser, Secure mutual authentication protocols for mobile multi-hop relay WiMAX networks against rogue base/relay stations, in *Communications (ICC)*, 2011, pp. 1–5.

17. D. Satish Kumar and N. Nagarajan, Relay technologies and technical issues in IEEE 802.16j Mobile Multi-hop Relay (MMR) networks, *Journal of Network and Computer Applications*, 36, 91–102, 2012.

18. D. Xinmin and X. Xiaoyao, Analysis and research of security mechanism in IEEE 802.16j, in *Anti-Counterfeiting Security and Identification in Communication (ASID)*, 2012, pp. 33–36.

19. P. J. F. Ruiz and A. F. G. Skarmeta, Providing security using IKEv2 in a vehicular network based on WiMAX technology, in *Consumer Communications and Networking Conference (CCNC), 2011 IEEE*, pp. 282–286.

20. V. Genc, S. Murphy, Y. Yang, and J. Murphy, IEEE 802.16 J relay-based wireless access networks: An overview, *Wireless Communications, IEEE*, 15, 56–63, 2008.

21. R. Yusoff, M. D. Baba, R. Abd Rahman, M. Ibrahim, and N. Mat Isa, Performance analysis of transparent and non-transparent relays in MMR WiMAX networks, in *Industrial Electronics and Applications (ISIEA)*, 2011, pp. 237–240.

22. A. S. Khan, Efficient distributed authentication key scheme for multi-hop relay in IEEE 802.16j networks, *International Journal of Engineering Science and Technology (IJEST)*, 2, 2192–2199, 2010.

23. N. F. A. S. Khan, Sharifah Kamilah, Sharifah Hafizah, Mazlina Esa, and M. Abbas., An efficient self-organized authentication and key management scheme for distributed multihop relay-based IEEE 802.16 networks, *International Journal of Computer Science and Information Security (IJCSIS)*, 9, 30–38, 2011.

24. L. Tie and Y. Yi, Extended security analysis of multi-hop ticket based handover authentication protocol in the 802.16j network, in *Wireless Communications, Networking and Mobile Computing (WiCOM)*, 2012, pp. 1–10.

25. A. K. M. N. Sakib and M. M. S. Kowsar, Shared key vulnerability in IEEE 802.16e: Analysis & solution, *13th International Conference on Computer and Information Technology (ICCIT 2010)*, pp. 600–605, 23–25 Dec. 2010, Dhaka, Bangladesh.
26. D. Xinmin and X. Xiaoyao, Analysis and research of security mechanism in IEEE 802.16j, in *Anti-Counterfeiting Security and Identification in Communication (ASID)*, 2010, pp. 33–36.
27. S. Taha, S. Cespedes, and S. Xuemin, EM³A: Efficient mutual multi-hop mobile authentication scheme for PMIP networks, in *Communications (ICC)*, 2012, pp. 873–877.
28. S. Sidharth and M. P. Sebastian, A Revised Secure Authentication Protocol for IEEE 802.16 (e), in *Advances in Computer Engineering (ACE)*, 2010, pp. 34–38.
29. D. C. Wyld, J. Zizka, D. Nagamalai, N. Kahya, N. Ghoualmi, and P. Lafourcade, Key management protocol in WIMAX revisited, in *Advances in Computer Science, Engineering & Applications*. 167: Springer, Berlin, 2012, pp. 853–862.
30. M. Shojaee, N. Movahhedinia, and B. T. Ladani, Traffic analysis for WiMAX network under DDoS attack, in *Circuits, Communications and System (PACCS)*, 2010, pp. 279–283.
31. S. Sidharth and M. P. Sebastian, A revised secure authentication protocol for IEEE 802.16 (e), in *Advances in Computer Engineering (ACE)*, 2010, pp. 34–38.
32. X. Sen, H. Chin-Tser, and M. M. Matthews, Modeling and analysis of IEEE 802.16 PKM Protocols using CasperFDR, in *Wireless Communication Systems. 2008. ISWCS '08*, 2008, pp. 653–657.
33. X. Sen and H. Chin-Tser, Attacks on PKM protocols of IEEE 802.16 and its later versions, *3rd International Symposium on Wireless Communication Systems (ISWCS 2006)*, pp. 185–189. 6–8 Sept. 2006, Valencia.
34. S. Capkun, L. Buttyan, and J. P. Hubaux, Self-organized public-key management for mobile ad hoc networks, *Mobile Computing, IEEE Transactions on*, 2, 52–64, 2003.

Chapter 4

Fighting against Black Hole Attacks in Mobile Ad Hoc Networks

Manuel D. Serrat-Olmos, Enrique Hernández-Orallo, Juan-Carlos Cano, Carlos T. Calafate, and Pietro Manzoni

Contents

4.1 Introduction

Mobile ad hoc network, usually known as MANET, consists of a set of wireless mobile nodes that function as a multihop mesh network in the absence of any kind of networking infrastructure and centralized administration. The most usual MANET deployment scenario is the community-built MANET, where there is no single owner of the nodes. In this case, each node belongs to a different user, which could pose an obvious security risk for the whole network. As a result of their deployment, MANETs rely on cooperation schemes between their nodes for a correct operation, that is, every network node involved in a data communications flow generates and sends its own packets, also forwarding packets on behalf of other nodes. Although there are not many MANET implementations in real testbed scenarios, these networks have been a popular research topic in the last decade, and some technologies developed for them are applicable to other types of networks, such as in vehicular ad hoc networks (VANET), which are a type of MANET formed by cars basically to implement intelligent transport system platforms [1].

In a typical MANET, a node is a wireless mobile computing device, like a smartphone or tablet, with the appropriate software installed, which allows the node to join the network, identify neighboring nodes, configure the routing protocol, and participate in forwarding activities. Due to their mobile nature, these nodes are battery operated, so energy saving is a key design parameter for this equipment, which could encourage the implementation of techniques aimed at saving individual node resources against network performance maximization.

MANET nodes could be classified [2] as:

■ *Well-behaved nodes*, if they cooperate with the MANET forwarding activities to achieve the community goals, or
■ *Misbehaved nodes*, if they act against those global goals. In this case, nodes are further classified into three classes:
  – Faulty nodes, if they do not cooperate due to a hardware or software malfunction.
  – Selfish nodes, if they drop all the packets whose destination node are not themselves, but they use other nodes to send their own packets, motivated by saving their own resources.
  – Malicious nodes, when they try to disturb the normal network behavior for their own profit, maybe using multiple potentially disturbing techniques.

In a MANET, there are basically two kinds of packet flows: data packets and route maintenance packets. However, not all misbehaved nodes have the same impact on network performance, due to the type of packet flows they affect. A really malicious node could damage the network by spoofing routes, flooding the wireless channel, or carrying out a man-in-the-middle attack. These are classical attacks that every network could suffer, and solutions have been already devised for them. It is clear that some of these classical attacks can be easily carried out in MANETs because of the nature of the wireless communications channel. However, we are interested in those potential attacks which are specific to MANETs or whose effects are significantly worse in this kind of network. Even if we can achieve a good protection level against certain types of MANET attacks, this kind of network is prone to suffer other attacks, for example, eavesdropping [3], denial of service [4], or Sybil [5] attacks, whose prevention or remediation are outside the scope of this text.

When a MANET is deployed, we have to assume that there could be a percentage of misbehaved nodes. Their number, type, position, and movement pattern are key issues which deeply impact the MANET performance [6], but they are a priori unknown. So, network performance could be dramatically reduced if nothing is done to cope with these threats, due to the decreasing packet delivery ratios triggered by the misbehaved nodes dropping packets. To this end, an effective protection against these types of MANET nodes will be mandatory to preserve the correct functionality of the network [7].

All types of misbehaved nodes—faulty, selfish, and malicious—have a common behavior; they do not participate in forwarding activities, a behavior which could be classified as a kind of denial of service attack [8]. We comprise all these misbehavior types using the term *black hole*. We define a black hole as a node that disrupts, intentionally or not, the communication within its neighborhood, dropping the packets received without forwarding them to their final destination. We also include in this definition the concept of *grey hole*, which is a node that selectively forward only some of the packets, but not all of them.

However, how can we decide whether a node is misbehaved or, more specifically, a black hole? There are several proposed approaches as we will see later in this chapter, based on monitoring the traffic heard by every node to detect misbehaved nodes, and then taking the appropriate actions to avoid the negative effects of that misbehavior [9]. So, what we want to do is to detect the existing black hole(s) in the network, and to avoid their participation in the communications between well-behaved nodes. The main problem that arises at this point is how to detect these black holes, while avoiding, as much as possible, wrong diagnostics, like false positives or false negatives. A *false positive* appears when the selected technique identifies a well-behaved node as a misbehaved one. A *false negative* appears when the technique cannot detect a misbehaved node, so the network believes that it is a normal node, with its potentially disruptive effects. Then, if the detection technique has a low accuracy level, this could lead to problems like network partition and low packet delivery ratio, as a result of the production of false positive and false negatives. The sooner and more accurate the black hole detection process is done, the less disturbing will be their effects. Therefore, we can state that accuracy and detection speed are critical issues when designing an approach for black hole detection in MANETs.

The rest of this chapter is organized as follows: Section 4.2 introduces the concept of watchdog and overviews proposed solutions in the literature for the black hole problem; Section 4.3 shows how Bayesian filtering will increase watchdog accuracy and speed. In Section 4.4, we will explain how the collaborative watchdog works. The performance of this watchdog is evaluated through simulation in Section 4.5. This simulation only evaluates the local performance, so we introduce an analytical model to evaluate the global effect of collaboration in Section 4.6. Finally, we present our future research directions and the main conclusions of this chapter.

4.2 Basic Building Block: The Watchdog

To perform the detection of black hole-related threats, an *intrusion detection system* (IDS) is commonly used. It is a software piece that collects and analyzes the network traffic generated by its neighboring nodes in order to detect attacks. In this context, IDSs aim at monitoring the activity of the nodes in the network in order to detect misbehavior [10] based on several criteria. To do so, it is a requisite that the nodes' wireless interface is able to work in a promiscuous mode, capturing all the packets that are sent within the reception range of the nodes' antenna. One of the most common components of IDS dedicated to that task is the *watchdog*. For example, a simple watchdog could overhear the packets transmitted and received by its neighbors, counting the packets that should be retransmitted, and it could compute a *trust level* for every neighbor as the ratio of "packets retransmitted" to "packets that should have been retransmitted." If a node retransmits all the packets that it should have retransmitted, it will have a trust level of 1. If a node has a trust level lower than the configured tolerance threshold, that node will be marked as malicious.

These concepts are illustrated in Figure 4.1. In this example, node D is running a watchdog to detect misbehaved nodes. Node A has a packet to be sent to node C, but since node C is outside the neighborhood of node A, the message has to be sent through a multihop route which includes node B. In this situation, node D will overhear node A sending the packet to node B, and node B sending the acknowledgment of that packet. So, node D knows that node B has to forward this packet because its destination is node C. If node B forward the packet and *node D overhears this forwarding activity*, it will maintain its trust level. Otherwise, its trust level will be reduced. If the trust level of node B falls below a tolerance threshold, node D will identify it as a black hole.

At this point, we must note that there are two aspects that affect the overhearing capacity of the node running the watchdog, thus influencing the detection process [10]. The first of them is *mobility*: nodes in a MANET move at variable speeds along unknown paths, so maybe node B from Figure 4.1 will be outside the reception range of node D when it forwards the packet from node A to node

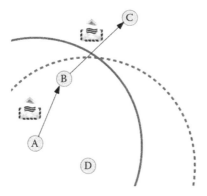

Figure 4.1 **Multihop transmission of a packet from nodes A to C.**

C. But node D has no way to know whether node B has actually sent the packet, and it reduces the trust level of node B. The second issue that may affect the detection process is the wireless physical channel, which is prone to *transmission errors* and interferences. Again, if an interference blocks the reception of these signals at node D when node B forward the packet, the trust level of node B will be reduced, although its behavior is correct. So, if the trust level of node B falls below a tolerance threshold, node D will wrongly identify node B as a black hole, generating a false positive detection. Now, let us assume that node B is really a black hole node. If nodes B and D are neighbors, there is not enough traffic to let node D characterize node B as black hole, causing a false negative detection to arise.

However, problems do not end when the black hole detection process is finished. Another issue to be addressed appears when a node acting as a black hole is detected: what should other nodes do to mitigate its effects? Basically, there are two approaches in the literature: isolation and incentivation.

Isolation methods are intended to keep the misbehaved nodes outside the network, excluding them from any ongoing communication and route maintenance processes. So, isolation is a response to the detection of a black hole that protects the working network, although it could lead to network partitioning if the detector component has a bad accuracy level or a very stringent trust requirement. On the other hand, isolation is the only suitable method for all classes of black holes.

Incentivation methods try to convince the misbehaving nodes to change their behavior, in order to become collaborative instead of selfish. Thus, these methods generally try to improve the MANET communication capabilities by increasing the number of collaborative nodes and, consequently, the global collaboration level. In this set of techniques, there is no distinction between the detection module and the response module, because they are combined in the basic functionality of the method. On the other hand, it is easy to deduce that incentivation is only useful for selfish nodes, and fails to prevent or mitigate the effects of other types of noncollaborative nodes.

Several solutions have been proposed for detecting and isolating or incentivating misbehaved nodes in MANETs. Marti et al. [11] proposed a Watchdog and a Pathrater over DSR protocol to detect nonforwarding nodes, maintaining a rating for every node and selecting routes with the highest average node rating. The response module of this technique only relieves misbehaved nodes from forwarding packets, but they continue getting their traffic forwarded across the network. Buchegger and Le Boudec [12] proposed the CONFIDANT protocol over DSR, which combines a watchdog, a reputation system, Bayesian filters, and information obtained from a node and its neighbors to accurately detect misbehaved nodes. The system's response is to isolate those nodes from the network, punishing them indefinitely. Also, there are some approaches designed for certain types of MANETs, like VANETs, whose aim is similar to one reported here [13].

Other approaches do not use reputation systems, in favor of incentivation. Buttyan and Hubaux [14,15] presented a method using a virtual currency called *nuglet*. Every node has a credit counter which will be increased when the node forward packets, and decreased when a node sends his own packets. When a node has no nuglets, it cannot send its packets anymore, and so it is a motivation for nodes to forward packets for the network benefit. Zhong et al. [16] proposed SPRITE, a credit-based system to incentivate participation of selfish nodes in MANET communication. It is based on a central clearance system, which charges or gives credit to nodes when they send or forward a message. So, if a node wants to send a message, it must have sufficient credit to do it. That credit is earned by forwarding messages for other nodes. The response module of this method is integrated into the incentivation method, so that if a node does not forward other nodes' messages, it will not have credit to send its own messages. However, incentivation methods proposed for MANETs present some basic weaknesses:

- They need some kind of infrastructure to maintain the accounting, so the MANET will lose its "infrastructureless" characteristic and its functionality will depend on additional elements, which could be affected by other types of vulnerabilities.
- They usually rely on some kind of tamper-proof hardware to store digital certificates or virtual currency amounts, which could be an unaffordable requirement.
- These techniques do not correctly mitigate the effects of other types of misbehaved nodes other than selfish ones, and it is a risky assumption to believe that this will be the only type of attacker in a MANET environment.

So, once incentivation methods have been discarded as a general solution for the black hole attacks, let us concentrate on the isolation techniques as a more general response to this problem. Once the response module of our IDS is clearly defined, we will deepen our study on the detection element of these methods, the watchdog, whose basic implementation has not yet been discussed. But is the reliability of this basic implementation suitable for MANETs, especially when nodes move at high speeds? Studies [10] available in the literature have shown that these kinds of watchdogs are characterized by a significant amount of false positives, basically due to mobility and signal noise over the wireless channel, and that they must be improved to become suitable for a wide range of MANET scenarios. Therefore, we can conclude that the basic watchdog technique is feasible, but unsuitable, for this kind of network in its actual form.

4.3 Bayesian Watchdog Approach

As we stated earlier, to detect misbehaved nodes, network monitoring is needed. Every node must be aware of its neighbors' behavior, and watchdogs are a popular component for IDSs dedicated to this task. However, they are not suitable if they simply compute the ratio of "packets retransmitted" to "packets that should have been retransmitted." Hortelano et al. [17] have evaluated a Bayesian watchdog over ad hoc on-demand distance vector (AODV) routing in MANETs. This Bayesian watchdog results from the aggregation of a Bayesian filter with a standard watchdog implementation.

The role of the Bayesian filter in the watchdog is to probabilistically estimate a system's state based on noisy observations [17]. The mathematical foundation of the Bayesian filter is the following: at time t, the state is estimated by a random variable θ, which is unknown, and this uncertainty is modeled by assuming that θ itself is drawn according to a distribution that is updated as new observations become available. It is commonly called *belief* or $\text{Bel}_t(\theta)$. To illustrate this, let us assume that there is a sequence of time-indexed observations $z_1, z_2,\ldots, z_n,\ldots, z_t$. The $\text{Bel}_t(\theta)$ is then defined by the posterior density over the random variable θ conditioned on all sensor data available at time t:

$$\text{Bel}_t(\theta) = p(\theta \mid z_1, z_2,\ldots,z_n,\ldots z_t) = \text{Beta}(\alpha_t,\beta_t,\theta) \qquad (4.1)$$

In this approach, the random variable θ belongs to the interval [0,1]. Bayesian filtering relies on the beta distribution, which is suitable to estimate the belief in this interval, as shown in expression 1; α and β represent the state of the system, and they are updated according to the Equations 4.2 and 4.3:

$$\alpha_{t+1} = \alpha_t + z_t \qquad (4.2)$$

$$\beta_{t+1} = \beta_t + z_t \qquad (4.3)$$

The beta function only requires two parameters that are continuously updated as observations are made or reported. In this approach, the observation z_t represents the information from the local watchdog obtained in time interval $[t, t + \Delta t]$

about the percentage of nonforwarded packets. The Bayesian watchdog uses three parameters. The first two parameters are α and β, which are handled over the beta function to obtain an estimation of the node's maliciousness. Among the isolation approaches, the *reputation* concept [12] is commonly used to improve the detection of black holes, just as reputation is used in human relations. Thus, we can say that α and β are the numeric representation of a node's reputation. The third parameter is γ, which represents the devaluation that old observations must suffer to adapt the watchdog's behavior to a continuously changing scenario without penalizing certain nodes forever. It is a mechanism to reintegrate nodes into the MANET if they change their behavior to a more cooperative one.

Hortelano et al. [17] simulated several MANET scenarios and, as a result of their work, they found that, compared to the standard one, the Bayesian watchdog reached a 20% accuracy gain, by reducing the false positives, and it presented a faster detection 95% of the time, thus demonstrating that this approach is a suitable technique to detect black hole attacks by an individual node.

4.4 Collaborative Bayesian Watchdog

In the previous section, we have claimed that the Bayesian watchdog technique is good enough to detect black hole attacks in MANETs. However, the obvious question is whether any further enhancement to this approach could increase its accuracy and detection speed even further. An answer to this question is to introduce a collaborative watchdog. Cooperation, or collaboration, is a trademark of the MANET environments, so why not combine individual watchdog results with information from other nodes to obtain a collaborative detection system? This is the basic concept behind the proposal of collaborative Bayesian watchdogs. Every single node running this kind of watchdog combines its direct observations with reputation information received from its neighboring nodes. The basic assumption in this approach is the *honest majority principle* [18], which assumes that the majority of nodes are likely to be well-behaved. It is arguable that this approach could be affected by other types of attacks and, as a message-passing technique it will generate a little amount of traffic overhead in the MANET. Nevertheless, this technique has shown to be an excellent solution for the black hole attack problem in these networks. This asseveration does not come only from simulation results [19]; it comes also from an analytical model [20] developed to evaluate the system performance, as we will show in this section.

Using the Bayesian watchdog as the building block, we want to implement a collaborative Bayesian watchdog, a technique based on a message-passing mechanism running in every individual node that allows publishing both self and neighbor reputations. Similarly to the Bayesian watchdog, the collaborative Bayesian watchdog overhears the network to collect information about the packets that its neighbors send and receive. Finally, it obtains the α and β values for its whole neighborhood, exactly in the same way as those obtained by the Bayesian watchdog. We call α and β "firsthand information" or "direct reputations." Periodically, the watchdog shares these data with its neighbors, for example, stuffing protocol HELLO messages with this information. We call this information "secondhand information" or "indirect reputations." Of course, indirect reputations must be modulated using a parameter δ, which represent the confidence degree that a node will put on other node's information about its neighborhood. Whenever required, every node running the collaborative Bayesian watchdog calculates, using Equations 4.4 and 4.5, the values of α' and β', which in this case are passed to the beta function to obtain an estimation of the maliciousness of a node.

$$\underset{j \in N_i}{\forall} \underset{k \in N_j}{\forall} \alpha(i)'_j = \frac{\alpha(i)_j + \delta \cdot \text{mean}(\alpha(i)^k_j)}{2} \tag{4.4}$$

$$\underset{j \in N_i}{\forall} \underset{k \in N_j}{\forall} \beta(i)'_j = \frac{\beta(i)_j + \delta \cdot \text{mean}(\beta(i)^k_j)}{2} \tag{4.5}$$

where
  i is the node which is performing detection,
  N_i is the neighborhood of node i,
  $\alpha(i)_j$ is the value of α calculated for every neighbor j of i, obtained from direct observations at i,
  $\beta(i)_j$ is the value of β calculated for every neighbor j of i, obtained from direct observations at i,
  $\alpha(i)^k_j$ is the value of α calculated for every neighbor j of i, obtained from observations of every neighbor k *of j*,
  $\beta(i)^k_j$ is the value of β calculated for every neighbor j of i, obtained from observations of every neighbors k *of j*, and
  δ represents the level of trust or the relative importance that a neighbor's observed reputations have for node i.

Algorithm 1: Black Hole Detector-processing algorithm
  Every observation_time **Do**
    For all Node_j which is a neighbor
      If (BayesianDetection() **or** CollaborativeDetection())
      Then Node_j is malicious
      EndIf
    EndFor
  EndEvery
  Function BayesianDetection()
  Obtain observations
  Compute α and β

If relationship between α and β exceeds tolerance Φ
Then return true
Else return false
EndIf
EndFunction
Function CollaborativeDetection()
Obtain neighborhood reputations
Compute α′ and β′
If relationship between α′ and β′ exceeds tolerance Φ
Then return true
Else return false
EndIf
EndFunction

When indirect reputations arrive at a node from one of its neighbors, it only processes those reputations for its own neighbors, because reputations about nodes that are not in its neighborhood are not very useful at that moment. Once the reputations have been obtained, and the adequate analysis has been done, the detection only needs a predefined tolerance threshold to identify whether a node is misbehaving.

Figure 4.2 shows the main components of the collaborative Bayesian watchdog. First, each individual watchdog overhears the network to make direct observations of its neighbors, thereby detecting black holes as the Bayesian watchdog does. Periodically, it receives reputation information from its neighbors and evaluates their behavior taking into account this secondhand information and its direct observations.

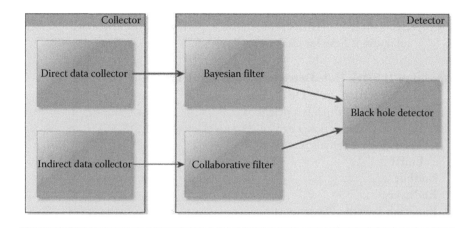

Figure 4.2   Main components of the collaborative Bayesian watchdog.

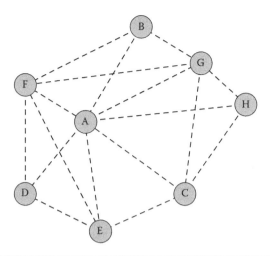

Figure 4.3   Example of a MANET.

The algorithm of the detector module is outlined in Algorithm 1. Basically, the Bayesian detection function performs analysis over direct observations, obtaining the values of α and β. If the relationship between α and β exceeds a predefined tolerance level Φ, the watchdog identifies that node as malicious. These values of α and β are also used in the collaborative detection function, according to Equations 4.4 and 4.5. This function operates similarly to the Bayesian detection function, although it uses secondhand information and other parameters to perform its task.

For the sake of clarity, Figure 4.3 shows an example of a MANET, where dashed lines represent neighbor relationships. Table 4.1 shows the secondhand information received by node A from its neighbors[*]

Node A combines data from Table 4.1 with the direct reputations obtained by it, and, for the sake of simplicity, it uses a δ value of 1; the tolerance threshold[†] Φ is set to 50. These operations are executed in every node running the collaborative Bayesian watchdog with its own received and produced data, but in this example we show in Table 4.1 only the values obtained at node A.

As Table 4.2 shows, node A will identify node H as a black hole using the Bayesian and collaborative versions of the watchdog, because both α and α' are 50 times bigger than β and β', respectively. Also, node C will be detected as malicious only by the collaborative version, reducing the false-negative ratio, thus improving the watchdog accuracy. In the next two sections we will evaluate the improvements of this approach: first using an analytical model, and second, through simulations.

---

[*] Information received about node A is discarded by the node, and it is not shown here.

[†] The tolerance threshold configured here raises a black hole detection alarm if α' is 50 times bigger than β'.

**Table 4.1   Second-Hand Information Received in
Node A**

Neighbor	Reputations Received ($\{\alpha(A)_j^k, \beta(A)_j^k\}$)
B	F: {5,1}, G:{11,1}
C	E:{1,4}, G:{18,1}, H:{1,1}
D	E:{1,2}, F:{7,1}
E	C:{34,1}, D:{1,6}, F:{15,1}
F	B:{1,1}, D:{1,4}, E:{1,3}, G:{13,1}
G	B:{1,2}, C:{52,1}, F:{27,1}, H:{1,6}
H	C:{21,2}, G:{2,13}

**Table 4.2   Values of Collaborative Reputations Calculated at Node A of the
Example**

Neighbor	Reputations		$\{\alpha(A)_j^r, \beta(A)_j^r\}$	Detection of Black Hole?	
	Direct	Indirect		Bayesian	Collaborative
B	{1, 2}	{1, 1.5}	{1, 1.75}	No	No
C	{43, 1}	{57, 1}	{50, 1}	No	Yes
D	{1, 4}	{1, 5}	{1, 4.5}	No	No
E	{1, 1}	{1, 3}	{1, 2}	No	No
F	{1, 4}	{14, 1}	{7.5, 2.5}	No	No
G	{3, 1}	{14, 1}	{8.5, 1}	No	No
H	{68, 1}	{44,1}	{56, 1}	Yes	Yes

4.5 Local Evaluation through Simulation

Having introduced the algorithms and the architecture, we now set the objectives
we are trying to achieve with this collaborative Bayesian watchdog. The goal of
this section is to evaluate the local improvement compared to previous versions of
the watchdog implementations. The global improvement and the effect of the collaboration will be evaluated in Section 4.6. The main objectives of the collaborative
Bayesian watchdog are shown graphically in Figure 4.4. In that figure, we show a
set of nodes, classified as well-behaved and misbehaved. Over them, we draw three

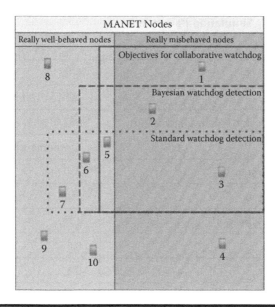

Figure 4.4 Graphical representation of the evaluation objectives for the collaborative Bayesian watchdog.

subsets representing which nodes are detected as misbehaved by the two types of noncollaborative watchdogs shown previously, and the expected results for our collaborative watchdog. Additionally, in Table 4.3 we summarize these expectations.

We have implemented our collaborative Bayesian watchdog as a network simulator 2 (NS-2) extension to the AODV routing protocol, although this implementation is protocol-agnostic. Once implemented, we have evaluated the impact that this approach has over the accuracy and the detection speed, comparing the results from the collaborative Bayesian watchdog with those obtained using the noncollaborative versions, both Bayesian and standard. Table 4.4 shows the characteristics of the scenarios we have selected for our performance evaluation.

Some of these parameters, like the area, the number of nodes, or the speed, are needed by NS-2 to execute the simulation. Others, like δ, γ, Φ, or *observation*

Table 4.3 Summary of Detection Results for the Three Types of Watchdogs in Figure 4.4

Node #	Positives	Negatives	False Positives	False Negatives
Standard	3	8,9,10	5,6,7	1,2,4
Bayesian	2,3	7,8,9,10	5,6	1,2
Collaborative	1,2,3	6,7,8,9,10	5	4

Table 4.4   Simulation Parameters

Parameter	Value
Nodes	50
Area	1000 × 1000 m
Wireless interface and bandwidth	802.11 at 54 Mbps
Antenna	Omnidirectional
Antenna range	250 m
Node speed	5, 10, 15, and 20 m/s
% of black holes	10%
δ	0.8
γ	0.85
Φ	50
Fading	1
Neighbor time	1 s
Observation time	0.2 s
UDP unicast traffic	Three flows
UDP broadcast traffic	Every 5 s
Simulation time	352 s
Scenarios	20

time, are needed by our code as input parameters. To obtain normalized results, we executed tests on the standard watchdog, the Bayesian watchdog, and the collaborative Bayesian watchdog with the same scenarios and parameters. For each test, we averaged the results of 20 independent simulations.

4.5.1 Detection Speed

Accuracy is a key issue when detecting black holes, but speed is also important. A watchdog that detects 100% of black holes but requires 10 minutes is a useless watchdog. So, it is crucial for accuracy and speed to be well balanced. In that sense, watchdog enhancements will target both speed and accuracy issues.

The collaborative Bayesian watchdog performed well in terms of speed. Table 4.5 shows that, on average, 7% of the times our approach detected black holes before the Bayesian watchdog, with the same traffic pattern. For the rest of the

Table 4.5 Percentage of Detections Where the Collaborative Bayesian Watchdog Detects the Black Holes before the Bayesian Watchdog Does

Node Speed (m/s)	Percentage of Earlier Detections
5	1.04
10	11.88
15	9.66
20	5.72

cases, it detects the malicious nodes at the same time. When a node B enters[*] node A's neighborhood, our approach allows node A to identify node B as a black hole with only a reputations-sharing phase with its common neighbors. This means that even if node B does not send or receive any data or routing packet when it enters node A's neighborhood, if it has been previously detected as black hole, node A will quickly mark it as a black hole, too.

In dense networks with traffic load equally balanced between malicious and well-behaved nodes, both watchdog versions will perform nearly equally, despite of the smaller number of packets that the collaborative Bayesian watchdog needs to perform detections. This is because the interval between packets is very short. Nevertheless, in networks with low traffic load and with black holes that transmit a very small amount of packets, the performance difference between the two approaches could be more significant in terms of time. A single packet would make the difference between detecting a black hole or not, and the collaborative Bayesian watchdog obtains better results in those cases.

Additionally, we can say that the collaborative Bayesian watchdog obtains the best results at a node speed of 10 m/s. In fact, when nodes move at 10 and 20 m/s, our approach introduces improvements of nearly 12% and 6%, respectively. These results lead to the conclusion that maybe the collaborative Bayesian watchdog could be suitable for VANET.

4.5.2 Accuracy

Now we present and evaluate the results about the accuracy of our approach. Tables 4.6 through 4.9 summarize the results of our simulations. The meanings of the different rows are the following:

- ■ "*(A)% of accuracy*" row shows the ratio of right detections with respect to the total number of detections. Note that 100-Accuracy is the percentage of false negatives.

---

[*] In this context, entering a node's neighborhood means that this node is within communication range and it announces its presence, for example, through a standard HELLO message.

■ *"(B)% of coverage"* denotes the percentage of real black holes present in the MANET that have correctly been detected as misbehaved.

■ *"(C)% of false positives"* indicates the percentage of detected black holes that are not real black holes.

■ *"(D)% only detector"* row shows the percentage of total detections (right or wrong) where the collaborative Bayesian watchdog has been the only one doing that detection.

The results show that the detection accuracy (row A in Tables 4.6 through 4.9) is also slightly better than for the noncollaborative Bayesian watchdog, which comes from the decreased level of false negatives (row C shows that there are no enhancements on the false positives side of the problem). The fact is that a small

Table 4.6   Simulation Results for Nodes Moving at 5 m/s

	Standard	Bayesian	Collaborative	Enhancement
(A)% of accuracy	61.19	91.15	92.23	1.17
(B)% of coverage	24.00	30.00	30.00	0.00
(C)% of false positives	64.00	17.00	17.00	0.00
(D)% only detector				0.78

Table 4.7   Simulation Results for Nodes Moving at 10 m/s

	Standard	Bayesian	Collaborative	Enhancement
(A)% of accuracy	57.27	96.88	97.39	0.53
(B)% of coverage	13.00	26.00	26.00	0.00
(C)% of false positives	37.00	20.00	20.00	0.00
(D)% only detector				3.77

Table 4.8   Simulation Results for Nodes Moving at 15 m/s

	Standard	Bayesian	Collaborative	Enhancement
(A)% of accuracy	55.45	95.41	96.06	0.67
(B)% of coverage	22.00	33.00	33.00	0.00
(C)% of false positives	42.00	18.00	18.00	0.00
(D)% only detector				0.78

Table 4.9 Simulation Results for Nodes Moving at 20 m/s

	Standard	*Bayesian*	*Collaborative*	*Enhancement*
(A)% of accuracy	40.45	91.57	92.25	0.74
(B)% of coverage	17.00	37.00	37.00	0.00
(C)% of false positives	42.00	29.00	29.00	0.00
(D)% only detector				0.55

amount of black holes, which are not detected by the Bayesian watchdog, are now detected by the collaborative Bayesian watchdog (row D). In fact, our approach is able to detect cases where a black hole enters and exits from the range of a watchdog quickly. Although there is not a big difference between them, the collaborative Bayesian watchdog performs better in terms of accuracy compared to the Bayesian watchdog, despite of the node speed. With respect to the standard watchdog, our approach clearly surpasses it in terms of detection accuracy.

4.5.3 Message Overhead

It is obvious that a message-passing technique introduces overhead in the network, because the information shared by the nodes running the collaborative Bayesian watchdog competes for network resources with data packets and protocol packets. We propose that this reputation information will be included in standard HELLO packets. In this case, if we compare the number of messages sent by a node using a noncollaborative watchdog and those sent by a node running our collaborative watchdog, there will be no difference between them. However, the key issue here is how to reduce the total amount of *bytes transferred* between nodes when they exchange reputation information, affecting as little as possible the protocol and the data packets' transmission. The amount of information that a node will send to its neighbors depends only on two dimensions: the size of the neighborhood and the interval established for the sharing process. The bigger its neighborhood is, and the shorter the interval is set, the greater the amount of data transferred. Also, the way individual watchdogs send this information could increase the total amount of data exchanged. In an *in-band* protocol, this information is attached to other protocol messages. In an *out-of-band* protocol, the reputation information will be sent using special-purpose packets. Since each packet introduces overhead due to headers and trailers introduced by the different network layers, we propose compressing the reputations data and inserting them in standard HELLO packets (in-band protocol). Let

- ■ S_H: Size of standard HELLO messages (bytes).
- ■ S_R: Size of a single node's reputation record (bytes).

- H: Size of overhead introduced in the network by protocols at lower layers for every packet transmitted (bytes).
- N: Total number of nodes in the MANET.
- n: Average number of neighboring nodes for every node during the life of the network or simulation time.
- t_h: Interval between two consecutive HELLO messages sent by a single node (seconds).
- t_m: Interval between two consecutive sharing reputation messages sent by a single node in an out-of-band protocol (seconds).
- r: Compression ratio (r:1). On average, a compression ratio of 2:1 reduces the size of data by 50%.
- TC_i: Total cost of an in-band protocol (bytes).
- TC_o: Total cost of an out-of-band protocol (bytes).

In order to compare the incurred overhead we need to introduce the cost of transmitting the "HELLO" messages and the special packets. Equations 4.6 and 4.7 analyze the generalized overhead inserted in the MANET traffic by an in-band protocol and out-of-band protocol for a time T of network life.

$$TC_i = \frac{H + S_H + (s \cdot n \cdot S_R/r)}{t_h} \cdot N \cdot T \qquad (4.6)$$

$$TC_o = \left(\frac{H + S_H}{t_h} + \frac{H + (s \cdot n \cdot S_R/r)}{t_m} \right) \cdot N \cdot T \qquad (4.7)$$

Note, that the out-of-band Equation 4.7 includes the cost of transmitting the "HELLO" message. Let us denote $((s \cdot n \cdot S_R)/(r))$ as S_C, the average size of compressed complete neighborhood reputations, and substitute it in Equations 4.6 and 4.7. To compare the costs of every alternative, let us subtract TC_i from TC_o, as

$$TC_o - TC_i = \left(\frac{H + S_H}{t_h} + \frac{H + S_C}{t_m} \right) \cdot N \cdot T - \frac{H + S_H + S_C}{t_h} \cdot N \cdot T \qquad (4.8)$$

and operating and simplifying Equation 4.8

$$TC_o - TC_i = \left(\frac{H + S_C}{t_m} - \frac{S_C}{t_h} \right) \cdot N \cdot T \qquad (4.9)$$

For similar accuracy and speed results between the in-band protocol and the out-of-band protocol, it is mandatory that $t_h = t_m = t$. Substituting in Equation 4.9, the difference between both approaches is

$$TC_o - TC_i = \frac{H}{t} \cdot N \cdot T \qquad (4.10)$$

This value is obviously greater than zero, and then we can say that using an out-of-band protocol to exchange reputation information is more expensive than using an in-band protocol, not only if we compare messages transmitted, but also in terms of bytes transmitted. Thus, we can conclude that our approach could *allow saving a significant amount of bandwidth* while achieving the results previously shown in the terms of black hole detection.

4.6 Global Evaluation through an Analytical Model

In the previous section, we focused on evaluating the local performance of the collaborative Bayesian watchdog. In order to evaluate the global behavior we found that simulation was not feasible. The complexity and time consumption of the network simulation under realistic scenarios was the main reason to develop an analytical model. Thus, the goal of this section is to model and evaluate the performance of our collaborative Bayesian watchdog, taking into account the effect of collaboration, false positives, and false negatives.

The network is modeled as a set of N wireless mobile nodes, with C collaborative nodes and one black hole node ($N = C + 1$). Our goal is to obtain the time required by all collaborative nodes to realize who the black hole node in the network is. For our model, we assume that the occurrence of contact between two nodes follows a Poisson distribution with rate λ. This has shown to be valid for both human and vehicle mobility patterns [21–23]. There is some controversy about whether this exponential distribution can reflect some real mobility patterns. Empirical results have shown that the aggregated intercontact times distribution follows a power-law and has a long tail [24]. In [25] it is shown that, in a bounded domain (such as the one selected along this paper), the intercontact distribution is exponential but, in an unbounded domain, it is a power law. The dichotomy of this distribution is described in [26]: a truncated power law with exponential decay appearing in its tail after some cut-off point. The work in [27] analyzed some popular mobility traces, and found that over 85% of the *individual pair distribution* fits an exponential distribution.

Therefore, we consider that using an exponential fit is a good choice to model intercontact times. Moreover, using exponential distributions we can formulate analytical models using Markov chains. We introduce a first model for evaluating the detection of black hole nodes. This model only takes into account the effect of false negatives. Thus, a second model is introduced that evaluates the impact of false positives. Finally, using these models, we evaluate the impact of false positives and negatives.

4.6.1 Modeling Bayesian and Collaborative Detection

The watchdog is modeled using three parameters: the probability of detection (p_d), the ratio of false positives (p_{fp}), and the ratio of false negatives (p_{fn}). The first parameter, p_d, reflects the probability that, when a node contacts another node, the

Bayesian watchdog has enough information to decide whether a node is a black hole or not (i.e., a positive or a negative). This value depends mainly on the observation time, and the transmission and mobility pattern of the nodes.

Furthermore, the watchdog can generate false positives and false negatives. In order to measure the performance of a watchdog, these values can be expressed as a ratio or probability: p_{fp} is the ratio of false positives generated when a node contacts a black hole node and p_{fn} is the ratio of false negatives generated when a node contacts a black hole node.

The collaboration detection is modeled using a function f_{cp}. This function reflects the probability that a node changes to positive when it contacts another collaborative node. As detailed in the previous section, the α and β values are updated using the mean of the α and β obtained from the neighbor nodes (see Equations 4.4 and 4.5). Thus, f_{cp} needs to reflect the probability that a new pair of α and β values obtained from the new contact node makes the detection positive. This function depends on the difference between nodes that have a positive and nodes that have negatives. When this difference is zero or negative, then the probability of change is zero, but when this difference is greater than zero the probability rises to one up to a given threshold C_t. Thus, function f_{cp} can be defined as

$$
f_{cp}(c_p, c_n) = \begin{cases} 0 & (c_p - c_n) \leq 0 \\ \delta(1 - p_{fn})\dfrac{\max[(c_p - c_n), C_t]}{C_t} & (c_p - c_n) > 0 \end{cases} \tag{4.11}
$$

where c_p is the number of collaborative nodes that have a positive, and c_n is the number of nodes that have a negative. Factor $(1 - p_{fn})$ reflects that only the true positives are taken into account and δ corresponds to the trust level. A similar function f_{cn} can be derived for false negatives interchanging c_p and c_n.

Using the previous parameters we can model the probability of generating a positive and a negative when a contact occurs:

- Positive: There are two possibilities: (i) the node contacts with the black hole node and the local watchdog detects it, with probability $p_d(1 - p_{fn})$; and (ii), the node contacts another node that has a positive about the black hole node with probability f_{cp}. Note that a false positive can also be generated with probability $p_d \cdot p_{fp}$.
- Negative: Two possibilities: (i) a contact with a nonblack hole node with probability $p_d(1 - p_{fp})$, and (ii) the node contacts another node that has a negative so the probability is f_{cn}. A false negative can also be generated when it contacts with the selfish node with probability $p_d \cdot p_{fn}$.

In the next subsection, we introduce a generic analytical model for evaluating the performance of the collaborative-watchdog approach. The goal is to obtain the detection time of a black hole node in a network.

4.6.2 Model for the Detection of Black Hole Nodes

This model takes into account the effect of false negatives. False positives do not affect the detection time of black hole nodes, so p_{fp} is not introduced in this model. The effect of false positives will be studied later.

Using λ we can model the network using a 2D continuous time Markov chain (2D-CTMC) with states (c_p, c_n), where c_p represents the number of collaborative nodes that have a positive about the black hole node at time t, and c_n represents the number of *collaborative* nodes that have a negative of the black hole node (note that, in this case, is a false negative). At the beginning all nodes have no information about the black hole node. Then, when a contact occurs, c_p and c_n can be increased by one. Note that c_p and c_n are not independent: $c_p + c_n \leq C$, so some states are not reachable. The final (absorbing) states are achieved when $c_p = C$. A 2D-CTMC model is used, with an initial state, $s_1 = (0, 0)$, $C(C + 1)$ transient states (from $s_1 = (0, 0)$ to $s_\tau = (C - 1, C)$ states), and $C + 1$ absorbing states (from $s_{\tau+1} = (C, 0)$ to $s_{\tau+\upsilon} = (C, C)$. We define τ as the number of transient states ($\tau = C(C + 1)$) and υ as the number of absorbing states ($\upsilon = (C + 1)$). This model can be expressed using the following transition matrix P in the canonical form

$$P = \begin{pmatrix} Q & R \\ 0 & I \end{pmatrix} \qquad (4.12)$$

where I is a $\upsilon \times \upsilon$ identity matrix, 0 is a $\upsilon \times \tau$ zero matrix, Q is a $\tau \times \tau$ matrix with elements p_{ij} denoting the transition rate from transient state s_i to transient state s_j, and R is a $\tau \times \upsilon$ matrix with elements p_{ij} denoting the transition rate from transient state s_i to the absorbing state s_j.

Now, we derive the transition rates p_{ij}. Given the state $s_i = (c_p, c_n)$, the following transitions can occur

- (c_p, c_n) to $(c_p + 1, c_n)$: A new collaborative node has a positive. The transition probability is $\lambda(p_d(1 - p_{fn}) + f_{cp}(c_p, c_n)) \cdot \max(C - c_p - c_n, 0)$. The term $p_d(1 - p_{fn})$ represents the probability of a positive from the watchdog and $f_{cp}(c_p, c_n)$ from collaboration. Finally, the factor $(C - c_p - c_n)$ represents the number of pending collaborative nodes. If there are no pending nodes, this value is 0.
- (c_p, c_n) to $(c_p, c_n + 1)$: A new collaborative node has a negative (a *false negative*). The transition probability is $\lambda(p_d p_{fn} + f_{cn}(c_p, c_n)) \cdot \max(C - c_p - c_n, 0)$.
- $(c_p + 1, c_n)$ to (c_p, c_n): A collaborative node that has a positive state changes to negative. So, the transition probability is similar to the new negative case: $\lambda(p_d p_{fn} + f_{cn}(c_p, c_n))c_p$.
- $(c_p, c_n + 1)$ to (c_p, c_n): A collaborative node that has a negative changes to positive. The transition probability is similar to the new positive case: $\lambda(p_d(1 - p_{fn}) + f_{cp}(c_p, c_n))c_n$.
- (c_p, c_n) to (c_p, c_n): This is the probability of no changes, and it is $1 - \sum_{j \neq i} p_{ij}$.

Summing up, the transition rates (p_{ij}) of the transition matrix P are

$$
p_{ij} = \begin{array}{ll}
\lambda \cdot (p_d(1 - p_{fn}) + f_{cp}(c_p,c_n)) \cdot C_{max} & (c_p \rightarrow c_p + 1) \\
\lambda \cdot (p_d p_{fn} + f_{cn}(c_p,c_n)) \cdot C_{max} & (c_n \rightarrow c_n + 1) \\
\lambda \cdot (p_d p_{fn} + f_{cn}(c_p,c_n)) \cdot c_p & (c_p \rightarrow c_p - 1) \\
\lambda \cdot (p_d(1 - p_{fn}) + f_{cp}(c_p,c_n)) \cdot c_n & (c_n \rightarrow c_n - 1)
\end{array}
\qquad (4.13)
$$

where $C_{max} = \max(C - c_p - c_n, 0)$. Finally, p_{ii} is $1 - \sum_{j \neq i} p_{ij}$.

Using the transition matrix P we can derive the detection time T_d. From the 2D-CTMC we can obtain how long will it take for the process to be absorbed. Using the fundamental matrix $N = (I - Q)^{-1}$, we can obtain a vector t of the expected time to absorption as $t = Nv$, where v is a column vector of ones ($v = [1, 1, ..., 1]^T$). Each entry t_i of t represents the expected time to absorption from state s_i. Since we only need the expected time from state $s_1 = (0, 0)$ to absorption (i.e., the expected time for all nodes to have a positive), the detection time T_d, is

$$
T_d = E[T] = v_1 N_v
\qquad (4.14)
$$

where T is a random variable denoting the detection time for all nodes and $v_1 = [1, 0, ..., 0]$.

Now, we study the effect of the false positives. When a node has a false positive, the problem is that, due to the diffusion of positives, this false positive can be quickly distributed in the network. A way to evaluate this diffusion is to obtain the time when all nodes have a false positive about a given node. Following the same process that in the false negatives' model, we have a 2D-CMTC with the same states (c_p, c_n), but in this case c_p represents the number of nodes with false positives, and c_n the number of nodes with a negative. The transition rates (p_{ij}) of the transition matrix P are

$$
p_{ij} = \begin{array}{ll}
\lambda \cdot ((p_d p_{fn}) + f_{cp}(c_p,c_n)) \cdot C_{max} & (c_p \rightarrow c_p + 1) \\
\lambda \cdot ((p_d(1 - p_{fn})) + f_{cn}(c_p,c_n)) \cdot C_{max} & (c_n \rightarrow c_n + 1) \\
\lambda \cdot ((p_d(1 - p_{fn})) + f_{cn}(c_p,c_n)) \cdot c_p & (c_p \rightarrow c_p - 1) \\
\lambda \cdot ((p_d p_{fn}) + f_{cp}(c_p,c_n)) \cdot c_n & (c_n \rightarrow c_n - 1)
\end{array}
\qquad (4.15)
$$

where $C_{max} = \max(C - c_p - c_n, 0)$. We can see that the transition rates are the same than in the false negative model by replacing $p_{fp} = 1 - p_{fn}$. Therefore, we can use the previous model for obtaining the detection time T_d.

4.6.3 Model Evaluation

Now, based on the previous model, we evaluate the effect of collaboration, false positives, and false negatives on the performance of the collaborative Bayesian watchdog using the MATLAB software package. The models allow an overall evaluation of the collaborative watchdog under a large number of scenarios. For the next experiments we use the following parameters that were obtained from the previous experimental evaluation: $p_d = 0.1$, $C_t = 5$, $\delta = 0.3$, $p_{fn} = 0.1$, and $\lambda = 0.02$.

The first experiment evaluates the improvement on the global detection time using our collaborative approach. We evaluate the time that all nodes (except the misbehaving node) have a positive about this misbehaving nodes depending on the number of nodes. The results are shown in Figure 4.5. The graph starts in $N = 2$, that is a black hole node and a collaborative node, so both approaches have the same detection time (there is no collaboration). However, when $N \geq 2$, we can see that, using our collaborative watchdog, the detection time is nearly the same, but when using only the collaborative watchdog the detection time increases exponentially.

The second experiment evaluates the impact of the false negatives comparing the results with a noncollaborative approach (i.e., depending only on the local watchdog) for a network of 10 nodes ($N = 10$). In this case, we expect that the diffusion of α and β can reduce the influence of false negatives. Figure 4.6 shows the detection time depending on p_{fn} for different values of N (network nodes). First, we can see that the detection time is greatly reduced using the collaborative watchdog in the absence of false negatives (i.e., for $p_{fn} = 0$). Second, when p_{fn} increases, the detection time increases with a very little slope while for the local watchdog the time increases exponentially. Note that the detection time refers to all the nodes in the network, so this value can be very high with no collaboration. Figure 4.7 shows the percentage of reduction of the detection time between the collaborative and the local watchdog for several values of N. This confirms that the reduction is

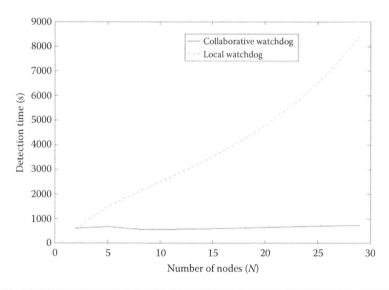

Figure 4.5   Detection time depending on number of nodes N.

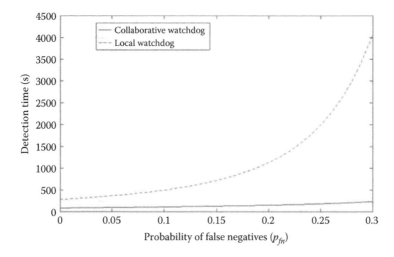

Figure 4.6 **Evaluation of the impact of false negatives on the detection time for** *N* = **10.**

very important in the absence of false negatives (from 65% to 75%), but it is even greater for higher values of p_{fn} and N.

The third experiment evaluates the impact of false positives. The model introduced obtains the detection time of a false negative. Therefore, a greater value of the detection time will imply a reduced impact of false positives. In this case, we

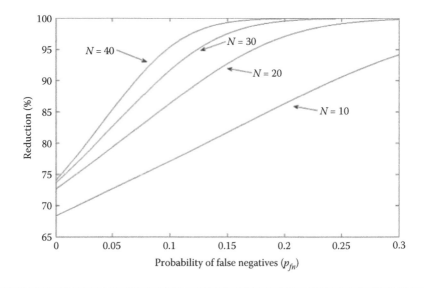

Figure 4.7 **Rate of reduction time for several values of** *N*.

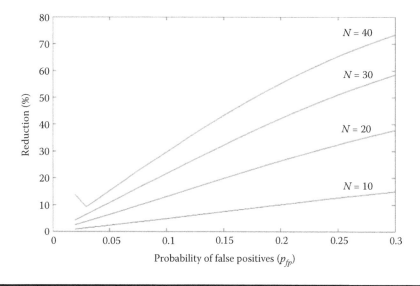

Figure 4.8   Reduction of the detection time depending on false positives.

expect that the collaborative watchdog can reduce the influence of false positives. This is shown in Figure 4.8. In this case, the reduction in not as high as in the false negatives case.

Two conclusions can be drawn from the performed analytical evaluation: our collaborative watchdog is able to reduce drastically the detection time of black hole nodes while also reducing the impact of false positives and false negatives.

4.7 Future Research Directions

There are two main future research directions: improving the watchdog and improving the analytical model in consequence. Improving the watchdog mechanism to deal with more misbehaving nodes requires, in a first stage, protecting the MANET against collaborative black hole attacks, and, in a second stage, avoiding the problems related to liar nodes during the reputation-sharing phase. Also, we aim at implementing this mechanism in a hardware testbed (Castadiva), while working on the fine tuning of the collaborative Bayesian watchdog to apply this technique on delay tolerant network environments. Finally, we want to explore other statistical functions to use them inside the collaborative detection module. Obviously, the analytical model proposed could be further developed to accommodate the evolution of the implemented watchdog, and to evaluate the performance of this watchdog when several parameters change, thereby addressing different collaboration scenarios.

4.8 Conclusions

This chapter discussed the problems related to the black hole attacks in MANETs, and some of the solutions proposed in the literature during the past few years to deal with them. The concept and functionality of a watchdog as basic IDS has also been introduced. The basic watchdog has been improved to enhance its accuracy and detection speed, using a Bayesian filter to remove from the collected data those measurements that are useless due to the nature of the wireless channel and mobility. Another enhancement to this technique, which aims at taking advantage of the cooperative essence of the MANET, has also been presented, allowing nodes to share neighbor reputations to improve both accuracy and detection speed. Overall, results show that the black hole detection problem is still a challenging issue, although different improvements can be introduced to mitigate its effects.

Acknowledgment

This work was partially supported by the Ministerio de Ciencia e Innovación, Spain, under Grant TIN2011-27543- C03-01.

References

1. S. Yousefi, M. S. Mousavi, and M. Fathy, Vehicular ad hoc networks (vanets): Challenges and perspectives, in *Proceedings of the 6th International Conference on ITS Telecommunications*, Chegndu, China, June 21–23, 2006, pp. 761–766.
2. C.-K. Toh, D. Kim, S. Oh, and H. Yoo, The controversy of selfish nodes in ad hoc networks, in *Proceedings of the Twelfth International Conference on Advanced Communication Technology (ICACT'10)*, Phoenix Park, Korea, February 7–10, 2010.
3. H. Kandavalli and M. Nagendra Nath, Minimizing malicious eavesdropping ability in wireless mesh networks using skems, *International Journal of Computer Science and Information Technologies*, 3(2), 3476–3478, 2012.
4. M. Raya and J.-P. Hubaux, Securing vehicular ad hoc networks, *Journal of Computer Security—Special Issue on Security of Ad-hoc and Sensor Networks*, 15(1), 39–68, 2007.
5. J. R. Douceur, The Sybil attack, *Revised Papers from the First International Workshop on Peer-to-Peer Systems (IPTPS'02)*, MIT Faculty Club, Cambridge, MA, USA, March 7–8, 2002.
6. T. Sundarajan and A. Shammugam, Modeling the behavior of selfish forwarding nodes to stimulate cooperation in manet, *International Journal of Network Security and its Applications (IJNSA)*, 2(2), 147–160, 2010.
7. F. Kargl, A. Klenk, S. Schlot, and M. Webber, Advanced detection of selfish or malicious nodes in ad hoc networks, in *Proceedings of the First European Conference on Security in Ad-Hoc and Sensor Networks (ESAS 2004)*, Heidelberg, Germany, August 6, 2004.
8. K. Sahadevaiah and P. Reddy, Impact of security attacks on a new security protocol for mobile ad hoc networks, *Network Protocols and Algorithms*, 3(4), 122–140, 2011.

9. L. Xu, Z. Lon, and A. Ye, Analysis and countermeasures of selfish node problem in mobile ad hoc networks, in *Proceedings of the Tenth International Conference on Computer Supported Cooperative Work in Design (CSCWD'06)*, Nanjing, China, May 3–5, 2006.

10. J. Hortelano, J.-C. Cano, C.-T. Calafate, and P. Manzoni, Watchdog intrusion detection systems: Are they feasible in manets?, in *XXI Jornadas de Paralelismo (CEDI'2010)*, Valencia, Spain, September 7–10, 2010.

11. S. Marti, T. Giuli, K. Lai, and M. Baker, Mitigating routing misbehavior in mobile ad hoc networks, in *Proceedings of the Sixth International Conference on Mobile Computing and Networking (MobiCom'00)*, Boston, MA, USA, August 6–11, 2000.

12. S. Buchegger and J.-Y. Le Boudec, Self-policing mobile ad hoc networks by reputation systems, *IEEE Communications Magazine*, 43(7), 101–107, 2005.

13. M. Ghosh, A. Varghese, A.-A. Kherani, and A. Gupta, Distributed misbehavior detection in VANETS, in *Proceeding of the IEEE Wireless Communications and Networking Conference (WCNC'09)*, Budapest, Hungary, April 5–8, 2009.

14. L. Buttyan and J.-P. Hubaux, Enforcing service availability in mobile adhoc wans, in *IEEE/ACM Workshop on Mobile Ad Hoc Networking and Computing (MobiHOC '2000)*, Boston, MA, USA, August 11, 2000.

15. L. Buttyan and J.-P. Hubaux, Stimulating cooperation in self-organizing mobile ad hoc networks, *Mobile Networks and Applications*, 8(5), 579–592, 2003.

16. S. Zhong, J. Chen, and Y. Yang, Sprite: A simple, cheat-proof, creditbased system for mobile ad-hoc networks, in *Proceedings of the Twenty-second Annual Joint Conference of the IEEE Computer And Communications Societies (INFOCOM'03)*, San Francisco, CA, USA, March 30–April 3, 2003.

17. J. Hortelano, C.-T. Calafate, J.-C. Cano, M. de Leoni, P. Manzoni, and M. Mecella, Black-hole attacks in p2p mobile networks discovered through Bayesian filters, in *Proceedings of OTM Workshops '2010*, Hersonissos, Crete, Greece, October 25–29, 2010, pp. 543–552.

18. B. Parno and A. Perrig, Challenges in securing vehicular networks, in *Proceedings of the Fourth Workshop on Hot Topics in Networks (HotNets-IV)*, College Park, MD, USA, November 14–15, 2005.

19. M. Serrat-Olmos, E. Hernandez-Orallo, J.-C. Cano, C. Calafate, and P. Manzoni, A collaborative Bayesian watchdog for detecting black holes in MANETS, in *Proceedings of the 6th International Symposium on Intelligent Distributed Computing (IDC/12)*, Calabria, Italy, September 24–26, 2012.

20. E. Hernandez-Orallo, M. Serrat-Olmos, J.-C. Cano, C. Calafate, and P. Manzoni, Improving selfish node detection in manets using a collaborative watchdog, *IEEE Communications Letters*, 16(5), 642–645, 2012.

21. R. Groenevelt, P. Nain, and G. Koole, The message delay in mobile ad hoc networks, *Performance Evaluation*, 62, 210–228, 2005.

22. H. Zhu, L. Fu, G. Xue, Y. Zhu, M. Li, and L. M. Ni, Recognizing exponential inter-contact time in VANETS, in *Proceedings of the 29th conference on Information communications (INFOCOM'10)*, San Diego, CA, USA, March 15–19, pp. 101–105.

23. Y. Li, G. Su, D. Wu, D. Jin, L. Su, and L. Zeng, The impact of node selfishness on multicasting in delay tolerant networks, *IEEE Transactions on Vehicular Technology*, 60(5), 2224–2238, 2011.

24. A. Chaintreau, P. Hui, J. Crowcroft, C. Diot, R. Gass, and J. Scott, Impact of human mobility on opportunistic forwarding algorithms, *IEEE Transactions on Mobile Computing*, 6, 606–620, 2007.

25. H. Cai and D. Y. Eun, Crossing over the bounded domain: From exponential to power-law intermeeting time in mobile ad hoc networks, *IEEE/ACM Transactions on Networking*, 17(5), 1578–1591, 2009.
26. T. Karagiannis, J.-Y. Le Boudec, and M. Vojnovic, Power law and exponential decay of inter contact times between mobile devices, in *Proceedings of the 13th Annual ACM International Conference on Mobile Computing and Networking (MobiCom'07)*, Montreal, Canada, September 9–14, pp. 183–194.
27. W. Gao, Q. Li, B. Zhao, and G. Cao, Multicasting in delay tolerant networks: A social network perspective, in *Proceedings of the Tenth ACM International Symposium on Mobile Ad Hoc Networking and Computing (MobiHoc'09)*, New Orleans, LO, May 18–21, pp. 299–308.

Chapter 5

Mutual Authentication in IP Mobility-Enabled Multihop Wireless Networks

Sanaa Taha, Sandra Céspedes,
and Xuemin (Sherman) Shen

Contents

5.1 Introduction

An important requirement for current mobile wireless networks, such as vehicular ad hoc networks (VANETs), is their ability to provide ubiquitous and seamless IP communications in a secure way. Moreover, these networks are envisioned to support multihop communications, in which intermediate nodes help relay packets between two peers in the network. The combination of the two capabilities, namely, the seamless support of IP communications and multihop communications, is denominated IP mobility-enabled multihop wireless networks.

In such an infrastructure-connected multihop mobile network, the connection from the mobile node (MN) to the point of attachment may traverse multiple hops (Figure 5.1). The reasons for relaying packets in infrastructure-connected mobile networks are twofold: (1) direct connection to the infrastructure may not always be available; thus, by using relayed communications, the network coverage can be extended, and its throughput and capacity can also be increased [1]; and (2) relay nodes (RNs) may benefit from offering their services as temporary relays; for example, different cooperation incentive schemes have shown that it is the best interest of each node to participate in multihop packet forwarding to earn credits that reward them per forwarded packets [2].

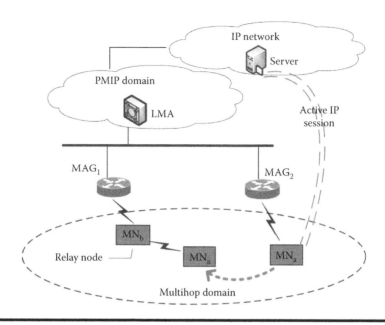

Figure 5.1 Infrastructure-connected multihop mobile network. MN$_a$ is roaming to a relayed communication through relay node MN$_b$.

In this chapter, we review the different approaches that help achieve authentication in IP mobility-enabled multihop wireless networks. Next, we introduce a new mutual authentication scheme, called *EM³A*, for authentication in multihop mobile networks that employ Proxy Mobile IPv6 (PMIP) to support IP mobility. *EM³A* thwarts authentication attacks, including denial of service (DoS), colluding, impersonating, replay, and man-in-the-middle (MITM) attacks. In addition, we present a key establishment scheme based on symmetric polynomials [3–5], which generates a shared secret key between an MN and an RN. Compared to existing authentication schemes, our proposed scheme achieves higher secrecy as well as lower computation and communication overheads. Using a symmetric polynomial of degree *t*, and for a domain with *n* mobile access gateways (MAGs), our scheme achieves $t \times 2^n$–secrecy, whereas existing symmetric polynomial-based authentication schemes achieve only *t*–secrecy. Extensive simulations are performed to show that our scheme can be used with seamless handover since it results in low authentication delay. In addition, the proposed key establishment scheme achieves lower revocation overhead than that achieved by existing symmetric polynomial-based schemes. We also present a case study that employs *EM³A* in the context of asymmetric VANET [6], in order to show the efficacy of *EM³A* to thwart authentication attacks when handovers occur through the infrastructure-to-vehicle-to-vehicle (I2V2V) communications while achieving reduced overhead.

5.2 Mutual Authentication Schemes for Multihop Wireless Networks

In this section, a brief overview of existing mutual authentication schemes for multihop wireless networks is presented. The authentication solutions are categorized according to two different approaches: (1) end-to-end authentication and (2) hop-by-hop authentication. We also discuss the key generation schemes based on symmetric polynomials.

5.2.1 End-to-End Authentication

In the end-to-end authentication approach, the RN is used only as the forwarder of the authentication credentials between the MN and the infrastructure. Public key-based authentication schemes are proposed in refs. [7,8] for end-to-end authentication. A trusted delegation entity and an MN's public key certificate are used in ref. [7] by the MN to authenticate itself to the foreign gateway. On the other hand, the scheme in ref. [8] uses a symmetric key for authenticating an MN to its home network, and a public key for mutual authentication between the home network and the foreign network. However, the expensive computation generally involved with public key operations may increase the end-to-end delay of these schemes.

Conversely, ref. [9] proposes a symmetric key-based authentication scheme for multihop mobile IP. In that work, an MN authenticates itself to its home authentication, authorization, and accounting (HAAA) server, using the extensible authentication protocol (EAP) [10]. After a successful authentication, the HAAA derives a group of keys to be used by the MN, including a shared master key, an extensible master key, and a foreign mobile IP (MIP) key. Despite the low computation and communication overheads, the symmetric key-based schemes cannot achieve as strong levels of authentication as those achieved by public key-based schemes. This is because the sharing of the secret key between the two peers increases the chances for adversaries to identify the shared key. Instead, public key-based schemes create a unique secret key for each user; hence, it is more difficult for adversaries to identify the keys.

5.2.2 Hop-by-Hop Authentication

In the hop-by-hop authentication approach, authentication algorithms are implemented between every two hops. Ref. [11] presents a mutual authentication that depends on both secret splitting and self-certified schemes. However, it has been shown that both schemes are prone to DoS attacks. In addition, Alpha, another scheme for hop-by-hop authentication, is presented in ref. [12]. In Alpha, the MN signs the messages using hash chain elements as different keys for signing, and then delays the key disclosure until receiving an acknowledgment from the intermediate node. Although Alpha protects the network from insider attacks, it suffers from a high end-to-end delay. A hybrid approach, the adaptive message authentication scheme (AMA), is proposed in ref. [13]. AMA adapts the strength of the security checks depending on the security conditions of the network at the moment of packet forwarding. AMA works under the assumption that different spots inside the entire network cannot be attacked simultaneously. In some spots, the adversary attacks all the transmitted messages, while in other spots, there is no attack at all. Consequently, AMA proposes two different modes: a relaxed mode, to be used as default mode, and a check-all mode, which is used when attacking is discovered.

5.2.3 Symmetric Polynomial-Based Key Generation Schemes

By definition, a symmetric polynomial is any polynomial of two or more variables that achieves the interchangeability property, that is, $f(x,y) = f(y,x)$. Such a type of mathematical function is often used by key generation schemes to establish a shared secret key between two entities. Centralized and decentralized key generation schemes are proposed.

In the centralized key generation schemes, a polynomial distributor, such as the access router (AR), securely generates a symmetric polynomial and evaluates this polynomial with each of its users' identities. For example, given three users identities 1, 2, and 3, and the symmetric polynomial $f(x,y) = x^2y^2 + xy + 10$, the resultant evaluation functions are $f(1,y) = y^2 + y + 10$, $f(2,y) = 4y^2 + 2y + 10$, and $f(3,y) = 9y^2 + 3y + 10$, respectively. Then, in addition to keeping the original polynomial secured, the polynomial distributor sends the evaluated polynomials to each user in a secure way. Afterwards, any two users can share a secret key between them by calculating the evaluation function for each other. Continuing with the previous example, if user 1 evaluates its function $f(1,y)$ for user 3, it obtains $f(1,3) = 22$. In the same way, user 3 evaluates the function $f(3,y)$ for user 1, it obtains $f(2,1) = 22$. Therefore, both users share a secret key, 22, without transmitting any additional messages to each other.

In addition, new decentralized key generation schemes are proposed in refs. [4,5] to generate a shared secret key between two arbitrary MNs that are located in two heterogeneous networks. These schemes achieve t–secrecy level, where t represents the degree of the generated polynomial. However, a scheme with a t–secrecy property can be broken if $t + 1$ users collude to reveal the secret polynomial. Moreover, for only one MN's revocation, the decentralized schemes require changing the entire system's keys, and hence leading to a high communication overhead.

5.3 IP Mobility-Enabled Multihop Wireless Networks

5.3.1 Network and Communication Model

Consider an infrastructure-connected multihop mobile network such as that depicted in Figure 5.1. The IP mobility support in MNs is provided by means of

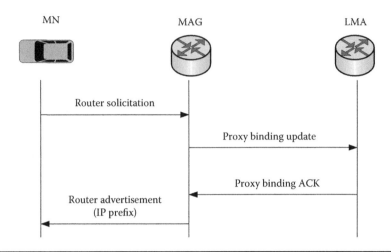

Figure 5.2 MN joins the PMIP domain.

an adapted version of Proxy Mobile IP for multihop domains [14]. In addition to the adaptive PMIP, we also add the strict requirement for the MN to first connect directly to an MAG, in order to obtain a valid IP prefix in the domain. As illustrated in Figure 5.2, once an MN, such as a vehicle, joins the domain for the first time, it sends router solicitation (RS) messages, which are employed by the MAG as a hint for detecting the new connection. After the PMIP signaling has been completed, the MAG announces the IP prefix in a unicast router advertisement (RA) message delivered to the MN over the one-hop connection. Afterward, the MN may eventually divert to use an RN to reach the fixed network. We assume that once nodes have been authenticated in the PMIP domain, such legitimate nodes faithfully follow the routing protocol when they are selected to provide their relay services for another MN in their surroundings.

In this network model, we focus on the multihop communications that are occurring between an MN and an RN, when the MN uses the RN to maintain a connection to the infrastructure. Applications of multihop mobile networks have been largely studied not only for vehicular communications networks but also for wireless personal area networks [15] and wireless local area networks [16].

5.3.2 Adversary and Trust Models

Two types of adversaries are defined: internal and external. Internal adversaries are legitimate users who exploit their legitimacy to harm other users. Since they have the same capabilities as the legitimate users, those adversaries obtain authorized credentials that can be used in the PMIP. Two types of internal adversaries are defined: impersonator and colluder. The former impersonates another MN's identity and sends neighbor discovery messages, such as RS, through the RN. The latter colludes with other domain users using their authorized credentials in order to identify the shared secret key between two legitimate users.

On the other hand, external adversaries are unauthorized users who aim at identifying the secret key and breaking the authentication scheme. Those adversaries' capabilities are unlimited in such a way that the adversaries have monitoring devices to eavesdrop messages transmitted between an MN and an RN. Moreover, the external adversary may inject their own messages and delete other authorized user's transmitted messages as well. We consider that external adversaries may execute replay, MITM, and DoS attacks. The goal of the MITM and replay attackers is to identify a shared key between two legitimate users, while the goal of a DoS attacker is to exhaust the system resources following a kind of irrational attack. DoS adversary can also be considered as internal adversaries when the attacker is one of the legitimate nodes.

For the trust model, the local mobile anchor (LMA) and all MAGs in the domain are considered to be trusted entities. An MN trusts its first attached MAG in such a way that this MAG does not reveal the MN's evaluated domain polynomial, which is used by the MN to create shared keys with RNs. In addition, the MN trusts the LMA that maintains the secret domain polynomial, which can be used to reveal the shared keys for all nodes in the network.

5.4 Efficient Mutual Multihop Mobile Authentication Scheme (*EM³A*)

Unlike the different approaches of mutual authentication schemes, for multihop wireless networks, presented in Section 5.2, in this section, a novel scheme, *EM³A* [17], is introduced. *EM³A* scheme guarantees the authenticity of an MN and an RN in a multihop-enabled PMIP network that follows the communication model described in Section 5.3. The scheme consists of four main phases: the key establishment, MN registration, authentication, and MN revocation. The key establishment phase generates distributing keys for entities in the PMIP domain. The MN registration phase is executed when the MN joins the PMIP domain for the first time. Finally, the authentication phase mutually authenticates the roaming MN and the RN.

5.4.1 Key Establishment Phase

Assuming a unique identity for each MAG inside the PMIP domain, the LMA maintains and distributes the list of such identities for all MAGs to all legitimate users in the domain. The size of MAGs' list is expected to depend on the number of MAGs in the domain. Assuming n MAGs in the domain, each legitimate MN requires $(n \times \log n)$ bits to store this list. Such storage space can be adequately found for MNs in mobile networks, such as vehicular networks. In addition, the LMA is responsible for replacing the identity of any MAG with another unique identity (this is especially useful for the management of MN's revocation, as it is illustrated in Section 5.4.4).

As the first step, each MAG in the domain generates a four-variable symmetric polynomial, $f(w,x,y,z)$, which we call the network polynomial. The MAG then sends this polynomial to the LMA in the domain. The LMA collects all network polynomials, $f_i(w,x,y,z), i = 1,2,\ldots,n$, from every MAGs ($MAG_1,MAG_2,\ldots,MAG_n$), and creates the domain polynomial, $F(w,x,y,z)$, as follows:

$$F(w,x,y,z) = \sum_{i \in_R n}^{l} f_i(w,x,y,z), \quad 2 \leq l \leq n \qquad (5.1)$$

where n is the number of MAGs in the domain. Out of the received n network polynomials, the LMA randomly selects and sums l polynomials in order to construct the domain polynomial. The reasons for not summing all the network polynomials are twofold: (1) increasing the secrecy of the scheme from t−secrecy to $t \times 2^n$−secrecy, and (2) decreasing the revocation overhead at the time of the MN's revocation. The LMA then evaluates the created domain polynomial $F(w,x,y,z)$, for each MAGs identity, ID_{MAG}, individually. The LMA then securely sends to each MAG its corresponding evaluated polynomial. Later on, the evaluated polynomials, $F(ID_{MAGi},x,y,z)$, with $i = 1,2,\ldots,n$, are used to generate shared secret keys among arbitrary nodes in the domain.

5.4.2 MN Registration Phase

This phase is executed when an MN first connects to the PMIP domain. The MN initially authenticates itself to the MAG, to which it is directly connected, by any available authentication schemes, such as the RSA. This MAG, also called MN's first-attached MAG, checks the MN's credentials, and securely responds by evaluating its domain polynomial, $F(ID_{MAG},x,y,z)$, using the MN's identity to obtain $F(ID_{MAG},ID_{MN},y,z)$. In addition, when the domain's LMA guarantees the MN's legitimacy through its first-attached MAG, the MN also receives and stores the list of current MAGs's identities along with the identity of its first-attached MAG (ID_{FMAG}) from the LMA. Consequently, any two MNs, a and b, located in the same PMIP domain can establish a shared secret key with each other. The MN a evaluates its received polynomial, for MN b to get $F(ID_{FMAGa},ID_a,ID_{FMAGb},ID_b)$. Similarly, b evaluates its received polynomial, $F(ID_{FMAGb},ID_b,y,z)$, for MN a to obtain $F(ID_{FMAGb},ID_b,ID_{FMAGa},ID_a)$. Since the domain polynomial, F, is a symmetric polynomial, the two evaluated polynomials result in the same value and they represent the shared secret key, K_{a-b}, between MNs a and b.

5.4.3 Authentication Phase

The authentication between an MN and an RN, depicted in Figure 5.3, is executed when the MN roams to a relayed connection in order to communicate with a new MAG, rather than its first-attached MAG. Therefore, as a part of the multihop-enabled PMIP scheme, the MN needs to detect its movement by sending neighbor discovery messages, which go through an RN. The goal of the authentication phase is to support mutual authentication between the roaming MN and the RN. In other words, this mutual authentication proves to each entity, the MN and the RN, that the other entity is a legitimate user in the PMIP domain. The authentication phase is composed of the three stages described as follows.

5.4.3.1 MN Initialization

When detecting its movement, the MN reacts by broadcasting a RS or neighbor solicitation message that includes its identity and its first attached MAG's identity, $ID_{FMAG-MN}$. Therefore, as a first step, the intended RN checks its stored MAGs list to see if $ID_{FMAG-MN}$ is currently a valid identity. If there is no identity equals to $ID_{FMAG-MN}$, the RN rejects the MN and assumes it is a revocated or malicious node. Otherwise, if $ID_{FMAG-MN}$ is a valid identity, the RN continues with the next step to check the MN's authenticity.

5.4.3.2 Challenge Generation

Employing the MN's identity and $ID_{FMAG-MN}$, the RN generates the shared key K_{MN-RN}, as described in the registration phase. The RN then constructs a challenge message, which includes its own identity, ID_{RN}, the MN's identity, a random number $Nonce_{RN}$, and a time stamp t_{RN}. Finally, the RN encrypts the challenge message

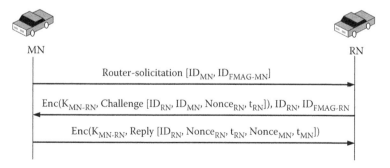

Figure 5.3 *EM³A* authentication phase. (From S. Taha, S. Céspedes, and X. Shen, **EM3A: Efficient mutual multihop mobile authentication scheme for PMIP networks**, in *Proceedings of IEEE ICC 2012*, Ottawa, Canada, June 10–15 2012.)

using the shared key, K_{MN-RN}, and sends the encrypted message, along with ID_{RN} and its first attached MAG's identity, $ID_{FMAG-RN}$, to the MN.

5.4.3.3 Response Generation

When the MN receives the RN's encrypted challenge message, it first uses its stored MAGs' identities list to check the received $ID_{FMAG-RN}$ validity. The MN then reconstructs the shared key with this RN, by using the RN's identity and $ID_{FMAG-RN}$, and decrypts the received challenge message. The MN accepts the RN as a legitimate relay if the RN's decrypted identity is the same as the identity received with the challenge message, that is, ID_{RN}. The MN then constructs a reply message, which includes RN's identity, $Nonce_{RN}$, t_{RN}, a new random number $Nonce_{MN}$, and a time stamp t_{MN}. The MN encrypts the reply message using the shared key, and sends it to the RN, which decrypts the message and accepts the MN as legitimate user if the decrypted $Nonce_{RN}$ equals the original random number that the RN sent in the challenge message.

Once the authentication phase is completed, the neighbor discovery messages are properly forwarded toward the MAG, which allows for the multihop-enabled PMIP to continue its operation as described in ref. [14] and maintain seamless communications. In Figure 5.3, *Enc(K, M)* represents an encryption operation of a message M using a key K. In addition, the RS, challenge, and reply are the three messages transmitted between the MN and the RN.

5.4.4 Mobile Node Revocation

EM^3A aims at achieving backward secrecy, which means that a revoked MN will not be able to use any of its previous shared keys to deceive the RN. However, achieving this goal in symmetric polynomial-based schemes costs high revocation overhead, in which all legitimate MN's and the secret domain polynomial need to be changed. Therefore, in EM^3A scheme, we decrease this revocation overhead while achieving the backward secrecy.

When an MN is revoked from the PMIP domain, the LMA replaces this MN's first-attached MAG's identity, $ID_{FMAG-MN}$, with another unique identity, ID_{NFMAG}, and sends the new identity to all legitimate nodes in the domain. Subsequently, each legitimate node updates its stored MAGs list by replacing the old identity with the new one. The LMA also sends a message to each MAG in the domain, which includes a list of the MNs that have ID_{NFMAG} as their first-attached MAG's identity, along with an evaluated polynomial, $F(ID_{NFMAG},x,y,z)$, for the FMAG's new identity. Afterwards, the MAGs send the evaluated polynomial for those MNs that are in the received list and under MAGs' coverage areas. Eventually, each MN, in the MNs list, receives a new evaluated polynomial, $F(ID_{NMAG},ID_{MN},y,z)$, for both its identity and the new first-attached MAG's identity. Therefore, instead of changing the entire domain keys, only the MNs that share the same $ID_{FMAG-MN}$ need to change their evaluated polynomials and keys.

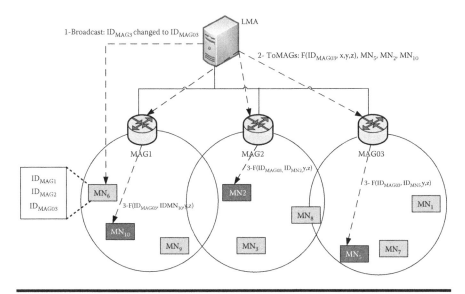

Figure 5.4 *EM³A* **MN revocation.**

Figure 5.4 shows an example of the revocation operation. Consider a revoked MN_4 with a first-attached MAG as MAG_3. Three main messages are transmitted after revocation. First, the LMA changes the MAG_3 identity to MAG_{03} and broadcasts the new identity to all MNs in the domain. The second message is transmitted from the LMA to all MAGs, and it includes the new evaluated polynomial with the new identity, $F(ID_{MAG03},x,y,z)$, along with the MNs that share the MAG_3 as their first-attached MAG. In Figure 5.4, MN_2, MN_5, and MN_{10} are those intended nodes that share MAG_3 as their first attached MAG. Note that nodes sharing the first-attached MAG may not be located under the same MAG's coverage area; therefore, the LMA has to send the second message to all MAGs. Finally, the last message is transmitted from the MAGs to those intended nodes, MN_2, MN_5, and MN_{10}.

5.5 Security Analysis

In Section 5.3.2, the internal and external adversaries defined in our network model were presented. Therefore, in the following subsections, we prove that the proposed EM^3A scheme thwarts those adversaries by achieving high security level.

5.5.1 Internal Adversaries

5.5.1.1 Impersonation Attacks

By using a shared secret key, which is only known by the two communicating entities, the MN and the RN, the proposed EM^3A authentication scheme thwarts the impersonation attacks. For instance, consider an adversary A, which aims at impersonating an MN in order to join a new MAG through an RN, and illegally benefit from the domain services. First, A sends an RS message and attaches the MN's identity, ID_{MN}. The RN replies with a challenge message, which is encrypted by the shared key K_{MN-RN}. In order to pass the authentication check, A needs to decrypt the challenge message and identify the RN's random number, $Nonce_{RN}$, which is included in the encrypted challenge message. However, A cannot reconstruct the shared key by using only the identities of the MN and the RN. In addition to the identities, the adversary needs to know one of the evaluated polynomials, $F(FMAG_{MN}, ID_{MN}, y, z)$ or $F(FMAG_{RN}, ID_{RN}, y, z)$. Since the evaluated polynomials are secret, it is impossible for an impersonation adversary to break EM^3A.

5.5.1.2 Collusion Attacks

By increasing the secrecy of the proposed key establishment scheme, we argue that the EM^3A scheme mitigates the collusion attacking impact. Generally, a key establishment scheme used a symmetric polynomial of degree $t-$ allows for a $t-$secrecy level, which means that $t + 1$ colluders are able to identify the secret polynomial and reconstruct the whole system's keys. However, in EM^3A, the domain polynomial is constructed as in Equation (5.1), where the LMA randomly selects a group of the network polynomials to calculate the domain polynomial. Therefore, by using the following theorem, we prove that at least $t \times 2^n + 1$ colluders should collude to break our authentication scheme.

Theorem 5.1

Using a symmetric polynomial of degree t, and having n different MAGs in the PMIP domain, the proposed key establishment in EM^3A scheme achieves $t \times 2^n$ secrecy level.

Proof
Since t is the degree of the used symmetric polynomials, the secrecy level of each MAG's network polynomial is also t. Hence, the secrecy s of the domain polynomial can be computed as follows:

$$
\begin{aligned}
s &= \sum_{k=2}^{n} \binom{n}{k} \times t \\
&= t \times \sum_{k=0}^{n} \binom{n}{k} - \left[\binom{n}{0} + \binom{n}{1} \right] \\
&= t \times [2^n - (1 + n)] \\
&\approx t \times 2^n
\end{aligned}
\tag{5.2}
$$

Since the secrecy increases from t to $t \times 2^n$, the number of colluders that can break the scheme also increases from $t + 1$ to $(t \times 2^n) + 1$. Therefore, in practice, t is chosen to be a large number, and n should be preferably large as ways to harden the colluder attacks.

5.5.2 External Adversaries

5.5.2.1 DoS Attacks

The EM^3A scheme thwarts the DoS attackers, which trigger forged RS messages in order to exhaust the RN and MAG resources. Without using our proposed scheme, the RN forward all RS messages to the MAG and facilitates the DoS attack. However, using EM^3A, the adversary, A, needs to know a valid shared key, K_{MNi-RN}, in order for the RN to forward the A's RS messages. However, A is an external adversary, which means A is an unauthorized user. Therefore, A cannot construct any key, even if this adversary knows the identity of a legitimate MN.

5.5.2.2 Replay and MITM Attacks

In addition, EM^3A thwarts the replay attacks, which repeat one of the RS messages that have been previously transmitted by a legitimated user. By appending both time stamps and random nonces at the end of each transmitted message between the MN and the RN, the proposed scheme prevents those replay attackers' impact. Finally, the MITM attack that impersonates an MN or an RN is thwarted by our EM^3A, because both the challenge and reply messages are encrypted, and the replay adversary, A, cannot replace the MN or the RN identities. Once more, A would need to know the shared key first in order to perform such attack.

5.6 EM^3A Performance Evaluation

5.6.1 Computation and Communication Overheads

In this section, we aim at evaluating the EM^3A scheme compared to previous multihop authentication schemes. Illustrating the computation and communication overheads for EM^3A comparisons, Tables 5.1 and 5.2 use T and B to represent the required time and the transmitted bytes for an operation, respectively. In these comparisons, we use Crypto++ benchmark [18] to calculate the time and bytes needed for each scheme. The AES and RSA 1024 schemes are used as symmetric and public key operations, respectively, in order to calculate the computation time required by the different schemes. The round trip time (RTT) considered between the MN and RN is 5 ms.

Table 5.1 Computation Overhead

Scheme	Computation Overhead	Time (ms)
AMA [13]	$T_s + T_v \times Pr_{check}$	2.55
GMSP [8]	$T_s + T_v + T_c$	2.60
Multihop MIP [9]	$T_c + T_{EAP}$	0.0194
ALPHA [12]	$T_c + T_{disclose}$	7.5094
EM^3A	$2 \times T_c$	0.0194

As depicted in Table 5.1, EM^3A has the smallest computation overhead among other schemes, because EM^3A requires only two symmetric-key encryption operations $(2 \times T_c)$. On the other hand, the AMA scheme [13] and the generalized multihope security protocol (GMSP) [8] consume much time for signing and verifying signatures (T_s, T_v); hence their computation overheads are higher than that in EM^3A. Similar to our proposed scheme, the multihop MIP scheme [9] consumes small time computation; however, it requires high communication overhead to exchange a large number of keys. Moreover, ALPHA [12] requires an extra time $(T_{disclose})$ to delay the disclosure of the secret key.

For the communication overheads, depicted in Table 5.2, we observe that AMA, GMSP, and multihop MIP require transmitting a sender's certificate in each transmitted message. Instead, the $EM^3 A$ scheme exchanges the list of MAGs only once at the key establishment phase, and the challenge/response messages $(B_{CHL-RESP})$ during handovers. To illustrate this, consider the average length of the sender certificate is 3500 bytes, while the list of MAGs has a length of $n\log_2 n$ bits, where n is the number of MAGs in the PMIP domain. In order for EM^3A to have a higher communication overhead than that in the other schemes, it would require to satisfy the condition $n\log_2 n \geq 28000 bits \times m$, where m is the number of transmitted messages in the certificate-based schemes. Consequently, n should be at least $236.64\sqrt{m}$ to satisfy such a condition. However, since n, number of MAGs in the domain, is

Table 5.2 Communication Overhead

Scheme	Communication Overhead
AMA [13]	B_{cert}
GMSP [8]	B_{cert}
Multihop MIP [9]	$B_{EAP} + B_{key-exchange}$
ALPHA [12]	$B_{ACK} + B_{disclose}$
EM^3A	$B_{FMAGs-list}$

a fixed value, and *m*, number of transmitted messages during an active session, increases over time with the length of active sessions, *n* becomes much smaller than *m* with time. Therefore, the condition cannot be satisfied and the EM^3A commutation overhead is lower, compared with the certificate-based scheme.

5.6.2 Simulation Results

To evaluate the overall performance of the network when using the EM^3A scheme, experiments were conducted through simulations using the OMNeT++ network simulator. We focus on the MN, which experiences handovers that involve the use of RNs. The RTT between LMA and MAGs is fixed to 10 ms. A server for the downloading of data traffic is located in an external network, so that RTT between server and LMA is 20 ms. In our simulation, because the MN is moving at different speeds, the frequency of handovers vary from one every 10 s to one every 50 s (i.e., highly dynamic and slow changing scenarios). We also consider the worst-case scenario in which every time the MN roams to a new MAG, it first connects to an RN, so that EM^3A authentication is required before the exchange of neighbor discovery packets and PMIP signaling may happen. Other details of the simulation parameters are provided in Table 5.3.

Figure 5.5a shows the average throughput achieved for the multihop-enabled PMIP when the EM^3A scheme is de-activated and activated, respectively. It can be observed that the authentication scheme does not negatively impact the communications and the achieved performance is almost equivalent to that achieved when no authentication has been activated. Thanks to the registration phase, which is executed when every node first joins the PMIP domain, EM^3A requires only one RTT between an MN and an RN before allowing for the continuation of normal handover signaling (i.e., the forwarding of RS from the MN to the MAG, the PMIP signaling between MAG and LMA, and the RA message sent back to the MN). The downside of such a registration phase is the overhead and storage required for sending and maintaining the list of current identities for all the MAGs in the domain.

To better illustrate the impact of EM^3A, we provide the details for the handover delay obtained during highly dynamic and slowly changing scenarios

Table 5.3   Simulation Parameters

PHY layer	2.4 GHz, 5.5 Mbps, 100 mW Tx power, −110 dBm sensitivity
MAC layer	802.11 ad hoc mode, 150 m radio range
Traffic type/rates	UDP/VBR video (mean 600 Kbps)
	VBR audio (mean 320 Kbps), CBR best effort 100 Kbps
Session time	~3 min

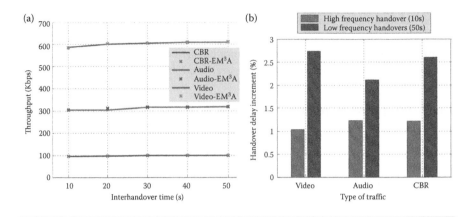

Figure 5.5 Comparison of performance between *EM³A* and nonsecure multihop-enabled PMIP. (From S. Taha, S. Céspedes, and X. Shen, EM3A: Efficient mutual multihop mobile authentication scheme for PMIP networks, in *Proceedings of IEEE ICC 2012*, Ottawa, Canada, June 10–15 2012.)

in Figure 5.5b. When the *EM³A* has been activated, the delay increases by ~1.1% and ~2.5% in each scenario. Consequently, the low computation overhead of the symmetric key encryption/decryption operations makes the authentication process a lightweight mechanism for securely using multihop communications in PMIP domains.

Figure 5.6 illustrates the performance of the network in terms of packet losses for real time (audio and video) and best effort traffic. In general, the use of the authentication scheme does not present a major impact compared to the nonsecure multihop PMIP. In the most demanding scenario, where handovers occur every 10 s, a low 0.03% average increment among the three types of traffic results due to the delay caused by the processing of *EM³A* traffic. In the case of medium-to-slow changing scenarios, packet losses remain as low as 1%, and *EM³A* accounts only for a 0.01% increment.

5.7 Case Study: MA-PMIP for Asymmetric VANET

A multihop authenticated PMIP (MA-PMIP) for asymmetric VANETs [6] is a new proposed scheme, which demonstrates that multihop paths are a useful tool for improving the performance of infotainment applications and Internet access in vehicular environments. MA-PMIP also handles the asymmetric links commonly found in VANET scenarios. More importantly, MA-PMIP integrates the mutual authentication scheme *EM³A* in order to thwart authentication attacks when handovers occur through I2V2V communications.

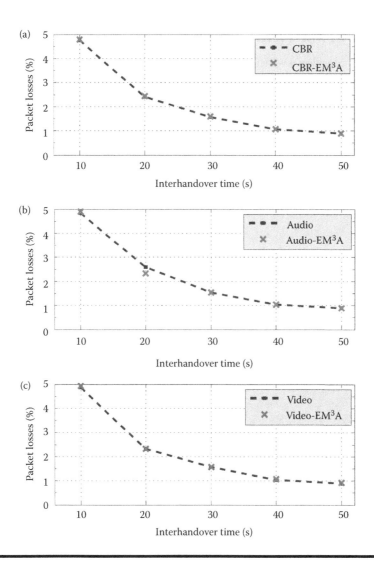

Figure 5.6 Average packet losses obtained by *EM³A* **compared to nonse-cure multihop-enabled PMIP. (From S. Taha, S. Céspedes, and X. Shen, EM3A: Efficient mutual multihop mobile authentication scheme for PMIP networks, in** *Proceedings of IEEE ICC 2012***, Ottawa, Canada, June 10–15 2012.)**

In the following subsections, we first explain the network model for the MA-PMIP scheme. Then, we briefly explain the handover operation through I2V2V communications in MA-PMIP. Afterward, we present a comparison evaluation of the computation overhead of *EM³A* when implemented with MA-PMIP and compared to previous schemes. For more details about the MA-PMIP scheme, the reader is referred to ref. [6].

5.7.1 MA-PMIP Network Model

MA-PMIP considers a vehicular communications network such as the one shown in Figure 5.7. Connections to the infrastructure are enabled by means of road-side ARs, each one is in charge of a different wireless access network. Vehicles are equipped with wireless interfaces, as well as GPS systems that feed a satellite service from which the vehicle location is obtained. Beacon messages are employed by vehicles to relay their location, direction, speed, acceleration, and traffic events to their neighbors.

The ARs consume a higher transmission power than the one consumed by vehicles. Therefore, the presence of asymmetric links in the VANET is considered. The delivery of packets is assisted by a geographic routing protocol. To serve this protocol, a location server stores the location of vehicles, and is available for providing

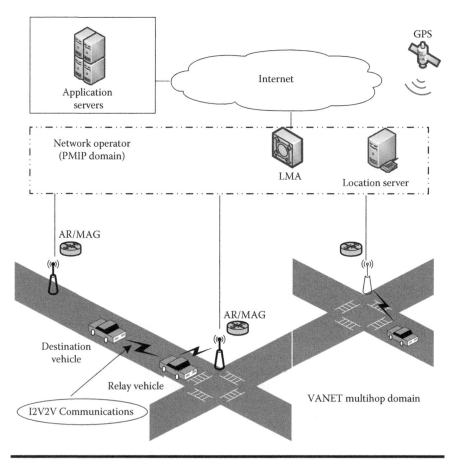

Figure 5.7 MA-PMIP network model. (From S. Céspedes, S. Taha, and X. Shen, A multihop authenticated proxy mobile IP scheme for asymmetric VANET, *IEEE Transactions on Vehicular Technology,* **62(7), 3271–3286, 2013.)**

updated responses to queries made by the nodes' geo-networking layer. In order to forward packets within the multihop VANET, a virtual link between AR and vehicle is created [19]. That means that a geo-routing header is appended to each packet, where the location and geo-identifier of the recipient are indicated. In this way, the geo-routing layer is in charge of the hop-by-hop forwarding through multihop paths, with no need of processing the IP headers at the intermediate vehicles.

The scheme only considers IP-based applications accessed from the VANET. Such applications are hosted in external networks that may be private (for dedicated content), or public, such as the Internet. Since PMIP is selected for handling the IP mobility in the network, the communication model is consistent with the one described in Section 5.3.

5.7.2 MA-PMIP Handover Operation

Figure 5.8 shows the basic MA-PMIP signaling employed when a vehicle experiences a handover through a relay and it has moved to the service area of a new MAG. On the other hand, if the vehicle losses one-hop connection toward the old MAG, but it is still inside the same service area, then no IP mobility signaling is required and packets are forwarded by means of the geo-routing protocol.

In Figure 5.8, after movement detection occurs, the RS message is an indicator for others (i.e., relay vehicle and MAG) of the vehicle's intention to reestablish a connection in the PMIP domain. Thus, an authentication is required to ensure that

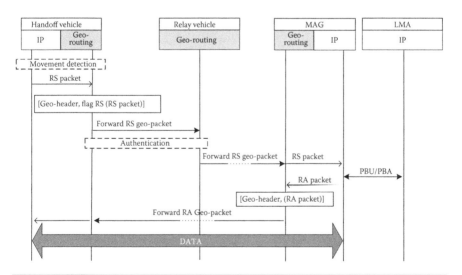

Figure 5.8 MA-PMIP handover through I2V2V communications. (From S. Céspedes, S. Taha, and X. Shen, A multihop authenticated proxy mobile IP scheme for asymmetric VANET, *IEEE Transactions on Vehicular Technology*, 62(7), 3271–3286, 2013.)

both mobile router and relay are legitimate and are not performing any of the attacks described in Section 5.3.2. Therefore, the *EM³A* scheme is employed to implement the authentication required in MA-PMIP. Once the nodes are authenticated, the RS packet is forwarded until it reaches the MAG, and the PMIP signaling is completed in order to maintain the IP assignment at the vehicle's new location.

5.7.3 Authentication Evaluation in MA-PMIP

To measure and compare the impact of the MA-PMIP authentication mechanism, an implementation of AMA [13] has been integrated with a simplified version of a multihop PMIP scheme (i.e., MA-PMIP with *EM³A* disabled). Figure 5.9 evaluates the MA-PMIP authentication mechanism, compared to previous multihop authentication schemes, in order to calculate the computation time required by the different schemes.

Figure 5.10 shows the authentication delay when the vehicle moves at different average velocities. Figure 5.11 depicts the comparison in terms of authentication overhead to payload ratio. As shown in both figures, MA-PMIP not only requires smaller delay and communication overhead than multihop PMIP and AMA, but also has almost fixed impact for different velocities. On the other hand, multihop PMIP and AMA have authentication delay and communication overheads that increase almost linearly with velocity. Compared with multihop PMIP and AMA, MA-PMIP achieves 99.6% and 96.8% reductions in authentication delay and communication overhead, respectively. The reason for these reductions is the high

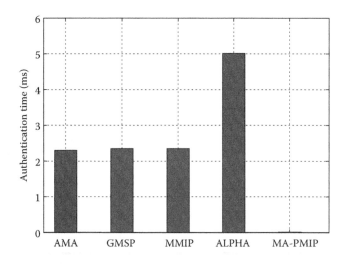

Figure 5.9 Comparison of computation time. (From S. Céspedes, S. Taha, and X. Shen, A multihop authenticated proxy mobile IP scheme for asymmetric VANET, *IEEE Transactions on Vehicular Technology,* **62(7), 3271–3286, 2013.)**

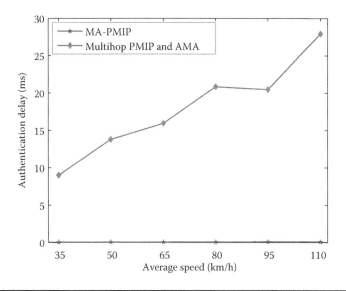

Figure 5.10 MA-PMIP authentication delay.

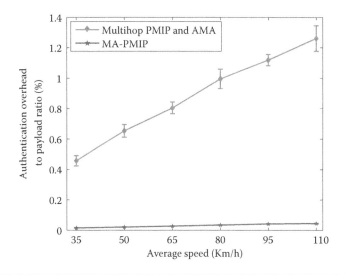

Figure 5.11 MA-PMIP communication overhead. (From S. Céspedes, S. Taha, and X. Shen, A multihop authenticated proxy mobile IP scheme for asymmetric VANET, *IEEE Transactions on Vehicular Technology,* 62(7), 3271–3286, 2013.)

computation and communication efficiency achieved by our proposed authentication scheme. Therefore, unlike multihop PMIP and AMA, MA-PMIP can be used with seamless mobile applications, such as VoIP and video streaming.

5.8 Summary

In this chapter, different approaches to achieve authentication in IP mobility-enabled multihop wireless networks have been reviewed. In addition, an efficient multihop authentication scheme, EM^3A, has been introduced to be employed between an MN and an RN in a multihop-enabled Proxy Mobile IP (PMIP) mobile network. With EM^3A, both the MN and the RN guarantee the legitimacy of each other, and construct a shared key using a novel proposed symmetric polynomial-based key establishment scheme.

The multihop communications studied in this chapter are those occurring between an MN and an RN, when the MN intends to maintain a connection to the infrastructure. In such a multihop communication, both the MN and the RN belong to the same PMIP domain while there is no presecurity association between them. Therefore, we employ the symmetric polynomials to generate a shared key between those arbitrary nodes. Unlike previous symmetric polynomial schemes that are used to generating a shared key between two arbitrary nodes, our proposed scheme increases the secrecy level achieved in the key establishment from t–secrecy to $t \times 2^n$–secrecy. Hence, colluder attacks require at least $t \times 2^n + 1$ colluders, instead of only $t + 1$ colluders, in order to reveal the shared key and break the system's security.

Furthermore, we have mitigated the problem of MN's revocation by proposing new security steps that also achieve MN backward secrecy. Unlike traditional revocation methods that change the whole system keys, our revocation scheme changes only the keys of the MNs, which share the same first-attached MAG. However, the cost of achieving the EM^3A scheme is the storage space required to store the MAG list of identities in each legitimated MN.

From the security perspective, we have showed that EM^3A thwarts internal adversaries, including impersonation and collusion attacks, and external authentication adversaries, including replay, MITM, and DoS attacks. In addition, we have shown that our proposed scheme achieves a higher secrecy level than that achieved by other symmetric polynomial authentication schemes, and lower computation and communication overheads than those achieved by multihop authentication schemes. By means of simulations, we have showed that EM^3A results in a low delay and allows for seamless communications even in highly mobile/highly traffic-demanding scenarios.

To show the efficacy of the EM^3A scheme, we have also presented the case study of MA-PMIP, a multihop vehicular network scheme that supports IP mobility and implements the EM^3A authentication scheme. Compared to other authentication schemes, the implementation of EM^3A in MA-PMIP achieves 99.6% and 96.8% reductions in authentication delay and communication overhead, respectively.

5.9 Future Research Directions

In this section, we address the future directions related to the multihop authentication for wireless networks, namely, EM^3A for arbitrary MNs, and mutual authentication for sensor networks.

5.9.1 EM³A for Arbitrary Mobile Nodes

In EM^3A, a MN should first directly connect to one of the MAGs before using a multihop authenticated communication with as RN. The reason of this condition is because we employ the identity of the MNs' first attached MAG in the evaluated symmetric polynomial, as illustrated in Section 5.4.3. Therefore, each MN needs to add its first attached MAG's identity to calculate a shared key with the RN. However, the direct communication between an MN and a MAG reduces the MN's mobility, and hence presents an impractical scenario for multihop wireless networks such as VANETs. Furthermore, deleting the MAGs' identities from the evaluated symmetric polynomials would change EM^3A to a pure centralized symmetric polynomial-based with a low secrecy level and high revocation overhead. Thus, a new direction is required to alleviate the trade-off between the security level and the network performance.

5.9.2 Mutual Authentication for Sensor Networks

Mutual authentication schemes are envisioned for multihop communications among sensors to securely relay transmitted data in sensor networks. Employing the proposed EM^3A scheme in such multihop communications is a challenge due to the limited storage space in sensors. However, EM^3A requires storage space to maintain the list of MAGs' identities at each sensor. One alternative is to divide the sensor networks into clusters, in which only the cluster head, that is, one sensor, needs to store the identities list while the other cluster members trust the information stored in their cluster head.

References

1. M. Grossglauser and D. Tse, Mobility increases the capacity of ad hoc wireless networks, *IEEE/ACM Transactions on Networks*, 10(4), 477–486, 2002.
2. M. E. Mahmoud and X. Shen, PIS: A practical incentive system for multihop wireless networks, *IEEE Transactions on Vehicular Technology*, 59(8), 4012–4025, 2010.
3. R. Blom, An optimal class of symmetric key generation systems, in *Proceedings of the 84th Workshop on the Theory and Application of Cryptographic Techniques, EUROCRYPT'85*, Linz, Austria, 1985, pp. 335–338.
4. A. Gupta, A. Mukherjee, B. Xie, and D. P. Agrawal, Decentralized key generation scheme for cellular-based heterogeneous wireless ad hoc networks, *Journal of Parallel and Distributed Computing*, 67(9), 981–991, 2007.

5. K. Pillai and M. Sebastain, A hierarchical and decentralized key establishment scheme for end-to-end security in heterogeneous networks, in *Proceedings of the IEEE International Conference on Internet Multimedia Services Architecture and Applications, IMSAA 2009*, Bangalore, December 9–11, 2009, pp. 1–6.

6. S. Céspedes, S. Taha, and X. Shen, A multi-hop authenticated proxy mobile IP scheme for asymmetric VANET, *IEEE Transactions on Vehicular Technology*, 62(7), 3271–3286, 2013.

7. C. Tang and D. Wu, An efficient mobile authentication scheme for wireless networks, *IEEE Transactions on Wireless Communications*, 7(4), 1408–1416, 2008.

8. B. Xie, A. Srinivasan, and D. Agrawal, GMSP: A generalized multi-hop security protocol for heterogeneous multi-hop wireless network, in *Proceedings of IEEE Wireless Communications and Networking Conference, WCNC 2006*, vol. 2, Las Vegas, USA, April 3–6, 2006, pp. 634–639.

9. A. Al Shidhani and V. C. M. Leung, Secure and efficient multi-hop mobile IP registration scheme for MANET-internet integrated architecture, in *Proceedings of IEEE Wireless Communications and Networking Conference, WCNC 2010*, Sydney, Australia, April 18–21, 2006, pp. 1–6.

10. M. Catur Bhakti, A. Abdullah, and L. Jung, EAP-based authentication with EAP method selection mechanism, in *Proceedings of the International Conference on Intelligent & Advanced Systems, ICIAS 2007*, Kuala Lumpur, Malaysia, November 25–28, 2007, pp. 393–396.

11. Y. Jiang, C. Lin, X. Shen, and M. Shi, Mutual authentication and key exchange protocols for roaming services in wireless mobile networks, *IEEE Transactions on Wireless Communications*, 5(9), 2569–2577, 2006.

12. T. Heer, S. Götz, O. G. Morchon, and K. Wehrle, Alpha: An adaptive and lightweight protocol for hop-by-hop authentication, in *Proceedings of the 4th ACM International Conference on Emerging Networking Experiments and Technologies, ACM CoNEXT'08*, Madrid, Spain, December 9–12, 2008, pp. 23:1–23:12.

13. N. Ristanovic, P. Papadimitratos, G. Theodorakopoulos, J.-P. Hubaux, and J.-Y. Le Boudec, Adaptive message authentication for multi-hop networks, in *Proceedings of 8th International Conference on Wireless On-Demand Network Systems and Services, WONS 2011*, Bardonecchia, Italy, January 26–28, 2011, pp. 96–103.

14. M. Asefi, S. Cespedes, X. Shen, and J. W. Mark, A seamless quality-driven multi-hop data delivery scheme for video streaming in urban VANET scenarios, in *Proceedings of IEEE ICC 2011*, Kyoto, Japan, June 5–9, 2011, pp. 1–5.

15. M. J. Lee, R. Zhang, J. Zheng, G.-s. Ahn, C. Zhu, T. R. Park, and S. R. Cho, IEEE 802.15.5 WPAN mesh standard-low rate part: Meshing the wireless sensor networks, *IEEE Journal on Selected Areas in Communications*, 28(7), 973–983, 2010.

16. G. Hiertz, Y. Zang, S. Max, T. Junge, E. Weiss, B. Wolz, D. Denteneer, L. Berlemann, and S. Mangold, IEEE 802.11s: WLAN mesh standardization and high performance extensions, *IEEE Network*, 22(3), 12–19, 2008.

17. S. Taha, S. Céspedes, and X. Shen, "EM3A: Efficient mutual multi-hop mobile authentication scheme for PMIP networks," in *Proceedings of IEEE ICC 2012*, Ottawa, Canada, June 10-15 2012.

18. W. Dai, Crypto++ 5.6. 0 benchmarks, http://www.cryptopp.com/benchmarks.html.

19. J. Choi, Y. Khaled, M. Tsukada, and T. Ernst, IPv6 support for VANET with geographical routing, in *Proceedings of 8th International Conference on ITS Telecommunications, ITST 2008*. Phuket, Thailand: IEEE, October 24–24, 2008, pp. 222–227.

Chapter 6

Detection of Misbehaving Nodes in Vehicular Ad Hoc Network

Rajesh P. Barnwal and Soumya K. Ghosh

Contents

6.1 Introduction

In the era of information and communication technology, the convergence of computing and telecommunications (fixed and mobile) paves the way for development of new technologies, services, and applications in all aspects of life. Vehicular Ad hoc Network (VANET) is one such promising technology, envisioned to appear in a big way in the near future. The facts from the US National Highway Traffic Safety Administration (NHTSA) reveal that a total of 33,808 people died in traffic accidents in 2009 [1]; however, the situation is more disastrous in a country like India, where in the same year, a total of 1,25,660 people died in 4,86,384 road accidents [2]. These statistics demand that technology like VANET should be rolled out as soon as possible to put a brake on increasing road fatalities.

Considering the phenomenal growth in road traffic activities worldwide, it becomes more important to make the driving safer and easier with the help of new technologies. In spite of various innovations in manufacturing technology and improvement in vehicle design, road accidents and the safety of the passengers continues to be a matter of major concern. The most effective solution lies in getting timely and accurate information about the forthcoming danger, and this is where VANET comes into play. In addition to safety issues, it is also predicted that, if VANET technology can be successful be put into work, it may help in fulfilling the dream of ubiquitous and pervasive computing. Evidently, past decades have witnessed a number of projects initiated to standardize the VANET technology. Before successful implementation of these technologies, making it feasible and acceptable in real environments, there are many issues and challenges that need to be addressed by the researchers working in the area of information and communication technology.

Privacy and security are the two challenging issues to overcome before rolling out the VANET in real scenarios [3]. Privacy concerns have been handled to a great extent with the introduction of pseudonyms (the dummy short-life identifiers) for the VANET nodes. However, security is still a matter of concern. The multihop nature of VANET makes it more vulnerable to security attacks in comparison to other single-hop wireless networks [4]. For this reason, security in VANET attracted the interest of researchers in recent years. In past decades, a number of articles [5–9] stressed upon the need of security and identified its challenges for the specific case of VANET. The CAMP report [10,11] describes various futuristic VANET applications. Security, particularly in case of safety applications, is of more concern due to the involvement of human lives.

Out of several security concerns, the misbehaving vehicles inside the network are the major threats to the VANET and participating nodes. The unknown/unauthenticated attacker can be detected and filtered out with the help of IEEE 1609.2 standards [12] based Public Key Infrastructure (PKI) security mechanism. However, the main problem arises when an authenticated node does not behave properly in the network. This type of node is known as a misbehaving node.

Nodes misbehave either unintentionally, due to malfunctioning of its sensing equipment or intentionally, for taking illegal benefits or creating trouble in the peer-to-peer network. Irrespective of the reasons, misbehaving nodes pose a big threat to VANET. Moreover, it is unsafe for the communicating vehicles to blindly rely on the received messages that are originating from such misbehaving nodes.

This chapter discusses the misbehavior-detection problem and available Misbehavior Detection Scheme (MDS) for the VANET environment. Section 6.2 gives a brief introduction of VANET and its applications. Section 6.3 discusses various potential misbehavior problems in VANET. Section 6.4 gives an idea of available security layer in the VANET communication standard and its inability to detect the misbehavior detection problem in VANET. Section 6.6 enumerates various issues and challenges of designing a good MDS for VANET. Section 6.7 presents currently available propositions for misbehavior detection in VANET. Section 6.8 discusses future research direction in the area of misbehavior detection in VANET, and Section 6.9 concludes the chapter.

6.2 Vehicular Ad Hoc Network

VANET is the short-range wireless communication technology that enables dynamic creation of the Mobile Ad hoc Network (MANET) for exchange of information messages among moving vehicles and between vehicles and roadside infrastructures. VANET is a special kind of MANET, where communication nodes are vehicles with sensing, processing, and communication capabilities. From the perspective of mobility, the movement of nodes in VANET are guided by various infrastructural and temporal factors like road structure and traffic conditions; however, the nodes movement in MANET can move in any possible directions. Figure 6.1 depicts a typical VANET.

The major motivation behind the VANET is to increase the safety of road travelers and traffic optimization. Thousands of people are killed worldwide due to road accidents yearly and many more are getting seriously injured. In addition to this, congestion/traffic jams are major concerns in ever-increasing road traffic scenarios. Both these problems can be solved or mitigated to a great extent by giving timely information to the drivers using cooperative intervehicular communication for vehicle safety. In addition to this, traffic advisory, driving assistance, infotainment, and so on, are some of the applications which are compelling major

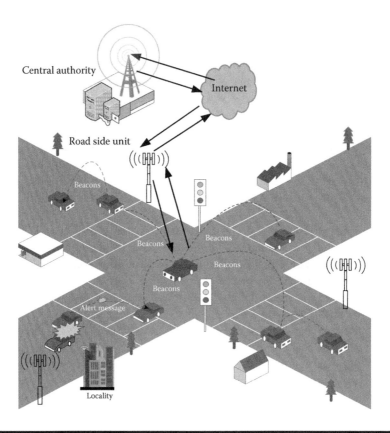

Figure 6.1   A typical vehicular ad hoc network.

car-manufacturing giants to shift their research focus toward telematic, crash avoidance, and collision mitigation in vehicles.

The typical components of a VANET-enabled vehicle are comprised of computer-controlled devices and radio transceivers for exchange of messages. The standardized protocol for communication in VANET is Dedicated Short Range Communication (DSRC), which is for working in a transmission range of 100–1000 m [10]. In addition, the roadside base stations are there to assist the drivers by providing the necessary information throughout the journey. The computer-controlled device on the vehicle is also popularly known as On-Board Unit (OBU), which is a computation and storage device with two-way short-range communication capability. OBU is powered by the car battery and, thus, battery power is normally not a constraint in VANET. Additionally, the OBU gets its input from other onboard sensors such as the Global Positioning System (GPS), speed sensors, tilt sensors, steering angle sensors, heading sensors, and so on. However, the fixed infrastructure of VANET is comprised of a Road Side Unit (RSU), which is also supposed to have built-in computation and storage capability. RSU is meant to provide a link between the nearby vehicles to backend infrastructure through wired/wireless connectivity.

6.2.1 Characteristics of VANET

VANET has some unique characteristics, which makes it different from other communication networks including MANETs. The typical characteristics of VANET are as follows:

■ High mobility: The mobility of the nodes is very high in comparison to any other MANET. The high mobility of nodes poses specific challenges, namely very dynamic network topologies, problems of frequent network partitioning, and ephemeral connectivity among nodes and unknown neighbors.

■ Absence of infrastructure: VANET are supposed to work in the absence of a fixed infrastructure. The communication between nodes, if not accessible through direct radio links, take place in multihops through intermediate nodes.

■ Security and privacy requirements: Multihop communications in the absence of infrastructure make VANET a potential target for diverse types of attacks such as eavesdropping, packet dropping, denial of service, modification of messages, and so on. Moreover, privacy of the user is a major issue and poses a challenge for implementation of security schemes.

■ Restricted mobility zone: The mobility of nodes in the vehicular network is restricted by the road topology and thus can be very much predicted. The vehicles move on existing roads, which enables the system to take advantage of road information for decision making.

■ Sufficient power supply: VANET nodes are less concerned about energy as the vehicles have sufficient battery power supply.

6.2.2 VANET Applications

VANET application areas can be classified [11] into three major categories:

1. Safety applications
2. Convenience applications
3. Commercial applications

Safety applications are supposed to assist the drivers in achieving safe and comfortable driving. These include the systems which can monitor surrounding roads, nearby vehicles, and other environmental features to avoid road hazards for enhancing the traveler's safety. Cooperative Collision Warning (CCW) and Slow Vehicle Adviser (SVA) are two such safety class applications.

Convenience applications include all those applications that are for the purpose of better traffic management to facilitate the traffic authority and enhance the convenience of the drivers. Congested Road Notification (CRN) and Parking Availability Notification (PAN) are the examples of this class of VANET applications, which detects and notifies about road congestions and availability of parking slots, respectively.

Commercial applications are targeted to provide various infotainment services such as onboard entertainment, web access, audio–video streaming, and so on. Applications for Remote vehicle personalizations/ diagnostics are examples of commercial VANET applications. Content Map Database Download (CMDD) is another application in this category, which acts as a portal for receiving valuable information from mobile hotspots or home stations.

In spite of various potential applications of VANET, the major motivation behind VANET is to enhance road safety. Some of VANET safety-related applications [10] are as follows:

- Cooperative collision warning: This application warns a vehicle if it is attempting to collide with other vehicle. The collision avoidance can be achieved by use of information gathered through OBU sensors and other received messages from other vehicles in the vicinity.
- Slow vehicle advisor: The slow vehicle advisor application is meant for raising alert notifications in the event of possible danger due to increased speed of the vehicle. The alert triggers, based on information such as current vehicle speed, brake, throttle position, steering angle, and so on, and the neighborhood information such as road curve, approaching vehicle on the same lane, and so on.
- Public safety: This application gives warning to driver about the approaching emergency vehicle to avoid unnecessary safety hazards. The alert is based on compilation of the information received from the messages of emergency vehicles. The broadcast message includes lane information, speed, and intended path.
- Postcrash warning: This safety application is intended to avoid postcrash danger and inconvenience. The application warns the approaching vehicle about the approaching accident site. This kind of information may extend its message reach to several kilometers depending upon the availability of traffic density and roadside infrastructure.
- Sign extension: This application extends the roadside signs to the approaching vehicles to reduce the potential dangers due to carelessness of drivers. The application includes school zone warnings, animal crossing zone warnings, road characteristic information, and so on.
- Emergency electronic brake light: This is an accident/collision avoidance warning that may arise due to sudden and hard brake engagement by the leading vehicle in the same lane. This application sends the message to the

approaching vehicle about the sudden and hard engagement of brakes by the driver. The emergency electronic brake light is very useful in limited visibility situations.

■ Lane change warning: This application warns approaching vehicles about intended lane-changing activity by the leading vehicle in the lane. Based upon periodic messages about position, heading, and speed of surrounding vehicles, the applications predict the possibility of the danger that may arise due to a reduction in the gap between the leading and approaching vehicle during the lane-changing process.

From the above discussions, it is clear that the objective of achieving vehicle safety has now become dependent on safety applications. This, in turn, is dependent on proper information exchange among the vehicles. For instance, a car can send a crash alert even if there is no crash or the sensing module of the vehicle is faulty, therefore giving wrong information in sent messages. These nodes, irrespective of reasons, are not conforming to the expected behavior of a genuine peer and are thus categorized as misbehaving nodes.

6.3 Misbehavior Problems in VANET

Misbehavior, as a word, refers to the behavior of an individual or community that breaches the rule or causes discomfort or annoyance to others. In general, the word defines the improper behavior of living beings. However, in past decades the term misbehavior has been extensively used by communication system security researchers to define the unexpected activities of communication entities in the network.

Among other security concerns, misbehavior detection in VANET is one of the most challenging problems [13–18]. An authenticated VANET node is termed as a misbehaved node if it deviates from the expectation of the authority and its peers in the vehicular network. Contrary to different security attacks from illegitimate outsiders, the malicious behavior of a legitimate insider is more difficult to discover in VANET. Moreover, the inherent trustworthiness of an authenticated node makes it more immune from suspicion of the peers and, thus, may have much larger devastating effects on the network [14]. This is a situation where the insider node becomes untrustworthy even after passing through the highest authentication system of the network. Misbehavior is thus a very difficult problem to detect and mitigation of such a problem in a privacy-preserving mode is more challenging by nature. Therefore, it is of utmost necessity that a vehicle should be able to detect the false position or kinematics information of beacons/alert messages and identify the nodes responsible for the same.

Vehicles in a vehicular network exchange their *position, speed, heading, steering angle,* and other such information by transmitting periodic beacon messages [10].

Most of the safety applications rely on the position and kinematics information contained in beacon and alert messages received from other vehicles in the network. Moreover, it is also imperative that routing decisions in VANET should also be based on the position of destination and other relay nodes in the vicinity. It is highly possible in MANET, like VANET, that a node can manipulate its own position and kinematics information or disseminate incorrect values due to faulty sensing devices. Irrespective of the cause, the false information is always prone to create an accident or, at bare minimum, may result into the failed purpose behind the implementation of VANET. In case of position-based routing, incorrect position or kinematics information severely degrades the throughput of the network [19].

Misbehavior in VANET can be of a different kind: *denial of routing, dropping of relay messages, dissemination of false information, identity spoofing* by an authenticated node, and so on. Though all of these misbehavior areas affect the vehicular network and its stake-holders in one way or another, the dissemination of false positional information by an insider (authenticated node) is a big threat to the VANET users. Thus, a robust misbehavior-detection framework is needed to detect such misbehaving nodes in VANET. It is also required to devise a scheme/method to detect and identify the continuously misbehaving nodes and take proper action against them. Such a scheme is termed as Misbehavior Detection Scheme (MDS).

Based upon the motives, the misbehaviors in VANET can be broadly classified into two classes:

1. Intentional misbehavior
2. Unintentional misbehavior

6.3.1 Intentional Misbehavior

This class of misbehavior results from the wrong intention of networked nodes to get the unnecessary benefits by not cooperating with their peers. Some examples of intentional misbehavior in VANET, for instance, are raising bogus alert messages, dropping/delaying of routing packets, un-cooperative behavior during collective decision making, denial for message relay, spoofing of identity, dissemination of false information to the peers, and so on. Intentional misbehavior may be of several types. In vehicular networks, each node is supposed to be cooperative during route discovery and in the message-routing process. But in adverse situations, a node in the network may start denying the routing request of others or simply starts dropping the received routing packets. This is a case of intentional misbehavior, which can be exhibited by the node either behaving just for fun or due to selfish reasons, like saving of computation resources and energy conservations. Another example of intentional misbehavior can be demonstrated by the real malicious node who wants to attack and collapse the vehicular network by infusing false information in the messages. This kind of misbehaving node is more detrimental to the success of vehicular networks. Intentional misbehavior mainly results from human involvement, which in turn affects the behavior of a vehicular node in VANET. For instance, drivers on the road misbehave due to common selfish reasons. A greedy driver can disseminate false crash information to the peers to avoid further congestion on the road. Due to false crash notifications, other vehicles will take diversions. This makes the road clear for greedy drivers and allows them to drive faster on a congestion-free road. Identification, isolation, and penalization of this category of nodes are more important steps to make the VANET more secure from misbehaved nodes.

6.3.2 Unintentional Misbehavior

This class of misbehavior is normally due to reasons that are not in control of the participating users and purely unintentional. The users of misbehaving nodes of this class are usually not aware of his undesired behavior pattern. For instance, generating wrong alert messages due to a faulty OBU, false positional information due to malfunction of onboard sensors, and malicious behaviors by the compromised nodes are some examples of unintentional misbehavior in VANET. Unintentional misbehavior is sometimes common in a real scenario. This can be appreciated by an example. Sensors attached with a vehicle need periodic checking and calibrations to ensure their accuracy. But in practice, the owners reluctantly miss the testing and calibration schedule and due to this, the sensing devices start malfunctioning. The vehicle's OBU, relying on the output of these faulty sensing units, starts disseminating false information in the network. In this situation, though intention is not to infuse false information, but unknowingly incorrect information is inculcated by the participating vehicle. Sometimes, like other computer-based networked systems, some nodes in the network get compromised and become the victim of the attackers. In this scenario, the attacked nodes start working as zombie machines to fulfill the bad intention of the attackers. Though the compromised nodes are innocent, they appear as misbehaving ones in the vehicular network. The identification of such misbehaving nodes and rectification of their problems are of the utmost necessity to avoid inconvenience of other genuinely behaving nodes and enhance the security of the vehicular network.

6.4 Trust and Misbehavior Detection

Trust is another aspect, which can be used for misbehavior detection in a VANET. The verification of correctness of the received data also depends upon the amount of trust placed by the receiver or the sender or originator of the message. Sometimes it may not be possible for detecting misbehavior of the node accurately due to time constraints or paucity of supporting and related information. However, in certain situations, quick decision making is required, even with less amount of information and evidence of correctness of the data received from the other vehicle. In this scenario, conflicting information received from different sources confuses a receiver. Due to false information, it is hard to decide on whom to believe and this is the time when the notion of trust comes into play. The objective behind incorporating trust in VANET is to provide an opportunity to each peer to identify potentially misbehaving nodes in the network so as to discard information originating from them [20].

Trust establishment between communicating vehicles in a highly dynamic scenario like VANET poses major challenges [21]. The vehicle in the network receiving communication from other peers has to ensure the authenticity and trustworthiness of the received messages before reacting to them. However, it is very difficult to say that the vehicle encountered now will also be encountered in the future. Moreover, for the sake of privacy, the provision of pseudonyms made the trust management even more challenging and forced the security researchers to think how the concept of trust can be used in the ephemeral network scenario like VANETs. In the recent past, a number of researchers across the globe proposed various schemes for establishing the trustworthiness of the vehicle in the VANETs, but all have their own strengths and limitations. In the normal situation, the trust used to be entity-centric but some of the recent papers argue that data-centric trust is more suitable in case of highly dynamic situations like VANETs.

6.4.1 Trust Establishment in VANET

In VANET, trust can be established in two ways: infrastructure based or static trust and self-organization based or dynamic trust [22].

- Infrastructure based or static trust: In infrastructure-based static option, the trust establishment process relies mainly on common, global, trusted, and well-known system parameters received from a central certification authority or such, which can be used for message authentication. This kind of trust remains static over time.
- Self-organization based or dynamic trust: In dynamic option, the trust establishment process lacks the global knowledge and single point of control. Thus, in absence of any online connection to a security infrastructure, decision regarding trust to other nodes must be made autonomously only based on partial information that is collected from unknown nodes during a short period of time only. Further, it may not be possible to always have the presence of infrastructure on the road, especially in the situations like highway. In this situation, self-organizing trust mechanism may be helpful.

Self-organizing trust establishment can be further classified into three categories:

- Direct trust: Direct trust establishment mechanisms mainly depend upon the self-observations method, where a car takes trust decisions according to its own past/present experience. This is based on mutual communication with other nodes.
- Indirect trust: In case of indirect method of trust establishment, the node takes trust decisions based on feedback from other neighbors and their trust relationships. This mechanism mainly depends upon the amount and quality of information received from other nodes about their experience with the subject node.
- Hybrid trust: Hybrid mechanism provides the advantages of both, direct as well as indirect mechanisms, where the trust decision taken by a node is not only based on its self-observations but also used to influence by the feedback from the neighboring nodes.

A number of works have been reported in the recent past on trust establishment in VANET. Schmidt et al. [34] proposed a trust calculation scheme based on various data values obtained from the sensors of observing vehicles. The scheme requires continuous exchange of information between the vehicles, which are used by observing the vehicle to analyze the behavior of its neighbor nodes and calculation of their trustworthiness. After obtaining the value of trust for a vehicle, the scheme uses it to identify a misbehaving vehicle.

M. Raya et al. [13] introduced data-centric trust establishment in ephemeral ad hoc networks such as VANET. The paper argued that node-specific reputation-based models would not be of much use due to the provision of pseudonyms. The scheme assigns default values of trustworthiness based on the type of node. For instance, police vehicles have a higher default trust level than private vehicles. Similarly, event- or task-specific trustworthiness is also assigned depending upon the type of event. Finally, trust ranks are proposed to be calculated using the weighted average of node-specific trust, event-specific trust, and security-status binary value. The authors suggested the use of *Bayesian inference* and *Dempster–Shafer Theory* to aggregate these values. However, the implementation of the system is not discussed.

Kim et al. [35] proposed a message-filtering model using multiple complimentary sources of information to filter out the malicious messages coming from the misbehaving vehicles. The model is based on two main components: a threshold curve and *certainty of event* (CoE) curve. An alert is only triggered by the OBU when the event certainty surpasses a predefined threshold. The model works based on assumption that the majority of vehicles are not malicious and multiple vehicles observe the same event. In paper, the suggested framework has been applied for misbehavior detection using EEBL messages.

Zhang [20] discussed various challenges of trust management in VANET and enumerated the desired properties of an effective trust management. The author presented a survey of various existing trust models in VANETs and finally concluded that none of the trust models satisfy all the properties of an effective trust management scheme in VANET.

6.4.2 Desired Properties of a Good Trust Model

A good *trust model* in VANET to detect misbehaving nodes (selfish as well as malicious nodes) is supposed to pose the following properties [20]:

- Decentralized trust establishment: Trust establishment should be fully decentralized to be applicable to the highly dynamic and distributed environment of VANETs. For any trust-related decisions, a node in VANET should be capable to handle decentralized information sharing and message passing.
- Coping with sparse information: In a VANET scenario, due to absence of sufficient number of participating nodes, there may be lack of information for trust estimation. In this situation, an effective trust model should be able to cope up with the information scarcity problem.
- Event/task and location/time specific: Highly dynamic environments of the peers in VANET is a matter of serious concern during the trust assessment. A good trust model should introduce certain dynamic trust metrics, which can allow capturing this dynamism in terms of event/task or location, as well as time.
- Scalable: A trust model should be capable of handling the scalability issue to handle critical situations; when during a very large size of VANET, each peer should only accept or consult a limited amount of trusted information to make very quick decisions in real time.
- Integrated confidence measure: Uncertainty of any trust model increases with incompleteness of information about the peers in the vicinity. Thus, it is very important to include a confidence measure in trust management to capture the uncertainty of the environment.
- System level security: Security mechanisms at the system level deal with protocols that, among other things, allow peers to authenticate themselves, that is, prove their identity. This is important because most of the trust-building models assume that a peer can be uniquely identified.
- Sensitive to privacy concerns: Privacy is an important concern in a VANET environment. In this environment, revealing a vehicle owner's identity may allow a possibly malicious party to cause damage to the owner. Thus, the trust model should be concerned about privacy and avoid such situations and protocol where risk of breach of privacy is possible.
- Robustness: Sometimes trust management itself may be a target of attacks and can be compromised. Thus, a good trust model should also be resistant to attacks like the Sybil attack, newcomer attack, betrayal attack, inconsistency attack, badmouthing/ballot stuffing attack, and collusion attack.

6.5 Security Layer in VANET

The security of VANETs is more crucial than any other network due to the involvement of critical life-threatening situations. There may be a case when vehicle movement and position-related vital information are manipulated by a malicious node/person intentionally or unintentionally. This may result in serious danger during the travel, not only for the receiving vehicle but also for all other vehicles in the vicinity. In this situation, it is very much desired that there must be some mechanism by which the system can mitigate or help in mitigating various types of life and property-threatening attacks. Vehicles in VANET mainly communicate by exchange of messages using wireless media. Wireless communication makes the communication most vulnerable to attacks. Thus, most of the attacks in the VANET are mainly performed by targeting these messages. Based on the nature and intention behind attacks, these VANET attackers may be classified into three broad categories, namely, *insider or outsider* attackers, *malicious* or *rational* attackers, and *active* or *passive* attackers [23]. However, various possible categories of message attacks [23] are as follows:

■ Bogus information: In this situation, the attackers may diffuse the false information in the network to affect the behavior of other vehicle (e.g., to divert traffic from a given road and thus free it for themselves). More malicious attackers could impersonate other vehicles or roadside infrastructure to trigger safety hazards. Vehicles could reduce this threat by creating networks of trusted nodes and ignoring, or at least distrusting, information from untrusted senders.

■ Cheating with positioning information: Vehicles are solely responsible for providing their location and positioning information. Attackers can use this data to alter their perceived position, speed, direction in order to escape liability, notably in the case of accident. Unsecured communication can allow attackers to modify or falsify their own position information to other vehicles, create additional vehicle identifiers (also known as a Sybil attack) or block vehicles from receiving vital safety messages. In other situations, an attacker may modify the message exchanged in vehicle-to-vehicle or vehicle-to-RSU communication in order to falsify transaction application requests or to forge responses.

■ Disclosure of ID: In this scenario, the global observer can monitor the trajectories of targeted vehicles and use this data for various commercial or illegal purposes. Location privacy and anonymity are important issues for vehicle users, which involve protecting users' privacy by obscuring the users' exact location in space and time. And thus it is very much required to mitigate the collection of information through eavesdropping.

■ Denial of service: This situation can arise when attackers may want to bring down the VANET or even cause an accident. A denial of service attack may be launched by an aggressive injection of dummy messages into the channel. Additionally, introduction of malware, such as viruses or worms, into VANETs has the potential to cause serious disruption to its operation. These types of attacks are more likely to be carried out by a rogue insider node rather than an outsider and may be introduced into the network, when the OBUs and RSUs receive software and firmware updates. The presence of spam is also a difficult case to control in absence of a basic infrastructure and centralized administration.

■ Masquerading: Masquerading attacks are easy to perform on VANET, as all that is required for an attacker to join the network is a functioning OBU. By posing as legitimate vehicles in the network, outsiders can conduct a variety of attacks such as forming black holes or producing false messages.

■ Tempering of sensors: Malicious nodes may temper with the sensing units of the vehicle. Attacker can also intentionally retrofit the faulty sensors in the vehicles to disseminate false information in the VANET. These types of attacks are easy to launch but are difficult to discover.

The intelligent transportation systems committee of the IEEE Vehicular Technology Society defined a set of standards for wireless access in vehicular environments (WAVE) under the name *IEEE 1609.x* [12]. The main *WAVE* standard is comprised of four substandards as follows:

■ IEEE 1609.1
■ IEEE 1609.2
■ IEEE 1609.3
■ IEEE 1609.4

Out of these four substandards, the *IEEE 1609.2* is dedicated for security service applications and management messages and is the topic for this chapter. Figure 6.2 depicts the schematic showing the relative position of the security services layer in a *WAVE* communication stack.

The security layer defined by *IEEE 1609.2* has the provisions for mitigating the effects of threats arising during data transmission over vehicular communication network. *IEEE 1609.2* defines security services for *confidentiality* (encrypting a message for a specific recipient such that only that recipient can use it), *authenticity* (confirmation of origin of the message), and *integrity* (confirmation that the message has not been altered in transit). In addition to this security services, the *WAVE* system adds an additional requirement, that of *anonymity*, for end users.

6.5.1 IEEE 1609.2 Security Services

The IEEE 1609.2 standard provides specifications for several security services for the VANET environment. These security services include definitions for message formats, message processing for authenticating WAVE management messages,

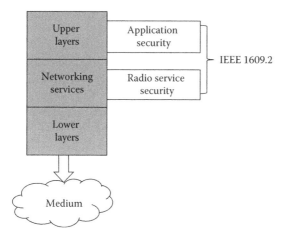

Figure 6.2 Schematic showing WAVE security layer. (From s.l. : IEEE Std 1609.2-2006, 2006. *IEEE trial-use standard for wireless access in vehicular environments—security services for applications and management messages.***)**

vehicle application messages, and encryption of messages intended for known recipients [24]. According to the standard, the entities that provide or use IEEE 1609.2 security services are classified into two broad categories: Certificate Authority (CA) and end entities (user). The CA issue certificates and Certificate Revocation Lists (CRL). However, the user or end entities comprise of all other stakeholders, for example vehicles, RSU, application servers, applications, and so on, which can use IEEE 1609.2 certificates, but is not entrusted to issue certificates or CRLs. The end entities obtain their certificates from the CAs.

A certificate is an electronic document in a cryptographic system that uses a digital signature to bind one or more public keys with a real or virtual identity. The certificate is intended for use by the message receiver to verify that a public key belongs to the claimed identity. The certificate also enlists the permission associated with that public key [24].

The IEEE 1609.2 defines four types of CAs namely *root CA*, *message CA*, *WAVE Services Advertisements* (WSA) CA, and CRL signers.

- *Root CAs* are entrusted to issue certificates to the other three CAs as well as end entities.
- *Message CAs* are responsible for issuing certificates to end entities that send application messages to other entities.
- *WSA CAs* are for issuing certificates to those end entities that send *WAVE Service Advertisements* for the services that are provided by the respective end entities.
- *CRL Signers* issues CRLs to all other entities.

The standard defines three types of end entities named *identified, identified not localized*, and *WSA signers*.

- *Identified and identified not localized* end entities are those entities that are supposed to send secured application messages to other entities. These end entities have to obtain their certificates from the message CAs.
- *WSA Signers* are comprised of those entities that are supposed to send signed WSAs. WSA signers receives their certificates from the WSA CAs.

An IEEE 1609.2 standard permits the encryption of a message before sending the same to the recipient(s). The message sender uses symmetric key algorithm to encrypt each message with a fresh symmetric key. The symmetric key is then encrypted with the public key of the recipient(s) separately before sending the same to intended receivers. The message receiver, upon receiving the encrypted message, first decrypts the symmetric key using its own private key and then uses the decrypted symmetric key for decrypting the actual message.

6.5.2 IEEE 1609.2 and Misbehavior Detection

The IEEE 1609.2 standard provides robust mechanism for authenticity, confidentiality, and integrity of the communications. However, there is no standard mechanism provided for handling the issue of misbehavior that may arise during the communications in VANETs. Moreover, it is always imperative to identify those insiders (authenticated node), which can exhibit improper behavior during the message creation and communication process. The misbehaving entity can be compromised RSU, on-road vehicles, application servers, or even the WSA signers. The detection of static misbehaving entities, for example, RSU, application servers, and WSA signers are relatively easy to detect and can be detected by the CAs eventually. But the misbehavior detections of highly dynamic end entities like vehicles are difficult for CAs without the feedback of the peer vehicles. Moreover, the problem becomes more difficult if an entity demonstrates Byzantine fault [25] and misbehaves intermittently. The centralized misbehavior detection mechanism may also fail in the case when the misbehavior-reporting entity itself is malicious. Thus, detection of misbehavior in a VANET, even in the presence of a security layer in IEEE 1609.2 standard, is a nontrivial task. Therefore, the responsibilities of detecting misbehavior and misbehaving nodes need to be handled separately based on local misbehavior-detection capabilities of the individual nodes.

6.6 Design Issues of MDS

The nodes in VANET generally comply with the standard protocol, but those nodes that deviate from their expected behavior are termed as misbehaving nodes. Misbehavior detection in VANET is a challenging and open problem. Due to the unique characteristics of VANET, misbehavior detection in this kind of MANET has several research challenges. However, contrary to this, VANET also poses some special strengths, which are normally absent in other kinds of MANETs.

6.6.1 Challenges

The challenges [3,26] in designing an MDS of highly mobile, ad hoc, and ephemeral network like VANET are as follows:

- Authentication: Authentication of a communicating entity in VANET is first and foremost security requirement. The entities like central authority, RSUs, or OBUs must be authenticated prior to trusting the information originating from them.
- Availability and fast response time: Critical safety applications warrant all-time availability of network and other communication services. Moreover, the information in a vehicular network is highly temporal in nature and a delay of even just a few seconds can make the information useless. Thus, a fast response time of any security system for VANET is another challenge.
- Efficiency: MDS in VANET is required to be efficient. This requires the scheme to be capable of handling a large number of vehicles in a very short duration of time. Also, it should not impose extra overhead on communication resources.
- Robustness: Due to frequent make and break links among highly mobile vehicles, message loss is more likely to occur in the network. At the same time, the system is also vulnerable to various types of outsider attacks like Sybil attack [27], collusion attack, illusion attack, and so on. The designed security system in VANET is supposed to be robust enough to work even in the presence of message loss, intermittent misbehavior, and have immunity to the types of attacks mentioned above.
- High node mobility and ephemeral network: Nodes in VANET are highly mobile and, thus, the network topology changes very fast, causing intermittent communication links. Due to high speed of the communicating vehicles, the communication links make and break very frequently. This results into very short interaction (sometimes in tune of few seconds) between two vehicles. Thus, the MDS needs to be capable of taking care of decision making in short interactions with limited information.
- Privacy: User's privacy is really a big concern in any public network. Vehicles used to participate in an open communication environment, and as per requirement of VANET, each vehicle has to send out beacons at certain intervals. The malicious attackers can use these periodic messages to track the source of the messages and its trajectory during longer period of observation. This is an undesired situation and needs to be avoided.
- Transmission range: Typically, a communication range of 100–1000 m can be achieved by the underlying wireless communication technology in VANET. Short communication range limits the capability of the vehicle in terms of

receiving direct messages from remote nodes. This in-turn limits the vehicle's observation capability of observing potential misbehavior by a particular node for a longer duration.

■ Key distribution: Use of a PKI mechanism is recommended for VANET to achieve authentication and integrity during communications among different entities. Therefore, secure key distribution is another issue to be resolved for any security system using PKI as a part of its operations.

■ Detection and eviction: When a vehicle misbehaves there should be some mechanism in place to detect, with high probability, the misbehaving node. At the same time, it is also necessary that a genuine vehicle should not be wrongly classified as a misbehaving one. Upon successful detection of a misbehaving vehicle, there should be some mechanism for isolation/eviction of those vehicle nodes from VANET. This process may use techniques such as revocation of certificates, imposing fine/penalty, and so on.

6.6.1.1 Strength

In spite of relatively more challenging environments for security system designers, VANET demonstrates some strength over other forms of MANET. From the perspective of a security system researcher, the strength of VANET architecture can be enumerated as follows:

■ Availability of road topology: Road topology is easily available in the form of digital onboard maps, which can be readily used for any of the security systems in VANET.

■ Availability of location information: Use of reasonably accurate GPS and other sensing systems helps the vehicle to readily know its own position to a certain precision.

■ No energy constraint: Unlike other classes of ad hoc networks, nodes in VANET have little or no energy constraints. Thus, high-energy consumption is not considered as an issue during implementation of a protocol or security system for VANET.

■ Better processing capability: Nodes in VANET have better processing power for executing relatively more complex algorithms which achieve better security in the network.

The presence of various unique challenges and strengths of the VANET architecture makes the design of MDSs a real interesting research problem.

6.7 Detecting Misbehaving Nodes in VANET

Misbehavior detection in VANET depends upon the type and nature of the misbehavior demonstrated by the insider nodes. All kinds of misbehaviors are affecting the security and performance of the vehicular network and its stake-holders. Sometimes, the misbehaviors of some nodes may result in serious devistation for the VANET users. Thus, a robust misbehavior-detection framework is needed to detect such misbehaving nodes in VANET. It is also required to devise a scheme/method to detect and identify the continuously misbehaving nodes and take proper action against them. A scheme meant for detecting misbehaving nodes in a networked environment is termed as Misbehavior Detection Scheme (MDS).

Based on targeted sources, MDSs can be classified as *data-centric* and *entity-centric* [13,18]. In *data-centric* misbehavior detection, the objective is to detect the false information rather than the source of the information. Contrary to this, in *entity-centric* misbehavior detection, the primary idea is to find the source of false information, so that the misbehaving nodes can be penalized or evicted from the VANET to help others. Both kinds of MDSs have their own advantages and disadvantages.

However, this classification of MDS is a bit hazy. The ultimate goal of any misbehavior detection is to find misbehaving nodes and isolate them to help the present and future participants in curbing the risk of misbehavior in the network. Therefore, the classification of MDSs needs to be explored from different perspectives. The classification of MDS can be better appreciated based on the mode of detecting misbehavior and misbehaving nodes in a network. Based on underlying modes for misbehavior detection, the MDS can be classified into two types:

1. Independent MDS
2. Cooperative MDS

6.7.1 Independent MDS

This class of MDS does not take the help of any external entity or agency for decision making about the existence of misbehavior in the network.

Sometimes observing the action of a reporting node after sending out the alert messages can also give strong clues about the peer's behavior [18,28]. In this type of MDS, the observer in VANET tries to detect the consistency of recent messages and new alerts with reported and estimated vehicle positions. The positive aspect of this MDS category is that the task of misbehavior detection is done independently and without taking help from any consensus or outside information.

One way to detect a falsely claimed position is by calculating the expected receive time of beacon messages based on its sending time stamp, assuming the speed of signal propagation is the same as the speed of light [18]. In this solution, when a vehicle receives an authentic fresh (lifetime not expired) event-related message M_A, it has been tested for the order of reporting vehicle (n_j), its own (n_j) location, and the reported location of the event E_x in the space. If the receiver finds that the order of the vehicle is either $n_i-n_j - E_x$ or $E_x-n_j - n_i$ (by calculating the relative distance among the E_x, n_i, and n_j), then it starts monitoring the beacon message from the same reporting vehicle for estimating the expected time for receiving the beacons. Suppose a node n_j sends a beacon at time t_1, and n_i (which is in the communication range of n_j) receives the same beacon message at time t_2. Then, the time t_2 is expected to be the same as calculated using the equation below:

$$t_2 = t_1 + \frac{dist(l_{it_2}, l_{jt_1})}{c}$$

where c is the speed of light, and $dist(l_{it_2}, l_{jt_1})$ is the distance between the position of node n_i at time t_2 and position of node n_j at time t_1. If the actual receiving time t_2 of the message is not within a threshold limit of the expected receiving time t_2, then one can infer that the reporting vehicle is a suspicious one.

Similarly, the physical behavior of a vehicle can also be compared with its disseminated information for detecting node misbehavior during communication. Short-term but continuous observation of successive beacon messages gives an idea of undesired behavior by an adversary node that disseminates false position information in the network [28]. In short-term MDS, the observing vehicle node V_o divides the region within its communication range into n equidistant (*interzone distance* = d) virtual zones as shown in Figure 6.3. As the virtual zones are associated with the communication range of V_o, these virtual zones always move with V_o and accordingly do not changes their relative positions with reference to V_o. Virtual zones are numbered from the left as Z_0, Z_1, ..., Z_n, where $n = (2R/d)$. For instance, in Figure 6.3, assuming the communication range of an observing vehicle V_o as $R = 300$ m, average speed of a vehicle as $S = 30$ m/s and beacon sampling rate as 1 Hz, the area covering the communication range is divided into 20 equidistant virtual zones, each of size $d = 30$ m.

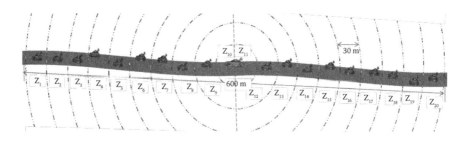

Figure 6.3 Virtual zone division for short-term misbehavior detection. (From Rajesh P. Barnwal, Soumya K. Ghosh. 2012, Heartbeat message based misbehavior detection scheme for vehicular ad hoc networks. s.l. : ACM/IEEE, *International Conference on Connected Vehicles and Expo (ICCVE-2012)*.)

In this scheme, whenever the observing vehicle receives continuous beacons from a reporting vehicle V_r, it calculates the current expected position of the reporting vehicle based on its last received information from the beacons of the same reporting vehicle V_r at time $t - \Delta t$. Here, beacons are assumed to carry information such as current position, speed, and steering angle of the reporting vehicle. Based on the calculation of the current expected position of the reporting vehicle V_r, the vehicle V_o calculates the expected zone for the V_r. Also, V_o estimates the observed zone of the vehicle V_r based on the current beacon information, received from V_r at time t. If the two zones (expected and observed zones) match, it is called a *match* otherwise the case is called a *mismatch*. After a number of such mismatches crosses a predefined threshold, the vehicle is declared as misbehaving.

An observing vehicle can also detect misbehaving nodes claiming false location by detecting the similar claim of positions by two or more vehicles at the same time with the help of Cooperative Awareness Messages (CAM) [29]. For instance, if two vehicles are on a road, there is always some gap between neighboring vehicles. This is also true in traffic jam situations. However, if a vehicle claims its position inside a footprint of another vehicle in the same time slot, then it can be detected as suspicious. For the purpose of computation, the vehicles are modeled as rectangles. Figure 6.4 illustrates the intersection of two vehicles V_A and V_B, each of length "l" and width "w."

In this approach, the positions of the two vehicles are taken from the CAM messages. The position information makes the system capable of computing whether two rectangles, representing two vehicles, are overlapping with each other or not. If two rectangles intersect each other, one may infer that either of two vehicles is disseminating false position information.

6.7.2 Cooperative MDS

This class of MDS uses the information collected from other sources to deduce the possibility of misbehavior in the network.

Detecting malicious data with the cooperation of other nodes, if possible, can be a very effective tool and the first step for detecting misbehaving nodes in VANET. In order to detect the correctness of the received information, the capabilities of onboard sensors can be leveraged [30]. This mechanism works based on four assumptions: (1) a node can bind observations of its local environment with the communication it receives, (2) a node can tell its neighbors apart locally, (3) the network is "*sufficiently*" dense, and (4) after coming in sufficiently close contact, nodes can authenticate their communication to one another. For example, to detect erroneous nodes (nodes sending false location information) in the vicinity, it is assumed that nodes are able to sense the precise location of its neighbors. This solution works based on a model of the VANET, which includes all possible events or sets of events. In this example, the model may contain location information of the neighbors. Here, nodes seek the help of other nodes for

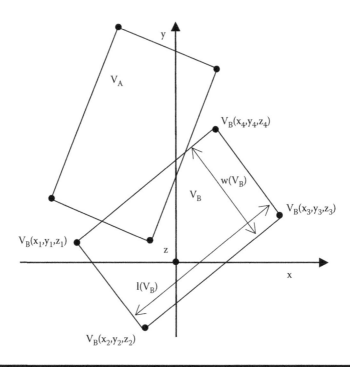

Figure 6.4 Intersection of two vehicles of length "l" and width "w." (From N. Bissmeyer, C. Stresing, and K.M. Bayarou. 2010, Intrusion detection in VANETs through verification of vehicle movement data. *IEEE Vehicular Networking Conference (VNC-2010).* **pp. 166–173.)**

creating the model of the VANET, assuming that majority of the nodes in the network are honest. Here, each node has its own database K, which consists of following tuples

$$K = \{< N_1, \vec{x}_1 > o_1, < N_2, \vec{x}_2 > o_2 \ldots\}$$

where the assertion $A_i = < N_i, \vec{x}_i > o_i$ means "*node O_i claims to have observed node N_i at location $\vec{x}_i$.*" If a node is out of sensing range, that node can be eliminated from the database. The scheme tries to compute the explanation for any missing observation in the model stored in the shared database. If this explanation supports some conflicts in the model, it can detect the attack. Figure 6.5 depicts a single malicious node, who tries to increase the evidence of its false presence by generating several spoof nodes. The *solid* arrow shows the observation and the *dashed* arrow shows missing observations. Figure 6.6 shows two possible explanations for the conflict. These explanations help the observer in detecting the possible malicious node(s) in the network.

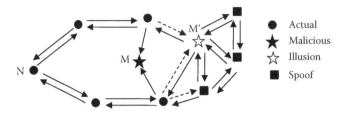

Figure 6.5 A single malicious node M spoofs to support a false location M'. Arrows shows observation in database, dashed arrows show missing observations. (From P. Golle, D. Greene, and J. Staddon. 2004, Detecting and correcting malicious data in VANETs. *1st ACM International Workshop on Vehicular Ad Hoc Networks (VANET-2004).* **pp. 29–37).**

Misbehavior in VANET can also be discovered by analyzing postnotification behavior of the notifying node or the other colocation nodes. For instance, in the case of a Post Crash Notification (PCN) alert, the driver's behavior after receiving the alert can give a strong clue about the correctness of the alert [15,17]. In case of a real PCN alert, the vehicles approaching the crash site are supposed to take necessary actions to change its lane near the crash site. On the other hand, if the driver of an approaching vehicle (very near to crash site) finds the PCN alert false, then he/she will continue to follow the same path/lane as earlier without any change in driving condition. Figure 6.7 shows the expected and actual trajectory in the case of a false PCN alert. The observing vehicle, which is distant from the crash site but

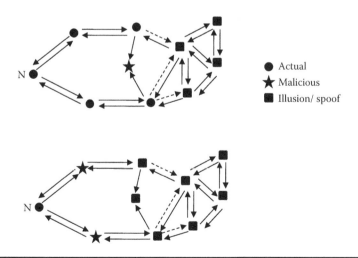

Figure 6.6 Possible explanations for conflicting observations. (From P. Golle, D. Greene, and J. Staddon. 2004, Detecting and correcting malicious data in VANETs. *1st ACM International Workshop on Vehicular Ad Hoc Networks (VANET-2004).* **pp. 29–37.)**

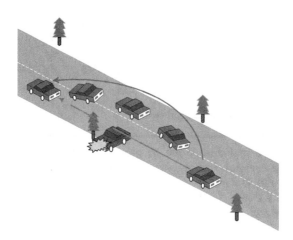

Figure 6.7 Expected and actual trajectory.

able to receive the PCN alert, can monitor the trajectory of other leading vehicles to detect deviation from expected trajectories in the presence of a real crash. If the difference of expected and actual trajectory of the crash-site-following vehicles is more than a threshold limit, a false alert can be detected.

Periodic beacons are the primary source of information in VANET. Misbehaving vehicles disseminating false position information can also be detected by monitoring the signal strength of their periodical beacons [31]. This can be understood with the help of Figure 6.8. A vehicle node S_1, who is claiming its position through broadcasted beacons, is observed by various witnessing nodes. A verifier node V_1 collects all signal strength measured by the neighboring nodes which have received the beacon. The Received Signal Strength (RSS) of the recently received beacon from a malicious node m (claiming as S_1) at the neighboring nodes can give an idea about its approximate position in the space.

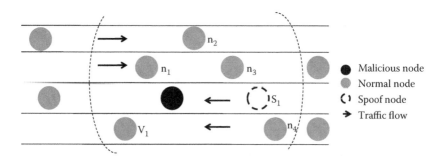

Figure 6.8 VANET scenario showing malicious, spoof, and normal nodes. (From B. Xiao, B. Yu, and C. Gao. 2006, Detection and localization of Sybil nodes in VANETs. Workshop DIWANS-2006, pp. 1–8.)

Other way of detecting misbehavior in VANET may also include different methods, which can be useful in deciding reliability of the received information. The reliability of received information can be determined based on the number of the peers sending the same information [32]. The reliability of the received information depends upon the fact that how many different vehicles are reporting and supporting the same event. Kounga et al. [32] suggested a time frame for which the vehicle should not be able to change their pseudonyms, which will restrict them in using more number of pseudonyms for sending supportive messages. The event report is considered reliable, when the number of supporting report crosses a given threshold. Further, the knowledge of the physical environments and facts can also help in deciding the reliability of the sender's information based on multiple criteria. The reliability of received information, thus in turn can be helpful in detecting the misbehaving nodes.

6.7.3 Independent and Cooperative MDS: Merits and Limitations

The various MDSs, proposed by different researchers, have their own merits and limitations. Independent MDS does not depend upon the response and help from other peers and thus is less prone to collusion attacks. At the same time, due to the paucity of a sufficient number of available evidences, it is more susceptible to misclassification. Moreover, the independent MDS is relatively faster in response compared to its cooperative counterpart. Independent misbehavior detection is suitable in time critical situations, when the response time matters a lot. In a ephemeral connectivity scenario like VANET, the independent MDS, in general, poses less response time than the cooperative scheme.

However, cooperative MDS has to depend upon the cooperation from other nodes for proper detection of misbehavior and misbehaving nodes. The dependency on unknown peers makes the cooperative MDS more open to collusion and Sybil attack. In general, the time complexity of this category of MDS is generally higher than the independent one. In spite of these limitations, the cooperative mode of misbehavior detection is more reliable and have better accuracy than the independent mode.

6.7.4 Hybrid Approach for Misbehavior Detection

The benefits of both independent and cooperative MDS can be reaped by adopting the hybrid approach for misbehavior detection in VANET. Figure 6.9 depicts a typical high-level schematic for hybrid approach-based MDS.

Hybrid misbehavior detection first relies upon the vehicle's independent MDS. The independent MDS first tries to detect the suspicious nodes in the network with the help of various onboard available data, for example, data from environmental sensors, received beacons, alert message, and so on. It enlists suspicious nodes based

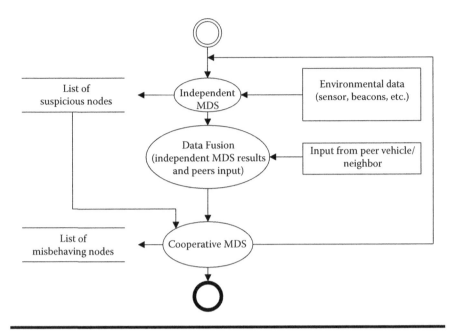

Figure 6.9 Hybrid misbehavior-detection approach.

on these preliminary observations. Furthermore, the independent MDS passes the compiled environment information into the *data fusion* module. The *data fusion* module takes the independent experience data and input from the experience of the other peer nodes for enrichment of the available information. Cooperative MDS with the help of available enriched information again cross-checks the behavior of the suspicious nodes from the global perspective and generates the list of misbehaving nodes. The misbehavior detection by this approach is supposed to be more reliable than the independent MDS and is also immune to collusion and Sybil attacks. Moreover, the list of suspicious nodes, generated in this process, can be used in a time critical situation, when response time is very urgent. However, the actual misbehaving nodes can be reliably identified from the list of misbehaving nodes generated after processing information by cooperative MDS. The identified misbehaving nodes can be isolated immediately by including their identity in CRL [24] or can be given some additional chance up to a certain threshold limit before revocation of their communication privileges [33].

6.8 Future Research Directions

There are several research challenges related to misbehavior-detection problem in VANET. Some of the research directions are discussed in this section.

Assignment of trust value to a participating VANET node is one of the major challenges. Identification of known misbehaving nodes can be done with the help of preassigned trust values of the nodes. The preassigned trust value may depend upon the intelligent fusion of multiple information and misbehavior reports received from different vehicular entities about the participating nodes. In the presence of required privacy, the distribution of these preassigned trusts, without distributing any extra information like certificate revocation list, is a nontrivial problem. Mechanisms for encoding and distributing preassigned trust through the pseudonyms itself may be an effective approach. A trust-based comprehensive framework for detection and isolation of misbehaving nodes from VANET communications can be a comprehensive approach to curb the effects of misbehavior to a great extent and restrain the misbehaving nodes from creating future damage to others.

Vehicular Cyber-Physical System (VCPS) is a more general abstraction of VANET. It is also an emerging area of research with the potential to become a pervasive and ubiquitous technology for the future. The dynamic and distributed nature of the VCPS makes it more vulnerable toward the false and incorrect information from other cyber-physical entities. In VCPS, it is imperative to model the behavior of cyber as well as physical components of the system to determine the robustness of the system from the effects of misbehaving entities. Unsupervised learning tools in VCPS can facilitate behavior modeling, which in-turn can help to identify previously unknown misbehaving node(s) in real time.

Future research also warrants proposals for designing and testing hybrid MDS to achieve better results in detecting the misbehaving nodes in VANETs. Moreover, designing a privacy-preserving framework for misbehavior detection and isolation of misbehaving nodes from the vehicular network is also a challenging and open problem.

6.9 Conclusions

VANET is a promising research area that may enhance the safety of the transportation system by many folds. However, there are many issues that must be addressed before deployment of the VANET in a real scenario. Misbehavior detection in VANET is one such issue. This is a challenging research problem and thus widely discussed among the contemporary research community in this field. Though in recent years, a few works have been reported in this direction, they have their own limitations and restricted applications. There is no fool proof security solution available for detection and isolation of misbehaving nodes in VANET. In this chapter, misbehavior problems in VANET have been discussed. Several research challenges in designing a good MDS for a vehicular network has also been enumerated. The effort has been made to provide the current state of research in this field. Several recent proposals have been discussed briefly to present a glimpse of several possibilities and approaches suggested by researchers worldwide. A hybrid approach for designing better MDS is proposed for eliminating the limitations of independent and cooperative MDSs. In the future, it is important to focus on the adoption of hybrid approaches for detection of misbehaving nodes in vehicular networks, so as to mitigate the effects of misbehavior in VANET.

References

1. *Fatality Analysis Reporting System (FARS) Encyclopedia*. [Online] September 2011. [Cited: January 15, 2013.] http://www-fars.nhtsa.dot.gov/main/index.aspx.
2. *Road accidents in India*. [Online] September 15, 2011. http://www.indiastat.com/.
3. H. T. Cheng, H. Shan, and W. Zhuang. 2011, Infotainment and road safety service support in vehicular networking: From a communication perspective. *Mechanical Systems and Signal Processing*, 25(6), 2020–2038.
4. N. A. Alrajeh, S. Khan, J. Lloret, and Jonathan Loo. 2013, Secure routing protocol using cross-layer design and energy harvesting in wireless sensor networks. *International Journal of Distributed Sensor Networks*, 2013, Article ID 374796, pp. 11.
5. M. Killat, M. Torrent Moreno, and H. Hartenstein. 2005, The challenges of robust inter-vehicle communications. *IEEE 62nd Vehicular Technology Conference (VTC 2005)*, Dallas, US. 1, 319–323.
6. V. Hubaux, P. Gligor, and J.P. Papadimitratos. 2006, Securing vehicular communications—Assumptions, requirements, and principles. *Workshop on Embedded Security in Cars (ESCAR)*, Berlin, Germany.
7. M. Raya, P. Papadimitratos, and J.P. Hubaux. 2006, Securing vehicular communications. *IEEE Wireless Communications Magazine (Special Issue on Inter-Vehicular Communications)*, 13, 8–15.
8. A. Festag, C. Harsch, T. Gerlach, G. Leinmuller, and M. Goldacker. 2007, Security architecture for vehicular communication. *5th International Workshop on Intelligent Transportation (WIT 2007)*. Hamburg, Germany.
9. H. Laberteaux, and K.P. Hartenstein. 2008, A tutorial survey on vehicular ad hoc networks. *IEEE Communications Magazine*, 46, 164–178.
10. *CAMP Vehicle Safety Communications Consortium, 2005. Vehicle safety communications project task 3 final report: Identify intelligent vehicle safety applications*. s.l. : National Highway Traffic Safety Administration, U.S. Department of Transportation.
11. Fan Bai, Hariharan Krishnan, Varsha Sadekar, Gavin Holl, and Tamer Elbatt. 2006, Towards characterizing and classifying communication-based automotive applications from a wireless networking perspective. *IEEE Workshop on Automotive Networking and Applications (AutoNet 2006)*. San Francisco, CA.
12. IEEE Std 1609.2-2006, 2006. *IEEE trial-use standard for wireless access in vehicular environments—Security services for applications and management messages*.
13. M. Raya, P. Papadimitratos, V. D. Gligor, and J. Hubaux. 2008, On datacentric trust establishment in ephemeral ad hoc networks. *IEEE INFOCOM-2008*. 25, 1238–1246.
14. J. Sun, and Y. Fang. 2009, Defense against misbehavior in anonymous vehicular ad hoc networks. *Ad Hoc Networks*, 7(8), 1515–1525.

15. M. Ghosh, A. Verghese, A. Kherani, and A. Gupta. 2009, Distributed misbehavior detection in VANETs. *IEEE Wireless Communication and Networking Conference (WCNC 2009)*, Budapest, Hungary.
16. Zhen Cao, Jiejun Kong, Mario Gerla, Zhong Chen, and Jianbin Hu. 2010, Filtering false data via authentic consensus in vehicle ad hoc networks. *International Journal of Autonomous and Adaptive Communications Systems*, 3, 217–235.
17. M. Ghosh, A. Varghese, A. Gupta, A.A. Kherani, and S.N. Muthaiah. 2010, Misbehavior detection scheme with integrated root cause detection in VANET. *Ad Hoc Networks*, 8, 778–790.
18. S. Ruj, M.A. Cavenaghi, Z. Huang, A. Nayak, and I. Stojmenovic. 2011, *Data-Centric Misbehavior Detection in VANETs*. Arxiv preprint arXiv:1103.2404, p. 12.
19. J.H. Song, V.W.S. Wong, and V.C.M. Leung. 2008, Secure location verification for vehicular ad hoc networks. *IEEE Global Telecommunications Conference (GLOBECOM 2008)*. New Orleans, LA.
20. J. Zhang. 2011, A survey on trust management for VANETs. *25th International Conference on Advanced Information Networking and Applications (AINA 2011)*. Biopolis, Singapore.
21. P. Wex, J. Breuer, A. Held, T. Leinmuller, and L. Delgrossi. 2008, Trust issues for vehicular ad hoc networks. s.l.: *IEEE Vehicular Technology Conference (VTC Spring 2008)*, pp. 2800–2804.
22. P. Wex, J. Breuer, A. Held, T. Leinmuller, and L. Delgrossi. 2008, *67th IEEE Vehicular Technology Conference (VTC Spring - 2008)*. pp. 2800–2804.
23. Hubaux, Maxim Raya and Jean-Pierre. 2007. Securing vehicular ad hoc networks. *Journal of Computer Security*, 15, 39–68.
24. T. Zhang, and L. Delgrossi. 2012, *Vehicle Safety Communications: Protocols, Security, and Privacy*. John Wiley & Sons, Hoboken, New Jersey.
25. M. Pease, R. Shostak, and L. Lamport. 1980, Reaching agreement in the presence of faults. *Journal of the ACM*, 27(2), 228–234.
26. B. Parno, and A. Perrig. 2005, Challenges in securing vehicular networks. *Proceedings of the Workshop on Hot Topics in Networks, HotNets-IV*. College Park, Maryland.
27. J. Douceur. 2002. The Sybil Attack. *First International Workshop on Peer-to-Peer Systems*. pp. 251–260.
28. R.P. Barnwal, and S.K. Ghosh. 2012, Heartbeat message based misbehavior detection scheme for vehicular ad hoc networks. *ACM/IEEE, International Conference on Connected Vehicles and Expo (ICCVE-2012)*. Beijing, China.
29. N. Bissmeyer, C. Stresing, and K.M. Bayarou. 2010, Intrusion detection in VANETs through verification of vehicle movement data. *IEEE Vehicular Networking Conference (VNC-2010)*, New Jersey, USA. pp. 166–173.
30. P. Golle, D. Greene, and J. Staddon. 2004, Detecting and correcting malicious data in VANETs. *1st ACM International Workshop on Vehicular ad hoc networks (VANET-2004)*. pp. 29–37.
31. B. Xiao, B. Yu, and C. Gao. 2006, Detection and localization of Sybil nodes in VANETs. *Workshop DIWANS-2006*. pp. 1–8.
32. Gina Kounga, Thomas Walter, and Sven Lachmund. 2009, Proving reliability of anonymous information in vanets. *IEEE Transactions on Vehicular Technology*, 58, 2977–2989.
33. J. Fang, and Y. Sun. 2008, A defense technique against misbehavior in VANETs based on threshold authentication. *Proceedings of IEEE Military Communications Conference.*

34. R.K. Schmidt, T. Leinmüller, E. Schoch, A. Held, and G. Schäfer. 2008, Vehicle behavior analysis to enhance security in vanets. *Proceedings of the 4th Workshop on Vehicle to Vehicle Communications (V2VCOM 2008)*. Eindhoven, Netherlands.

35. T.H.J. Kim, A. Studer, R. Dubey, X. Zhang, A. Perrig, F. Bai, B. Bellur, and A. Iyer. 2010, Vanet alert endorsement using multi-source filters. *Proceedings of the seventh ACM international workshop on VehiculAr InterNETworking*, Chicago, USA, pp. 51–60.

SECURITY IN WIRELESS SENSOR NETWORKS

Chapter 7

Security Architecture for Multihop Wireless Sensor Networks

Ismail Mansour, Gerard Chalhoub,
and Michel Misson

Contents

7.1 Introduction

Wireless sensor networks (WSNs) are more and more deployed for various applications including home monitoring, health, industrial, military, and so on. It is known that wireless networks are easy to attack because of the nature of the shared medium, which makes it relatively easy for intruders to eavesdrop, tamper, or inject data into the network. Sensor nodes are known to have limited computation, storage, and transmission capacities, but attackers are not necessarily using the same technology to launch their attacks.

Security techniques aim basically at offering the following proprieties: (1) data confidentiality, where only the entities concerned are able to decode information, (2) data integrity, when the destination is able to make sure that data sent by the source has not been tampered with by a third party, (3) data authentication, where the source of the data is authenticated, (4) entity authentication, which is to make sure that the entity is really who it is claiming to be. Our contribution is concerning the communication inside the WSN as depicted in Figure 7.1. Data is generated by the process equipment and passed through to wireless sensor nodes that route this information toward a sink that plays the role of a gateway that leads toward a wired or a wireless traditional network. Users communicate with the process by using a middleware layer and store the collected data in adequate servers.

The required security level might vary from one application to another according to the importance of the information that is being exchanged. In this chapter,

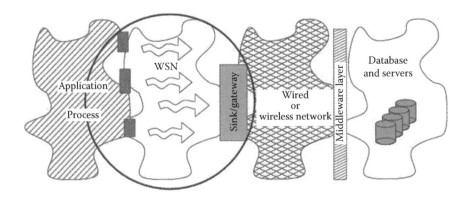

Figure 7.1 The missing link in the security process of applications using WSN.

we present a security mechanism using elliptic curve cryptography (ECC) [1] to establish a secured communication. We propose a secured system for critical applications with low mobility and where a very high level of security is required. We use an existing cryptographic system based on ECC that enables us to exchange secret session keys and encrypt exchanged data. We evaluate the additional cost of the cryptographic system during the initial phase of the network, that is, the join phase and the neighbor discovery phase. Evaluations are done through real measurements using TelosB motes. It should be noted that the main contribution of this chapter resides in the network protocols that are used to enable nodes to join the network in a secure manner and exchange messages in a secure manner as well. We adapt the cryptographic algorithms used, which are based on the ECC operations, such as the Diffie–Hellman method [2] with key predistribution.

The remainder of the chapter is organized as follows. In Section 7.2, we go through some of the most known attacks and countermeasures in WSNs followed by the related work in the field of cryptography and security in WSNs including the latest standardization efforts. In Section 7.3, we present our proposition that is based on public key management and symmetric key encryption, ending it with a security analysis. We analyze the results obtained through real measurements in Section 7.4. Finally, we give a quick perspective in Section 7.5 before concluding the chapter in Section 7.6.

7.2 State of the Art

7.2.1 Security in Wireless Sensor Networks

Most attacks in the literature concern wireless networks such as wireless local area network (WLAN), mobile ad hoc network (MANET), and WSN. The common vulnerability between them is the use of wireless communication. These attacks are divided into categories according to criteria or factors well-defined. We cite the best known of classifications:

- Communication protocol: The attacks are distinguished according to the layer where the attacks are made [3,4].
- Adversarial intent: Attacks are classified as primary intentions of adversaries such as collecting information, perturbing communications or data aggregation, and exhausting resources or physically capturing nodes of the network [5].
- Passive/active attacks: A passive attack is an attack that aims to listen and collect data that can be used later to start other types of attacks, while the active attack tries to modify, fabricate, or perturb data [6–8].
- Internal/external attacks: When a node in the network is involved in an attack, it will be considered an internal attack. This can be done using the nodes of the network that are physically captured. If the node participating

in the attack is not part of the network, the attack will be considered as an external attack [8].

■ Place of attack: The authors in [9] define a new model of attack suited for WSN based on the target of the attack. It depends on which entity of the network the attack is performed (i.e., sink, neighbor, or source node).

Each of these methods of classification helps to see the threats from different perspectives. In what follows, we present a list of some of the most important attacks and the basic countermeasures:

■ Eavesdropping or passive monitoring attack: This attack is one of the passive attacks where adversaries seek to monitor or collect information circulating in the network. The goal is to listen to traffic on the communication channels and intercept packets. Everything depends on the mechanisms used to secure communications. If they are not encrypted, the adversary can then immediately retrieve the content of the packets. Otherwise, this content needs time to be decrypted before exploited.

■ Sinkhole or blackhole attack: This attack is part of series of attacks that cause a denial-of-service of the network. It is an internal attack performed on the network level. The malicious node acts as a base station, attracts packets, and prevents them from continuing their ways. Authors in [10] proposed an energy-efficient end-to-end grouped acknowledgment security mechanism to detect such attacks and avoid sending packets to malicious nodes. One of the counterattack techniques that prevent such attacks from happening in the first place is entity authentication where nodes are authenticated before they are allowed to take part of the network and routing activity.

■ Sybil attack: A malicious node can claim to have multiple identities by using those of nodes targeted by the attack. It is an example of place of attack attacks. The goal of this attack is to degrade the integrity of data, the level of security, and resource utilization. The use of public key cryptography (PKC) can help to defend against this attack. The key is to verify the identities of neighboring nodes or sons.

7.2.2 Related Work

In this section, we go through the related work in the field of security in WSNs including key management and the standardization tendencies.

7.2.2.1 Key Management in WSNs

In order to provide security in WSNs, communication should be encrypted and authenticated. The main issue is how to set up secret keys between nodes to be used for the cryptographic operations, which are known as key agreements.

PKC has interesting properties for the security of wireless networks and has been applied for WSN in spite of the memory constraints of sensor nodes [11–13]. To avoid many attacks like Sybil or man-in-the-middle, the nodes of the network and all the public keys must be authenticated before the start of communications. Many authentication schemes are proposed in the literature in order to establish secure public key exchange [14–17].

The most common criticism on using PKC in sensor networks is its computational complexity and communication overhead. However, in [18–20], the authors show that PKC based on ECC is viable on small wireless devices. ECC is based on the problem of discrete logarithm. The main attraction of ECC over competing technologies such as Rivest, Shamir, and Adleman (RSA) [21] and Digital Signature Algorithm (DSA) [22], is the use of smaller parameters but with an equivalent level of security [23]. For example, 160-bit ECC key is equivalent to 1024-bit RSA key. The performance of ECC depends mainly on the efficiency of finite field computations and fast algorithms for elliptic scalar multiplications. The private key of the ECC is a random number d chosen in $[1, n-1]$, where n is a fixed parameter in the domain of the curve. The public key is then obtained by multiplying d by G, where G is the global point of the curve. ECC has three fundamental protocols: (i) elliptic curve Diffie–Hellman (ECDH) which is the elliptic version of the well-known Diffie–Hellman key agreement method, (ii) elliptic curve digital signature algorithm (ECDSA) which is the analog of the DSA, and (iii) elliptic curve authenticated encryption scheme (ECAES) which is a variant of public-key encryption. In [24], authors evaluated the energy consumption of the ECC cryptographic operations using 160-bit keys. In this chapter, we evaluate the time and energy consumption of these operations using an implementation on TelosB motes and compare them to symmetric cryptographic algorithms.

Symmetric key systems are known for their lightweight cryptographic operations compared to public key systems, thus they were the first to be considered for WSNs where computational resources are scarce. One of the most known symmetric key systems that were proposed for WSNs were security protocols for sensor networks (SPINS) [25], which uses a simplified version of timed, efficient, streaming, loss-tolerant authentication (TESLA) protocol [26]. In [27], authors used three types of symmetric keys: (i) a set of symmetric keys predistributed on every node, nodes with a common key are able to communicate; (ii) a wedge key shared by all nodes of the same wedge, and (iii) a path key established from source to destination. In [28], authors compared hardware implementation with a software implementation of the AES algorithm using MicaZ motes. Results show that the hardware implementation consumes less time but more energy. In our chapter, we considered a software implementation driven by cases where energy efficiency is more important than time efficiency. This is the case for most applications that use WSNs and aim at achieving a maximum network lifetime.

While symmetric key schemes are efficient in execution time, they require complicated key management, which consumes a lot of memory and overhead. In contrast, public key-based schemes have simple and adapted key management, but they are more complex to execute and thus consume more time.

The predistribution of symmetric or asymmetric keys is currently used almost widespread. It tries to replace the vacuum created by the absence of public key infrastructures in WSN. Other concepts to overcome the absence of certificate exist, such as the identity-based cryptography (IBC) [29], where every node in the network is able to compute the public key of another node based on the identity of that node. This method is used by [30], where authors proposed TinyPBC and showed that their implementation is less time-consuming and more optimized than that of [31]. The authors used the pairing operation in order to obtain the mutual secret key; they criticized the Diffie–Hellman method because of the lack of authentication and because of the messages that need to be exchanged between nodes before generating the mutual secret key. In our proposal, we used an authenticated no interaction Diffie–Hellman (NIDH) method to avoid the additional messages over the air as explained in Section 7.3.2.

Authors in [32] proposed a multihop key establishment between nodes called micro-PKI. Their method is based on the predistribution of the public key of the base station. Using this public key, every node is able to create a secret key with any other node of the network. The authentication process in this proposal is only dependent on the public key of the base station, if a node has this key, it is considered authenticated. This makes the procurement of the public key very critical and on which depends the whole security architecture. The evaluation presented in the chapter is based on results from [33] and does not take into account the fact the messages transit over multiple intermediate nodes. In addition, the authors did not implement their method, they only gave estimated results. In our proposal, we emphasized the authentication phase during a join process, and we evaluated the authentication cost over a multihop network based on implementation results.

7.2.2.2 Standardization Tendencies

In the ZigBee-2007 specifications [34], the protocol proposes a cryptographic mechanism based on AES in CounTeR mode (AES–CTR) with a 128-bit symmetric key to secure communications between devices. Three types of keys are used: (i) *link key* which is shared pairwise and used for unicast communications between peer entities, (ii) *network key* which is used for broadcast communication, and (iii) *master key* which is shared pairwise and used for exchanging keys. In each ZigBee network, there is a trust center application (configured on the ZigBee coordinator by default) that manages the key distribution, which is unique and known by all devices.

Other industrial standards that target monitoring and control [35] such as ISA100 [36] and WirelessHart [37] also use AES–CTR with 128-bit key and have a central security control entity that manages and authenticate the nodes. Symmetric encryption is known to be faster when it comes to computational complexity compared to an asymmetric encryption, but it lacks a digital signature and nonrepudiation. ZigBee published the smart energy profile [38] that includes public key-encryption mechanisms.

7.3 Security Architecture

As discussed in Section 7.2.1, a lot of attacks like sinkhole, black hole, and tampering can be avoided using public key-encryption infrastructure that offers the means to ensure entity authentication, source authentication, and data integrity and confidentiality. Our proposition is based on ECC encryption for the asymmetric operations and AES–CTR for the symmetric operations.

7.3.1 Network Topology

We consider a hierarchical multihop topology as depicted on Figure 7.2, where the node S is the sink of the topology and the default destination for the traffic. Hierarchical topologies are known to be more convenient for energy efficiency in WSNs as discussed in [10,39,40].

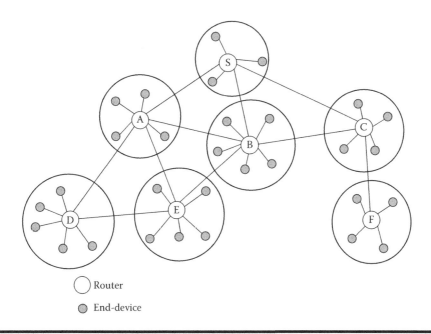

○ Router

● End-device

Figure 7.2 Network architecture. The network is organized in stars. Each star is under the control of one router. Sensors and actors are associated to a router and belong to only one star.

Nodes are grouped into stars; each star has one central node that we call router and several end-devices. Routers have more processing and memory capacities than end-devices and they constitute about 10% of the network devices. End-devices represent the sensors and actors of the networks. End-devices are only allowed to communicate with the router of the star to which they belong. The creation of stars is done during the network deployment phase. Routers are allowed to communicate with each other but not with end-devices that are not associated to them. This clustering technique is similar to the one used in IEEE 802.15.4 [41] and MaCARI [42].

The hierarchical aspect of the topology will serve to facilitate the key management as explained in the next section where the sink node will play an important role in authenticating newly arrived nodes.

7.3.2 Key Management

We use ECAES encryption for ensuring data integrity and authentication, and we use session keys (symmetric keys) for data encryption. Session keys are exchanged using ECAES encryption. Our key management is based on two phases: predeployment phase and deployment phase. The objective is to authenticate the nodes and generate session keys between the nodes and the sink, and between the nodes that need to communicate.

7.3.2.1 Predeployment Phase

In order to gain time at the deployment phase, public keys are generated and pre-distributed before deployment. Nodes in a WSN, where security is an issue, can often be configured before the deployment unlike the internet where nodes are very heterogeneous. This feature makes it easier to be able to distribute keys to nodes in a controlled manner.

For each node in the network (routers and end-devices) we generate a pair of keys (a public key and a private key). Keys are then distributed as follows: (i) we store in each node its own pair of public keys and the public key of the sink, (ii) in the sink we store its own pair of public keys and the list of all the public keys of the nodes of the network. Every time a new node has to join the network, a simple command to the sink to add its public key is done. This mechanism ensures the dynamic aspect of the network and the ability to add new nodes once the network is created. Nodes, routers, and end-devices, have thus the ability to be part of the network when they are in range of another node that is already part of the network. The dynamic aspect makes it possible to deal with topology changes and link failures.

The sink thus plays the role of a security manager that will authenticate the newly arriving nodes. A node that does not have a public key stored at the sink cannot be part of the network.

7.3.2.2 Deployment Phase

During the deployment phase, the sink is the first node to be activated. Then, when a new node is activated, it probes the medium in order to find either the sink or other previously joined nodes. When it finds a node in its communication range it proceeds to the join process as described in this section. This incremental creation of the network guarantees that when a node is activated it is going to join a secured network where nodes had already created session keys in order to secure the communications.

Once a node is activated, it should proceed to the join phase and authenticate with the sink, described as follows.

Join process: The first step after the activation of a node is the join phase. Let us consider a new node R that wants to join the network through the node F (see Figure 7.2).

R calculates a common key $DH_{RS} = K_R{}^*P_S$ according to Diffie–Hellman method without interaction (NIDH) between R and S, where K_R is the private key of R and P_S is the public key of S. No interaction is needed between R and S because P_S is predistributed in all the nodes of the network, thus R has it already and no need to exchange it over the air, and S has P_R as already explained in Section 7.3.2. In addition, the NIDH is an authenticated method because only R and S have P_R and P_S at the same time and are able to compute the DH key combine with their respective private keys.

A derivation of the calculated DH key is used as a symmetric key to encrypt the join request message (*joinReq*) that should be sent to the sink S. Note that the header of the message is not encrypted in order to avoid encryption and decryption operations on the intermediate node. Thus, address fields are sent in clear and the routing protocol can choose the next hop without any cryptographic operation.

When node F receives the join request, it forwards it toward the sink. Node C does the same. When the sink S receives the join request, it examines the join request and decides whether to accept it or not. The verification process is out of the scope of this chapter, it could be done using an external authentication server to which the sink has access, or the sink can be the authentication server if it has enough storage capacity.

In case node R is allowed to be part of the network, S creates locally the same key created by R $DH_{RS} = K_S * P_R$ (note that $DH_{RS} = K_R * P_S = K_R * (G * K_S) = (K_R * G) * K_S = P_R * K_S$). S sends back a positive join response (*joinRes*) to F (the node through which R is joining the network) encrypted using DH_{FS} (this key was previously created during the join phase of node F). This join response contains the public key P_R of R. Using this public key, F is able to encrypt and to exchange a symmetric session key SK_{FR} with R that will be used to encrypt communication between F and R. This join phase is depicted in Figure 7.3.

Using DHRS makes the join process an authenticated operation because S is the only node that is able to calculate DH_{RS} using its own private key K_S. The

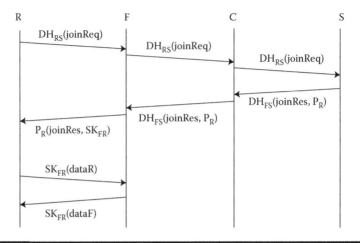

Figure 7.3 Join process. A multihop process guarantees that nodes are authenticated before taking part in the network. At the end of the join process, the new node R obtains a session key that enables it to communicate with the node through which it joined the network. (F, C, and S are the nodes that form the topology of Figure 7.2.)

fact that the public keys are predeployed, avoids any man-in-the-middle attack. All public keys are published by the sink using the authenticated keys DH. At the end of the join phase, the node R is authenticated and a session key between R and the sink S is established.

In case any of these operations do not succeed (the node does not figure in the allowed nodes list of the sink, or the DH key created by the new node is not compatible with the DH key created by the sink), the node is not allowed to join the network.

In the case of an end-device that wants to join the network, the same procedure is applied. The end-device will only need to join the network and be authenticated by the sink, and later on they only need to communicate with the parent router (the node through which they join the network). If an end-device changes position and is no longer in communication range with its parent, it can switch to another parent and rejoin the network.

The main reason behind the separation between end-devices and routers is related to the energy efficiency of communications. In an energy network, end-devices will spend most of the time in sleep mode to save energy, thus, routers will store messages destined to end-devices until they wake up and be able to receive them. This energy-efficient communication mode is adopted by IEEE 802.15.4 and Zigbee standards [34].

Neighbor discovery: The sink plays a critical role in the join phase. It needs to authenticate the public keys of every node that wants to join the network. When this authentication fails, it sends back a reject message to that node. Otherwise, it sends a positive join response. The sink saves all the public keys of the nodes in the network.

This way, whenever a node wants to obtain the public key of another node, it sends a public key request to the sink with the identifier of the other node (e.g., the identifier can be its physical address). The sink sends back the public key encrypted using its private key in order to ensure the authenticity of the sink.

The session key exchange is done with every neighbor with which the routing protocol needs to communicate. In the case of session key creations between neighboring nodes (e.g., nodes A and B in Figure 7.2), the node that creates the session key is the node that needs to send a message destined to the other.

Let us consider the case where A needs to send a message to B. A sends a public key request (*pkReq*) to the sink to obtain the public key of node B (P_B). The *pkReq* is encrypted using the Diffie–Hellman key DH_{AS} between A and S. S sends back a public key response (*pkRes*) to A containing the public key of node B (P_B) encrypted using the Diffie–Hellman key DH_{AS} between A and S. Then, A sends a session key establishment (*skEst*) message to B encrypted with P_B that includes a session key (SK_{AB}) and the public key of A (P_A). B replies with an acknowledgment (ACK) encrypted with the newly established session key to confirm that it was able to decrypt the key establishment message. This phase is represented on Figure 7.4.

Session keys are then updated periodically, where the periodicity is the maximum duration tolerated before compromising the session key. Concerning end-devices, they do not need to do network discovery. If the link between an end-device and its parent router is broken, then the end-device needs to find a new parent by doing a new join process.

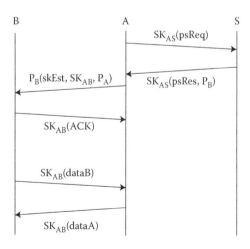

Figure 7.4 Neighbor discovery phase. The neighbor discovery enables the establishment of symmetric session keys between neighboring nodes.

Diffusion: In order to make encrypted diffusion possible using symmetric cryptography, the sink creates a global network session key. This session key is then propagated to all the nodes of the network in a hop-by-hop manner along the parent–child relationship starting from the sink. This session key is used to encrypt diffused messages such as synchronization beacons, for example. The global session key is updated periodically, where the periodicity is the maximum duration tolerated before the global session key is compromised.

7.3.3 Security Analysis

In what follows we go through the different security aspects that are taken into considerations in this architecture.

Confidentiality: During the join phase after the node activation, exchanges are encrypted using the ECAES protocol. Once pairwise session keys are established between nodes, data is exchanged and encrypted using a symmetric key using AES–CTR encryption.

Source authentication and integrity: Message integrity and source authentication can be ensured using ECDSA for creating the hash of the message and signing it using the private key of the sender.

Entity authentication: During the join process, the nodes are authenticated by the sink. When computing the DH key, the sink S and a node R are the only nodes that are able to calculate DH_{RS} using their private keys. The *joinReq* is encrypted by R and decrypted by S, and it contains information to verify the identity of R (i.e., physical address).

7.4 Results

In this section, we present, on one hand, the evaluation of the symmetric operations of AES algorithm (in CTR mode [43]), and on the other hand, we evaluate asymmetric operations which are: (i) the initialization phase of ECC, ECDH, and ECAES (ii) the generation of ECC private and public keys, (iii) the calculation of a common symmetric key between two nodes using ECDH, and (iv) the encryption/decryption using ECAES.

These evaluations are done in terms of execution time and energy consumption. We conclude the results with an analysis of the additional cost depending on the size of the network.

In what follows, we fixed the key size of AES–CTR algorithm to 128 bits, which is used with 10 rounds according to the standard recommendations, and the key size of ECC suite to 160 bits.

Results presented in this chapter are obtained through real measurements. We implemented the algorithms in nesC, a programming language for networked embedded systems [44], on TelosB motes [45]. These motes use a 16-bit 8 MHz TI MSP430 microprocessor.

It should be noted that our implementation of asymmetric algorithms is based on TinyECC [46], which is a configurable library for ECC operations in WSNs. It provides a number of optimization switches that can turn specific optimizations on or off. We used this library with all optimizations enabled to achieve a smaller execution time and less energy consumption with an acceptable cost of memory occupation.

The ECIES algorithm provided in [46] uses simple XOR operations for the symmetric encryption algorithm for ensuring the confidentiality of data exchanged between nodes. For better security, we integrated our implementation of the AES algorithm in the implementation of ECIES. The notations in this chapter for the use of ECIES with AES will be ECAES. The impact of this integration is evaluated in terms of execution time and energy consumption.

7.4.1 Computation of the Energy Consumption

The energy consumption evaluation that we propose here is the evaluation of the additional power consumption caused by the execution of the code of the different cryptographic algorithms. This evaluation is based on the measurements of the execution time of the algorithm components that have been introduced and justified previously. The global energy consumption of an IEEE 802.15.4 standard-based network is studied with more precision in [47].

In what follows, we have calculated the energy consumption using the following formula: U * I * t based on the execution time (t), the voltage (U), and current draw (I). For TelosB motes, we have fixed the voltage at 3 V, supposing that the batteries are all the time in full-charge state. The current draw in active mode (with radio off) is 1.8 mA [45]. The current draw of the radio in the receive or transmit state is 23 mA.

Note that this means that energy consumption is directly related to the execution time.

7.4.2 Elementary Values

In this section, we present previously published results [48] that constitute the elementary values for our evaluation. These results include the initialization phase of the different algorithms (AES, ECC, ECDH, ECAES) (Figure 7.5a), and the cost of encryption and decryption of 1 byte of data in terms of time and energy, using AES on one hand and ECAES on the other hand (Figure 7.5b). For more details and comments on these results please refer to [48].

In addition to the encryption and decryption, in order to establish secure links between nodes, other operations are necessary. To generate a private/public key, a

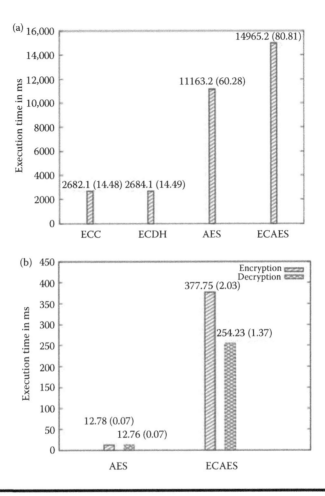

Figure 7.5 **Elementary values for the different cryptographic operations used by AES and ECAES. The energy consumption of these operations are presented between brackets and expressed in mJ. Each value represents the mean of the 10 measurements on TelosB motes. (a) Execution time of the initialization phase of the different algorithms. (b) Execution time for encrypting and decrypting 1 byte of data.**

node takes 2684.6 ms and consumes 14.49 mJ. To compute the secret key using the NIDH method, a node takes 3255 ms and consumes 17.57 mJ.

7.4.3 Time and Energy Consumption of the Join Process

In this section, we evaluate the join process presented in Section 7.3.2 in terms of time consumption. This evaluation is divided into two parts. The first part is the calculation of the processor's execution time at every node, and the second part is the time needed for sending and receiving packets between nodes. Figure 7.6 shows the decomposition of the process into periods of time Ti where $1 \leq i \leq 5$. In this example, we suppose that a node R wants to join the network through node F, and S is the sink of the network.

T1 denotes the time needed by R and S to calculate the DH_{RS} secret key. T2 denotes the time needed for encryption/decryption of the *joinReq* message (where only 16 bytes of the message is encrypted) using ES–CTR with the DH_{FS} key calculated in T1 which was already established between S and F before the arrival of R (during the join phase of node F).

T3 denotes the time needed for encryption/decryption of the *joinRes*, containing the public key of R (40 bytes of the message are encrypted), with AES–CTR using to DH_{FS} key.

T4 denotes the time needed for encrypting the symmetric key SK_{FR} (dedicated to encrypt the future communications between F and R) with ECAES. This operation costs 21, 92 mJ for a symmetric key of 16 bytes, which is a very similar result compared to the same operation presented in [32] that costs 22, 82 mJ.

Finally, T5 denotes the time needed for decrypting SK_{FR} with ECAES.

In Figure 7.7, we presented the time and energy consumption of these periods. Each value is the mean of 10 measurements of the same operation on the TelosB motes.

In addition of these time periods, the sending and receiving of packets must be taken into account. To send or receive a packet of 16 bytes it takes 23.95 ms which we denote as S16. While sending or receiving a packet of 40 bytes (*joinRes* message that contains the public key of the new node) take 25.75 ms which we denote as S40, and sending or receiving 80 bytes (*joinRes* message sent with 40 additional bytes in order to establish a symmetric key) takes 30, 89 ms. These values were also

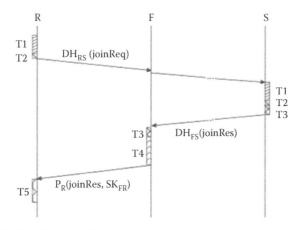

Figure 7.6 Decomposition of the join process into periods of execution time.

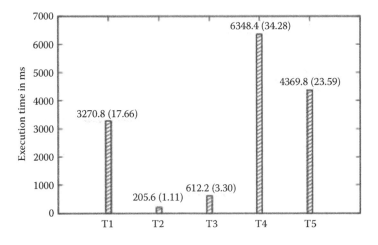

Figure 7.7 Execution time of the different periods of the join process. The energy consumption is presented between brackets and expressed in mJ.

obtained through real measurements on TelosB motes. We calculated the duration between the instant the mote is asked to send a message and the instant the message is sent on the medium.

It should be noted that the execution time of the send command takes more time than the duration of the transmission. In the case of sending or receiving 16 bytes, 40 bytes, and 80 bytes, only 0.512, 1.28, and 2.52 ms respectively, are spent by the radio activity (this calculation does not take into account the medium access delays).

TJoin denotes the total time needed for a node to finish its join process, TJoin is given in Equation 7.1. Note that Ti in Equation 7.1 denotes the execution times shown in the Figure 7.7 and n denotes the number of intermediate nodes on the route between R and S.

$$TJoin = 2 * (T1 + T2 + T3) + T4 + T5 + (n + 1) * S16 + n * S40 + S80 \quad (7.1)$$

The energy is calculated in the same manner by replacing Ti ($1 \le i \le 5$) in Equation 7.1 by the energy consumed in each of the time periods. S16 and S40 are replaced by their calculated energy.

7.4.4 Time and Energy Consumption of the Neighbor Discovery Phase

Figure 7.8 shows the decomposition of the neighbor discovery phase into time periods. These periods are the following:

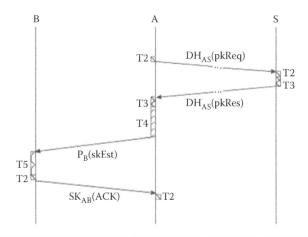

Figure 7.8 Decomposition of the neighbor discovery phase into periods of execution time.

T2 denotes the time needed for encryption/decryption of the *pkReq* message (where only 16 bytes of the message is encrypted) using AES–CTR with the DH_{AS} key.

T3 denotes the time needed for encryption/decryption of the *pkRes*, containing the public key of R (40 bytes of the message are encrypted), using AES–CTR with the DH_{AS} key.

T4 denotes the time needed for encrypting the *pkRes* message (containing a session key dedicated to encrypt the future communications between A and B) with ECAES. Finally, T5 denotes the time needed for decrypting the *pkRes* message with ECAES.

TDiscovery denotes the total time needed by a node to finish the neighbor discovery phase, TDiscovery is given in Equation 7.2. Where n is the number of intermediate nodes on the route between A and S.

$$TDiscovery = 4 * T2 + 2 * T3 + T4 + T5 + (n + 2) * S16$$
$$+ (n + 1) * S40 + S80 \tag{7.2}$$

The energy is calculated in the same way of execution time. We replace Ti $(2 \leq i \leq 5)$ in Equation 7.2 by the energy consumed. S16 and S40 are replaced by their calculated energy.

7.4.5 Network Size Effect

In this section, we study the effect of the network size in terms of number of hops on the energy and time consumption. When the number of nodes separating the newly arriving node and the sink is bigger the energy and time consumption is bigger as shown in Equation 7.1. The same can be noticed for the neighbor discovery process as shown in Equation 7.2.

In our proposal, the number of cryptographic operations remains the same regardless of the number of hops separating the nodes in the join phase and the neighbor discovery phase. Indeed, using the DH keys, only end-to-end encryption is needed and intermediate nodes only relay the message. Hence, when the network size gets bigger, only the S16 and S40 are affected, as both Equations 7.1 and 7.2 show.

Figure 7.9a and b shows the time consumption for the join process and the neighbor discovery process respectively when the number of intermediate nodes varies from 0 to 20.

It should be noted that during these processes, nodes are not necessarily awake during the entire process time, so the energy consumption varies according to the MAC protocol and the duty cycle of each node. We only presented the time

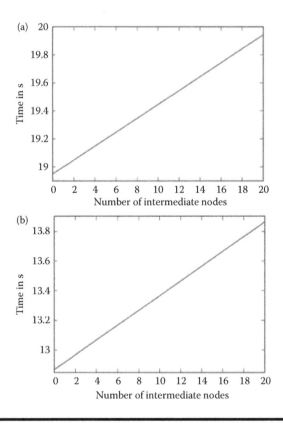

Figure 7.9 **Effect of the network size in terms of number of hops on the time consumption of the (a) join process and (b) neighbor discovery process.**

consumed before each process is completed. Which is the minimum and optimal delay consumed before each process is completed (we do not consider packet loss, repetitions nor medium access delays). This delay is very dependent on the MAC and routing protocols, as well as on the traffic charge of the network.

7.5 Future Research Directions

In our future work, we will evaluate the cost of key revocation and key updates. Public keys should be updated periodically in order to protect the system. This update generates additional control traffic between nodes, but helps enhance the robustness of the cryptographic keys. Key revocation and key updates help maintain backward and forward security in the network. When the network detects that a node is captured by a malicious intruder, the cryptographic keys that are known to that node need to be changed in order to protect the data that has been encrypted using those keys.

In this proposition, we based our test-bed measurements on ECC operations from the TinyECC library. Other cryptographic libraries for WSN exist; we will consider more optimized algorithms based on PBC and the use of more recent libraries such as RELIC [49].

We did not include any counter measurement against jamming attacks. Jamming consists of blocking the use of a certain communication channel by transmitting continuously on that channel in a way that prevents the nodes from receiving data frames. One of the techniques that can be used to avoid jamming is channel hopping. It has been proposed by Jones et al. in [50]. We plan on including a channel-hopping scheme to our proposition in order to enhance the robustness of the communications.

7.6 Conclusion

In this chapter, we presented dynamic lightweight security architecture for multihop WSNs based on predistributed ECC public keys before deployment. These public keys are then used to create session symmetric keys to encrypt the exchanged data and accelerate the cryptographic operations. A symmetric key can be established between any two nodes in the network in a secure manner. We proposed NIDH which is a Diffie–Hellman key exchange protocol that avoids message exchange in order to establish a common symmetric key. Our solution is dynamic for it allows nodes to join the network after the network creation, and, in case of a link failure, nodes can establish keys dynamically.

We are aware of the consequences that the encryption operations have on the network performance in terms of delay, memory usage, and processing. In order to evaluate our proposition, we implemented the different security algorithms on

TelosB motes and evaluated the time and energy consumption for the main phases: join phase and neighbor discovery phase. We showed how the network size does not affect the number of cryptographic operations needed for authenticated join process and authenticated neighbor discovery process.

Even though the computation time (and the energy consumed) can vary slightly according to the manner the code is implemented, this kind of optimal values shows that security in WSN remains a heavy option in terms of time and energy.

It should be noted that additional cost should be considered, compared to an unsecured network, for application traffic exchange, but this is out of the scope of this chapter. It is clear that the cost of cryptographic operations remains today very expensive on low capacity nodes such as wireless sensor nodes. Nevertheless, if the traffic generated by the application is low, cryptographic overhead can be tolerated.

References

1. Standards for efficient cryptography, sec 1: Elliptic curve cryptography, Certicom Research, Tech. Rep., September 2000.
2. W. Diffie and M. Hellman, New directions in cryptography, *IEEE Transactions on Information Theory*, 22, 644–654, 1976.
3. T. Kavitha and D. Sridharan, Security vulnerabilities in wireless sensor networks: A survey, *Journal of Information Assurance and Security*, 5, 31–34, 2010.
4. Y. Wang, G. Attebury, and B. Ramamurthy, Security vulnerabilities in wireless sensor networks: A survey, *CSE Journal Articles*, 5, 31–44, 2006.
5. M. Al and K. Yoshigoe, Security and attacks in wireless sensor networks, in *Network Security, Administration and Management: Advancing Technology and Practice*, 2011, pp. 183–216.
6. D. G. Padmavathi and M. D. Shanmugapriya, A survey of attacks, security mechanisms and challenges in wireless sensor networks, *International Journal of Computer Science and Information Security*, 4, 117–125, 2009.
7. W. Stallings, *Cryptography and Network Security: Principles and Practice*. Prentice-Hall, USA, 2006.
8. A. S. K. Pathan and C. S. Hong, Security attacks and challenges in wireless sensor networks, in *Encyclopedia on ad hoc and Ubiquitous Computing: Theory and Design of Wireless ad hoc, Sensor, and Mesh Networks*, World Scientific Publishing, USA, 397–426, 2009.
9. A. Uluagac, R. Lee, and J. Copeland, Designing secure protocols for wireless sensor networks, *Wireless Algorithms, Systems, and Applications*, 5258, 503–514, 2008.
10. Nabil Ali Alrajeh, S. Khan, J. Lloret, and J. Loo, Secure routing protocol using cross-layer design and energy harvesting in wireless sensor networks, in *International Journal of Distributed Sensor Networks*, 2013, 11 pages, 2013.
11. A. Wander, N. Gura, H. Eberle, V. Gupta, and S. Shantz, Energy analysis of public-key cryptography for wireless sensor networks, in *International Conference on Pervasive Computing and Communication*, Hawaii, USA, 2005.
12. H. Seo, S. Kim, and R. Ramakrishna, A new security protocol based on elliptic curve cryptosystems for securing wireless sensor networks, in *EUC Workshops*, Seoul, Korea, 2006.

13. P. Hong, Feasibility of PKC in resource-constrained wireless sensor networks, in *ICCIT*, December 2008.
14. K.-A. Shim, Y.-R. Lee, and C.-M. Park, An efficient identity-based broadcast authentication scheme in wireless sensor networks, *Ad Hoc Networks*, 11, 182–189, 2013.
15. A. K. Das, P. Sharma, S. Chatterjee, and J. K. Sing, A dynamic password-based user authentication scheme for hierarchical wireless sensor networks, *Journal of Network and Computer Applications*, 35, 1646–1656, 2012.
16. Y. Liu, J. Li, and M. Guizani, PKC based broadcast authentication using signature amortization for WSNs, *IEEE Transactions on Wireless Communications*, 11, 2106–2115, 2012.
17. C. Jiang, B. Li, and H. Xu, An efficient scheme for user authentication in wireless sensor networks, in *21st International Conference on Advanced Information Networking and Applications Workshop*, Niagara Falls, Canada, 1, 438–442, 2007.
18. L. Batina, N. Mentens, K. Sakiyama, B. Preneel, and I. Verbauwhede, Low-cost elliptic curve cryptography for wireless sensor networks, in *Security and Privacy in Ad-Hoc and Sensor Networks*, 4357, 6–17, 2006.
19. K. Piotrowski, P. Langendoerfer, and S. Peter, How public key cryptography influences wireless sensor node lifetime, in *Proceedings of the Fourth ACM Workshop on Security of Ad Hoc and Sensor Networks*, Alexandria, VA, USA, 2006, pp. 169–176.
20. N. Gura, A. Patel, A. W, H. Eberle, and S. C. Shantz, Comparing elliptic curve cryptography and RSA on 8-bit CPUs, 2004, pp. 119–132.
21. R. Rivest, A. Shamir, and L. Adleman, A method for obtaining digital signatures and public-key cryptosystems, *Communications of the ACM*, 21, 120–126, 1978.
22. S.-M. Yen and L. C.-S., Improved digital signature algorithm, *IEEE Transactions on Computers*, 44, 729–730, 1995.
23. J. Lopez and R. Dahab, An overview of elliptic curve cryptography, University of Campinas, Tech. Rep., May 2000.
24. G. Meulenaer, F. Gosset, F. Standaert, and O. Pereira, On the energy cost of communication and cryptography in wireless sensor networks, in *IEEE International Conference on Wireless and Mobile Computing*, Avignon, France, 2008.
25. A. Perrig, R. Szewczyk, J. Tygar, V. Wen, and D. Culler, SPINS: Security protocols for sensor networks, *Wireless Networks*, 8, 521–534, 2002.
26. A. Perrig, R. Canetti, J. Tygar, and D. Song, Efficient authentication and signing of multicast streams over lossy channels, in *IEEE Symposium on Security and Privacy*, April 2000.
27. K. Jones, A. Wadaa, S. Oladu, L. Wilson, and M. Eltoweissy, Towards a new paradigm for securing wireless sensor networks, in *New Security Paradigms Workshop*, Ascona, Switzerland, 2003.
28. F. Zhang, R. Dojen, and T. Coffey, Comparative performance and energy consumption analysis of different AES implementations on a wireless sensor network node, *International Journal of Sensor Networks*, 10, 192–201, 2012.
29. A. Shamir, Identity-based cryptosystems and signature schemes, in *Fourth Annual International Cryptology Conference*, pp. 47–53, Santa Barbara, California, 1984.
30. L. Oliveira, D. Aranha, P. Gouvea, M. Scott, D. Camara, J. Lopez, and R. Dahab, Tinypbc: Pairings for authenticated identity-based non-interactive key distribution in sensor networks, *Computer Communications*, 34, 485–493, 2011.
31. P. Szczechowiak, A. Kargl, M. Scott, and M. Collier, On the application of pairing based cryptography to wireless sensor networks, in *ACM Conference on Wireless Network Security*, pp. 1–12, Zurich, Switzerland, 2009.

32. E. Munivel and G. Ajit, Efficient public key infrastructure implementation in wireless sensor networks, in *International Conference on Wireless Communication and Sensor Computing*, Tamil Nadu, India, pp. 1–6, 2010.

33. A. Wander, N. Gura, H. Eberle, V. Gupta, and S. Shantz, Energy analysis of public-key cryptography for wireless sensor networks, in *International Conference on Pervasive Computing and Communication*, Hawaii, USA, 2005.

34. Zigbee, Zigbee Specification, ZigBee Standards Organization, Zigbee Standard 053474r17, January 2008.

35. P. Zand, C. Chatterjea, K. Das, and P. Havinga, Wireless industrial monitoring and control networks: The journey so far and the road ahead, *Computer Communications*, 1, 123–152, 2012.

36. International Society of Automation Std., ISA100.11a: 2009 wireless systems for industrial automation: Process control and related applications, Draft standard, in preparation, 2009.

37. HART Communication Foundation Std., HART field communication protocol specifications, Tech. Rep., 2008.

38. Zigbee, Zigbee Zigbee Smart Energy Specification, ZigBee Standards Organization, Zigbee Standard 075356r15, December 2008.

39. K. Akkaya and M. Younis, A survey on routing protocols for wireless sensor networks, *Ad hoc Networks*, 3, 325–349, 2005.

40. J. Ibriq and I. Mahgoub, Cluster-based routing in wireless sensor networks: Issues and challenges, in *SPECTS*, California, USA, 2004.

41. IEEE 802.15, Part 15.4: Wireless medium access control (MAC) and physical layer (PHY) specifications for low-rate wireless personal area networks (WPANs), ANSI/IEEE, Standard 802.15.4 R2006, 2006.

42. G. Chalhoub, A. Guitton, and M. Misson, MAC specifications for a WPAN allowing both energy saving and guaranteed delay—Part A: MaCARI: A synchronized tree-based mac protocol, in *IFIP WSAN*, 2008.

43. M. Dworkin, Recommendation for block cipher modes of operation, methods and techniques, NIST Special Publication 800-38A, December 2001.

44. D. Gay, P. Levis, R. von Behren, M. Welsh, E. Brewer, and D. Culler, The nesC language: A holistic approach to networked embedded systems, in *Programming Language Design and Implementation (PLDI)*, June 2003.

45. Crossbow, TelosB datasheet, document Part Number: 6020-0094-01 Rev B, 2004.

46. A. Liu and N. Ning, TinyECC: A configurable library for elliptic curve cryptography in wireless sensor networks, in *7th International Conference on Information Processing in Sensor Networks*, St. Louis, Missouri, April 2008, pp. 245–256.

47. N. Fourty, A. van den Bossche, and T. Val, An advanced study of energy consumption in an IEEE 802.15.4 based network: Everything but the truth on 802.15.4 node lifetime, *Computer Communications*, 35, 1759–1767, 2012.

48. I. Mansour and G. Chalhoub, Evaluation of different cryptographic algorithms on wireless sensor network nodes, in *International Conference on Wireless Communications in Unusual and Confined Areas*, Clermont-Ferrand, France, 2012.

49. D. Aranha and C. Gouv, RELIC is an efficient library for cryptography, http://code.google.com/p/relic-toolkit/. Last accessed on March 27th, 2013.

50. K. Jones, A. Wadaa, S. Oladu, L. Wilson, and M. Eltoweissy, Towards a new paradigm for securing wireless sensor networks, in *New Security Paradigms Workshop*, Ascona, Switzerland, 2003.

Chapter 8

Optimal Load Distribution in Wireless Sensor Networks
A Reliable Clustering Algorithm Approach

Dauda Ayanda

Contents

8.1 Introduction

Wireless sensor networks (WSNs) have recently emerged as an active research area due to advances in low-power wireless communications, microelectromechanical systems (MEMS), and nanotechnology (Chong and Kumar 2003). They are composed of sensor nodes which are tiny autonomous devices that combine sensing, computing, and wireless communication capabilities.

The wireless sensor nodes are deeply embedded into the physical surroundings where they gather and process information such as temperature, humidity, light characteristics, seismic activities or images, and sound samples from the physical world. Networked systems of such sensors are expected to be used in a variety of applications including habitat monitoring, precision agriculture, disaster recovery operations, health care, and supply chain management (Akyildiz et al. 2002; Kamal et al. 2005).

According to Chong and Kumar (2003), real-world sensor network deployments and prototypic implementations are still not commonplace. However, the experiences gained in such deployments are crucial for the sensor network research community. These results are needed to refine assumptions made when designing hardware, software, protocols, and mechanisms for WSNs.

Several studies were carried out in WSNs such as energy efficiency, load balancing, and reliability. However, these design goals are generally orthogonal to each other. For example, most of the load-balancing schemes are not robust—or resistive, enough to high-link failure rates (Woo et al. 2003). The existing load-balancing schemes are classified into two categories: local load balancing and global load balancing. These are also classified as hop-by-hop balancing and end-to-end balancing respectively.

On the other hand, reliable routing in WSNs is a challenging issue, first, because of the absence of global addressing schemes; second, because of the problem of data sources from multiple paths to a single source; and third, because of data redundancy and energy and computation constraints of the network. The performance of the existing routing algorithms, such as the shortest path routing algorithm for WSNs, varies from application to application because of the diverse demands for which they are deployed.

Puccinelli and Haenggi (2009) classified routing in WSNs into three major components: a link estimator that continuously assesses link quality, a routing engine that determines the address of the one-hop neighbor that provides the best progress toward the sink according to a given cost function based on the output of the link estimator, and a forwarding engine that injects its own traffic or relays upstream traffic by unicasting to the address determined by the routing engine. The rationale behind this partitioning is the separation of the data plane (forwarding engine) from the control plane (routing engine with the link estimator of choice). Connectivity discovery and route maintenance are carried out with the help of control beacons that diffuse global state used locally for route selection.

In this study, an enhanced load-balancing algorithm for reliable routing in WSNs is proposed. This algorithm aims to achieve both load balancing and reliability for large-scale WSNs. The use of a load-balancing scheme can be expected to provide significant lifetime benefits: rather than always using the nodes with the best channel, traffic is redistributed over a larger number of relays.

8.2 Historical Background of WSNs

According to Chong and Kumar (2003), the history of WSNs spanned four phases. First, the Cold-War era ushered in the development of extensive acoustic networks in the United States for submarine surveillance. The major impetus to research sensor networks took place in the early 1980s with the programs sponsored by the Defense Advanced Research Projects Agency (DARPA) on the distributed sensor networks (DSN) work aimed at determining if newly developed protocols (transmission control protocol internet protocol (TCP/IP)) and the ARPAnet's approach to communication could be used in the context of sensor networks. DSN postulated the existence of many low-cost spatially distributed sensing nodes that were designed to operate in a collaborative manner, yet be autonomous. Testbeds were custom built for tracking multiple targets in a distributed environment.

Moreover, based on the results generated by the DARPA–DSN research and the testbeds developed, military planners set out in the 1980s and 1990s to adopt sensor network technology, making it a key component of the network-centric warfare. An example of network-centric warfare includes the cooperative engagement capability, a system that consists of multiple radars collecting data on air targets. Other sensor networks in the military arena include acoustic sensor arrays for antisubmarine warfare, such as the fixed distributed system and the advanced deployable system, and autonomous ground sensor systems, such as the remote battlefield sensor system and the tactical remote sensor system.

The present day sensor network research evolves as a result of the advances in computing and communication that have taken place in the late 1990s and early 2000s that have resulted in a new generation of sensor network technology. These sensor networks represent a significant improvement over traditional sensors through a number of high-density technologies, including MEMS and nanoscale electromechanical systems (Akyildiz et al. 2002).

Advances in IEEE 802.11a/b/g-based wireless networking and other wireless systems such as Bluetooth, Zigbee, and WiMax are now facilitating reliable and ubiquitous connectivity. Also, inexpensive processors that have low power-consumption requirements make possible the deployment of sensors for a plethora of applications. Commercially focused efforts are now directed at defining mesh, peer-to-peer (P2P), and cluster-tree network topologies with data security features and interoperable application profiles.

The historical evidence shows that the essential function of a sensor network is to identify the state of an environment using multiple sensors. Other aspects of sensor network design, such as the communication, power, and computational resources, are evaluated by their ability to facilitate this function. Typically, a WSN is composed of a large number of sensor nodes, with processing, sensing, and radio communication capabilities, scattered throughout a certain geographical region, where the sensory data is routed in a multihop ad hoc fashion from the originator sensor node to a remote control station as shown in Figure 8.1.

As illustrated in the diagram, the basic elements of a WSN are sensor nodes. The sinks (or base stations [BSs]) and the gateways are usually more complex devices than the sensor nodes because of the functionalities they need to provide. The sensor node is the simplest device in the network, and in most applications the number of sensor nodes is much larger than the number of sinks. Therefore, their cost and size must be kept as low as possible.

Also, in most applications, the use of battery-powered devices is very convenient to make the deployment of such nodes easier. To let the network work under

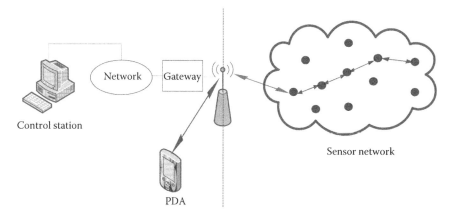

Figure 8.1 Topology of a WSN. (From Koubaa A, Alves M, 2005. A two-tiered architecture for real-time communications in large-scale wireless sensor networks: Research challenges, *Technical Report*, TR-050701, Version 1.0. Available at http://www.hurray.isep.ipp.pt, accessed on May 15, 2010.)

specified performance requirements for a sufficient time (denoted as network life-time), the nodes must be capable of playing their role for a sufficiently long period using the energy provided by their battery, which in many applications should not be renewed for a long time. Thus, energy efficiency of all tasks performed by a node is a must for the WSN design.

8.3 Features of WSNs

Wireless sensor nodes have distinguishing features that allow the nodes to self-organize into a noninfrastructure network with dynamic topology. Typically, the number of nodes in a sensor network is much higher than in an ad hoc network, and dense deployments are often desired to ensure coverage and connectivity (Puccinelli and Haenggi 2005). Hence, the sensor network hardware should be relatively cheap. Sensor network hardware should be power-efficient, small, inexpensive, and reliable in order to maximize network lifetime, add flexibility, facilitate data collection, and minimize the need for maintenance.

8.3.1 Network Lifetime

The energy consumption of the sensor nodes should be self-powering and is a factor that is dependent on the network lifetime. Network lifetime is important for most applications in WSNs. Many researchers (Goldsmith and Wicker 2002; Yuan and Qu 2004) have criticized the initial assumption that the transmit power associated with packet transmission accounts for the major share of power consumption. Rather, sensing, signal processing, and sensors' hardware operating in standby mode consume a consistent amount of power, while some applications require extra power for operation.

The authors also suggested that energy consumption can be favorably reduced by considering the existing interdependencies between individual layers in the network protocol stack. Energy-efficient routing should avoid the loss of nodes due to battery depletion. Schurgers et al. (2001) opined that lifetime benefits could be obtained with lower radio duty cycles and dynamic modulation scaling by varying the constellation size to minimize energy expenditure at the physical layer.

At the data-link layer, media access control (MAC) solutions exist and have a direct impact on energy consumption. Some of the primary causes of energy waste include collisions, control packet overhead, and idle listening, which are associated with data-link layer bottleneck. Energy-efficient routing should also avoid the waste of a node as a result of battery depletion. Some of the proposed routing protocols like ant colony optimization algorithm (ACOA) tend to minimize energy consumption on the forwarding paths, but if some nodes happen to be located on a forwarding path that is close to the BS, their lifetime will be reduced.

8.3.2 Flexibility

Sensor networks should be scalable and be able to dynamically adapt to changes in node density and topology. In surveillance applications, for instance, most nodes may remain quiescent as long as nothing interesting happens (Puccinelli and Haenggi 2005). However, they must be able to respond to special events that the network intends to study with some degree of granularity.

The sensor networks should be able to provide minimal delay service in critical applications such as actuator networks and thus an improved response time. In the same vein, sensor networks should be robust to changes in their topology due to the failure of individual nodes. Of particular importance is the connectivity and coverage that are needed in warfare and surveillance applications. Connectivity is achieved when there is a transmission agreement between the nodes and the BS while Meguerdichian et al. (2001) described coverage as a measure of quality of service in a sensor network. The authors went further by defining coverage as to how well a particular area can be observed by a network and also characterized the probability of detecting geographically constrained phenomena.

8.3.3 Maintenance

Sensor nodes should be updated because it forms the basis for the maintenance of the program code in the nodes over the wireless channel. In case of packet loss, it should be accounted for and reprogramming is carried out. Reijers and Loangendoen (2003) stated that the portion of the code always running in the node should guarantee reprogramming support with a small footprint, and updating procedures should only cause brief interruption of the normal operation of the node.

The functioning of the network as a whole should not be endangered by unavoidable failures of single nodes, which may occur for a number of reasons, from battery depletion to unpredictable external events, and may either be independent or spatially correlated (Ganesan et al. 2001).

8.3.4 Data Collection

Data collection only takes place when the connectivity has been guaranteed and the sensors are within the coverage area. Shah et al. (2003) proposed ubiquitous mobile agents named mobile ubiquitous LAN extensions (MULEs) that randomly move around to gather data-bridging sensor nodes and access points.

Data are directly relayed to a BS in a centralized form which may shorten the network lifetime. Relaying data to a data sink causes nonuniform power-consumption pattern that may overburden forwarding nodes (Haenggi 2004). This is particularly harsh on nodes providing end links to BSs, which may end up relaying traffic coming from all other nodes, thus forming a critical bottleneck for network throughput (Haenggi 2003).

Younis and Fahmy (2004) proposed an improved clustering technique for data collection. In clustering, nodes team up to form clusters and transmit their information to their cluster heads (CHs), which fuse the data and forward it to a sink. Fewer packets are transmitted, and a uniform energy-consumption pattern may be achieved by periodic reclustering. Also, as the aggregation process fuses strongly correlated measurements, data redundancy is reduced to the barest minimum.

8.4 Issues in Load-Balancing Clustering Algorithm for WSNs

Clustering in WSNs allows network model to be grouped into different clusters where CH is elected among the cluster member nodes that aggregate the sensed information from the member nodes and sends it to the BS. Despite the preference for clustering models in load-balancing algorithms, routing in WSNs based on clustering poses various challenges to the research community. Israr and Awan (2006b) highlighted some of the major challenges faced while clustering the WSN.

8.4.1 Network Deployment

Node deployment in WSNs is either fixed or random depending on the application. In fixed deployment, the network is deployed on predetermined locations whereas in random deployment, the resulting distribution can be uniform or nonuniform. In such a case, careful management of the network is necessary in order to ensure entire area coverage and also to ensure that the energy consumption is also uniform across the network.

8.4.2 A Heterogeneous Network

The WSNs are not always uniform. In some cases, a network is heterogeneous consisting of nodes with different energy levels. Some nodes are less energy constrained than others and the fraction of these sensor nodes are less. In this type of network, the less-energy constraint nodes are chosen as head of the cluster and the energy-constrained nodes are the worker nodes of the cluster. The problem arises in such a network when the network is deployed randomly and all CHs are concentrated in some particular part of the network, resulting in unbalanced cluster formation and making some portion of the network unreachable.

Also, if the resulting distribution of the CHs is uniform and multihop communication is used, the nodes which are close to the CH are under a heavy load as all the traffic is routed from different areas of the network to the CH via the neighbors of the CH. This will cause rapid dying of the nodes in the vicinity of the CHs resulting in gaps near the CHs, decreasing of the network size and increasing the network energy consumption. Heterogeneous sensor networks require careful management of the clusters in order to avoid the problems resulting from unbalanced CH distribution as well as to ensure that the energy consumption across the network is uniform.

8.4.3 Network Scalability

When a WSN is deployed, new nodes are sometimes added to the network in order to cover more area or to prolong the lifetime of the current node. In both cases, the clustering scheme should be able to adapt to changes in the topology of the network. This allows the algorithm to easily adapt to the topology change especially when it is local and dynamic.

8.4.4 Uniform Energy Consumption

The CHs which transmit data to the BS consume more energy compared to the rest of the nodes due to the fact that transmission in WSNs is more energy consuming as compared to sensing. Therefore, the clustering scheme should ensure that energy dissipation across the network is balanced and the CH should be rotated in order to balance the energy consumption of the network.

8.4.5 Multihop or Single-Hop Communication

The communication model that WSN uses is either single hop or multihop. Since energy consumption in wireless systems is directly proportional to the square of the distance, single-hop communication is expensive in terms of energy consumption. Most of the routing algorithms use the multihop-communication model since it is more energy efficient in terms of energy consumption. However, with multihop communication, the nodes that are closer to the CH are under heavy traffic and can create gaps near the CH when their energy terminates.

8.4.6 Attribute-Based Addressing

Due to the sheer number of nodes, it is not possible to assign identities (IDs) to the nodes in WSNs. Data is accessed from nodes via attributes, not by IDs. This makes intrusion into the system easier and implementing a security mechanism more difficult.

8.4.7 Cluster Dynamics

Cluster dynamics refers to how different parameters of the cluster are determined. For example, the number of clusters in a particular network might be preassigned in some cases, while in other cases, they are dynamic. The CH performs the function of compression as well as transmission of data. The distance between the CHs is a major issue that can be dynamic or can be set in accordance with some minimum value.

In the case of dynamics, there is a possibility of forming unbalanced clusters and, while limiting it by some preassigned value, minimum distance can be effective in some cases, but this is an open-research challenge. Also CH selection can either be centralized or decentralized and the number of clusters can either be fixed or dynamic. A fixed number of clusters cause less overhead in that the network will not have to repeatedly go through the setup phase when clusters are formed although it is poor in terms of scalability.

8.5 Routing in WSNs

Many routing and dissemination protocols have been studied specifically for WSNs, following all the essential design issues. In particular, the most important one is concerning energy consumption. Routing in WSNs is a very challenging task due to the inherent characteristics that distinguish these networks from other wireless networks like mobile ad hoc networks (MANETs).

According to Verdone et al. (2009), traditional IP-based protocols may not be applied to WSNs due to the large number of sensor nodes and because getting the data is often more important than knowing the specific identity of the source sending it. Second, almost all applications of sensor networks require the flow of sensed data from multiple sources to a particular BS through the sink. This makes sensor nodes to be constrained in terms of energy, processing, and storage capacities that require careful resource management. Third, sensor networks are strictly dependent from their application on sensor nodes; while the design requirements of a sensor network change with the application and the position awareness of the sensor nodes application in relation to the BS which is critical to data collection. Finally, sensor nodes are often correlated and contain a lot of redundancy as a result of data being collected by the sensors in WSN which are typically based on common phenomena. Such redundancy needs to be exploited by the routing protocols to improve energy and bandwidth utilization.

Routing protocols can be classified according to the network structure as flat, hierarchical, or location-based. In flat-based routing, all nodes are typically assigned equal roles or functionality. In hierarchical-based routing, nodes will play different roles in the network; for example, hierarchical protocols aim at clustering the nodes so that CHs can do some aggregation and reduction of data in order to save energy. Location-based routing exploits sensor nodes' positions to route data in the network.

8.5.1 Different Categories of Routing Models

There are three routing models in WSNs, which include: one-hop model; multihop model; and cluster-based model, which are depicted in Figure 8.2.

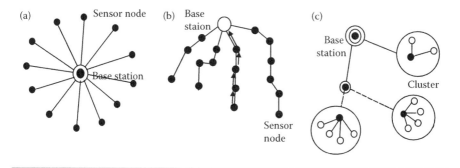

Figure 8.2   Routing models with (a) one-hop, (b) multihop, and (c) cluster-based model. (From Akhtar A, Minhas AA, Jabbar S, 2010. Energy aware intra cluster routing for wireless sensor networks, *in International Journal of Hybrid Information Technology,* **3(1).)**

The one-hop model is a simple model that uses direct data sending toward the BS. The multihop model is an energy-efficient model of routing that allows nodes to choose their neighbors to forward data toward the BS. Cluster-based models allow the network to be grouped into different clusters. Each cluster is composed of one CH and cluster member nodes. The CH gets the sensed data from the cluster member nodes, aggregates the sensed information, and then sends it to the BS.

8.5.2 Mechanisms of Data Relay in WSNs

According to Hedetniemi and Liestman (1988), the two classical data-relay mechanisms that do not require any routing algorithms and topology maintenance are flooding and gossiping. In flooding, each sensor receiving a data packet broadcasts it to all of its neighbors regardless of whether or not a neighbor has already received the data from another node. This process continues until the packet arrives at the destination.

Flooding is very easy to implement, however, it has several drawbacks that include implosion, which is caused by duplicated messages sent to the same node; overlapping that takes place when sensor nodes cover overlapped geographical areas and, therefore, collect overlapping pieces of data; and resource blindness as a result of nodes that do not modify their activities based on the amount of energy available to them at a given time.

On the other hand, gossiping is a slightly enhanced version of flooding whereby a sensor node that receives a packet randomly chooses one of its neighbors to which it forwards the packet. Although, gossiping does not solve the problem of overlap, the problem of implosion that is peculiar to flooding can be specifically avoided in gossiping by selecting a random next node rather than broadcasting. This can, however, cause delays in propagation of data.

8.6 Related Works

Cluster-based routing is an energy-efficient routing model in achieving load balancing as compared with direct routing and multihop routing protocol in WSNs, but there are some issues in cluster-based routing as well. Heinzelman et al. (2000) discussed the problem of load balancing in cluster-based routing and introduced a novel idea of rotation of CH role inside the cluster named low-energy adaptive clustering hierarchy (LEACH).

The LEACH operation is composed of two phases: a setup phase and a steady-state phase. The setup phase is needed in order to create the clusters inside the network and elect the CHs in each cluster. During the steady-state phase, the nodes inside each cluster sense data and transmit data to their CH. The CH collects all the data sent by the nodes in its cluster, aggregates it, and sends it to the sink as shown in the Figure 8.3. LEACH protocol assumes that all CHs can directly

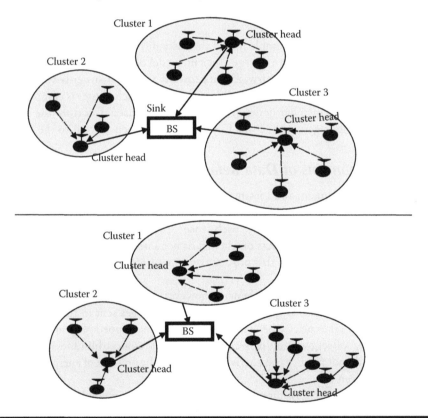

Figure 8.3 LEACH routing protocol showing the rotation of CHs among nodes. (From Heinzelmann H, Chandrakasan A, Balakrishnan H, 2000. Energy-efficient communication protocol for wireless microsensor networks *in Proceedings of the 33rd Annual Hawaii International Conference on System Sciences,* **Vol. 2.)**

communicate with the central BS of the network; therefore it is not applicable in large regions. The major drawback is that the resultant set of CHs may be unevenly distributed, which causes variable cluster sizes and higher intracluster communication costs. CHs far away from the BS have to transmit data over a long distance and suffer a high-energy consumption rate. In a large network, such a disparity will cause nodes in the far corners of the sensing area to die quickly.

Lindsey and Raghavendra (2002) proposed power-efficient gathering in sensor information systems (PEGASIS) that is an improvement of the LEACH approach. Instead of forming multiple clusters, PEGASIS forms chains of sensor nodes so each node communicates only with its closest neighbors and takes turns in transmitting data to the BS. The protocol assumes that all nodes are to communicate directly with the BS, so the role of the leader is rotated among all the nodes forming the chain in order to balance the energy consumption.

Considering data gathering in each round in PEGASIS, each node receives data from one neighbor, fuses it with its own data, and transmits to the other neighbor on the chain. The data gathered are transferred from one sensor node to another until it reaches the leader, which sends the data to the BS as shown in the Figure 8.4. In PEGASIS, a rotation scheme is used to share the cost of communication with the BS, however uneven energy dissipation still exists due to the difference in CH positions. In addition, this approach assumes global data aggregation, that is, sensor data from all nodes can be aggregated into a single packet. This is an assumption that is not always true. When it is not true, the cost of passing each packet along the entire chain will cause a very short network lifetime.

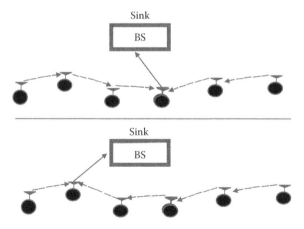

Figure 8.4 PEGASIS showing the rotation of leaders among the nodes composing the chain. (From Lindsey S, Raghavendra C, 2002. PEGASIS: Power-efficient gathering in sensor information systems. In *Proceedings of the IEEE Aerospace Conference*, 3, 1125–1130.)

These major drawbacks were addressed in a centralized optimization approach proposed by Ding and Liu (2004). In this chapter, the authors presented an AntChain method where ACOA is applied as a centralized optimization tool to form the near lowest cost chain for a particular area. Unlike the PEGASIS, sensors nodes do not have to have the prior global knowledge. The ACO approach is designed to deal with the dynamics of the sensor nodes, which can ensure the algorithm response to any network changes in a time.

The similarity between the ant colony and a web of energy-constraint sensor nodes inspired the need for the swarm intelligence ACO algorithm which the authors applied to their WSN data-gathering scheme. In Figure 8.5, each sensor node keeps data-gathering operations until this period ends. In each round, the chain tail will send the fused data and an integer n for which indicates how many nodes this data presents. All nodes wake up and expect to receive messages from BS when this period ends. Then the BS either sends another chain information according to the request it gets from the end user. If a node died in the AntChain, the information of the nodes that are at the left side of the dead node is lost until the next period begins. To make the network more robust and less information loss in case of any change happening during the data-gathering process, the bi-directional AntChain scheme was introduced as shown in Figure 8.6.

The ACO algorithm allows all the complicate computation, optimization, and set up overhead to be performed by BS which is assumed to have unlimited energy resource, much stronger communication, and computation ability. Unlike most other sensor nodes routing algorithm, ACO algorithm allows sensor nodes to communicate with each other in order to set up a transmission route while sensor nodes only receive useful information through the BS's broadcast. The drawback of this protocol is that the BS requires information about all the nodes in a network before the selection of CHs. In a larger network, this approach would not work well since it uses a centralized approach for the management of the clusters.

Younis and Fahmy (2004) proposed a hybrid energy-efficient distributed (HEED) clustering that periodically selects CHs according to a hybrid of their residual energy and a secondary parameter, such as node proximity to its neighbors

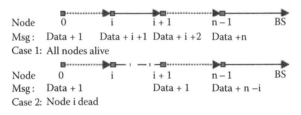

Figure 8.5   Data-gathering process in the unidirectional AntChain. (From Ding N, Liu PS, 2004. Data gathering communication in wireless sensor networks using ant colony optimization, *Proceedings of the 2004 IEEE International Conference on Robotics and Biomimetics*, Shenyang, China, pp 822–827.)

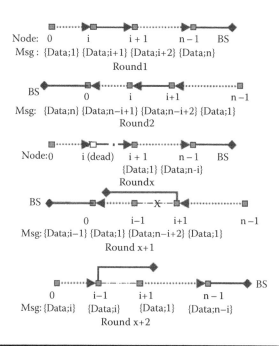

Figure 8.6 Data-gathering process in the bi-directional AntChain. (From Ding N, Liu PS, 2004. Data gathering communication in wireless sensor networks using ant colony optimization, *Proceedings of the 2004 IEEE International Conference on Robotics and Biomimetics,* **Shenyang, China, pp 822–827.)**

or node degree. In this approach, a probabilistic algorithm is employed to form a dominating set in a fixed number of rounds, with a penalty of slightly large dominating set size. This scheme builds higher quality clusters than LEACH and PEGASIS that uses random selections and single chain, which results in a longer network lifetime. In HEED, all CHs send aggregated data to the BS via the shortest path. These shortest paths form the shortest path tree as shown in Figure 8.7.

This scheme minimizes the total energy consumption. However, the energy consumption is still unbalanced since neighbors of the BS are responsible to relay all packets to the BS and have higher load. A hot spot is formed in the area surrounding the BS, which is congested with data traffic and consumes energy much faster than other areas of the network.

A multilayer multihop routing algorithm for intercluster communication was presented by Israr and Awan (2006a). The algorithm worked on the principle of divide and conquers and performed better in terms of load balancing and energy efficiency than LEACH.

The algorithm was aimed at exploiting the redundancy property of the WSNs. It selects a small percent of nodes from the network and marks them as temporary CHs and uses these nodes to make the intercluster communication multihop. The problem

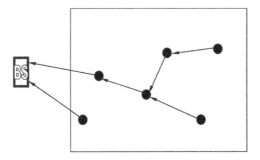

Figure 8.7 Shortest path routing. (From Younis O, Fahmy S, 2004. HEED: A hybrid, energy-efficient, distributed clustering approach for ad hoc sensor networks. In *IEEE Transactions on Mobile Computing*, 3, 366–379.)

with the algorithm was that it was selecting the temporary CHs randomly thus compromising occasionally on the area coverage of the network which it is monitoring.

A data-aggregation algorithm for WSNs was proposed in Mozumdar et al. (2009). The algorithm selects a cluster leader that can perform data aggregation in partially connected sensor networks. The algorithm reduces the traffic flow inside the network by adaptively selecting the shortest route for packet routing to the cluster leader. The proposed scheme is only limited to partially connected network within an intracluster network. Also, the algorithm did not address the time synchronization that is needed at the start of the algorithm.

Akhtar et al. (2010) proposed adaptive intracluster routing in order to address the problem of load balancing associated with the existing single hop and multihop routing protocol. The authors presented energy aware intracluster routing (EAICR) algorithm that keeps the scope of WSNs to intracluster communication where each node is not identical to other for routing the data.

As shown in Figure 8.8, some nodes are considered in close region and they perform direct routing and outside the region nodes adopt multihop routing. In this way, the closer nodes are not having an extra load on them, unlike HEED where the closer nodes to the BS exhaust energy very quickly because they perform the task of sensing their own data and routing the data of the other nodes. Thus, closer nodes have extra loads with them and there is no concept of load balancing as well.

The algorithm works in three phases. First, CH is selected, then with the collaboration of BS, clusters are formed and finally intracluster routing is carried out. Second, each selected CH broadcasts its status as CH, the nodes that get the high received signal strength indicator (RSSI) value respond to this beacon message and show their will to join the cluster. The node sends join request to the respective CH and, in response, join confirmed alert is sent by the CH and clusters are formed with one CH each. Finally, as the routing phase starts, the nodes first check their location if their location is inside the close region then the mode of routing for these nodes is direct routing; if the nodes are out of the close region, their mode

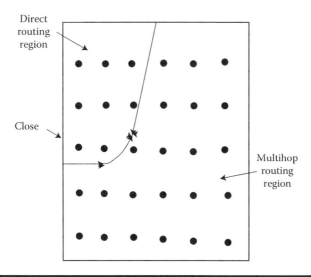

Figure 8.8 Intracluster routing topology using direct and multihop routing. (From Akhtar A, Minhas AA, Jabbar S, 2010. Energy aware intra cluster routing for wireless sensor networks, *in International Journal of Hybrid Information Technology*, 3(1).)

of routing is multihop. The major drawback of this algorithm is that the scope is only limited to an intracluster routing algorithm and not adaptive to an intercluster routing algorithm. Also, the load is still unbalanced due to hot spots that exist at the CH close to the BS as a result of data relay through multiple hops.

Alrajeh et al. (2013) presented a secure routing protocol for WSN, which is based on cross-layer design and energy-harvesting techniques. Although the study uses a cluster-based approach to group together nodes of two-hop neighbors, there exist a drawback as a result of the reliability of data packet due to packet loss.

8.7 Shortcomings of the Existing Load-Balancing Algorithms for WSNs

The major drawbacks of the existing load-balancing algorithms have been identified. The direct connection that was mostly employed by the LEACH approach allows the CH to directly communicate with the BS. As a result, the energy consumption is unevenly distributed coupled with the shorter network lifetime of the sensor nodes at the far end from the BS.

The multihop algorithm seems to be an improvement over the single-hop algorithm that uses direct connection. In this approach, instead of the nodes to send data directly to the BS, the nodes from the far end of the BS form a chain with the nearest neighbor and transmit through the shortest path to the BS. This scheme minimizes energy consumption compared to the single-hop scheme; however, the load is still unbalanced. This is due to the traffic load that leads to hot spots around the nodes closest to the BS since neighbors of the nodes close to the BS are responsible for relaying data to the BS. There is, therefore, the need for an appropriate scheme that improves the network lifetime with subsequent decrease in the energy being consumed by the nodes to the barest minimum; hence, the formulation of an improved scalable cluster-based load-balancing algorithm.

8.8 Model Description

Modeling is a simple description of a system, used for explaining how something works or calculating what might happen. In other words, modeling refers to the art of forming a standard for imitation, comparison, representation, and construction of a copy of something. Thus, sensor network modeling is the art of forming a standard for imitating a real-life network. A good network model will enhance transparent network evaluations and, hence, produce more accurate and realistic results.

In this research study, WSNs will be the network of interest. Modeling a WSN requires taking into consideration certain elements that form the network. The lists of elements to be considered and modeled are discussed in the following section.

8.8.1 Network Topology

Topology refers to the ways sensor nodes are arranged or structured within a network. The arrangement of sensor nodes plays an important role in network analysis as they can affect the network's performance. Topology modeling refers to the imitation of the existing network architecture for the purpose of evaluation.

Network topologies are usually in forms of single-hop, multihop, star, networked stars, tree, and graph. For the purpose of this study, a networked star structure, which is otherwise called a clustered-based network, would be adopted.

8.8.2 Network Mobility

In real-life situations, WSNs consist of static and mobile sensor nodes. Mobile sensor nodes affect the topology of the sensor network. Mobility is an element of real-life wireless network and since the aim of any research using a simulation method is to imitate as close as possible the real-life scenario, modeling the state of the sensor node whether static or mobile is included as part of the simulation model.

8.8.3 Network Load

Modeling the network load entails deciding a reasonable time that sensor nodes on the network can initiate or receive traffics on the network before death—that is, when the lifetime of the sensor node is fully exhausted. Network modeling will be responsible for handling which nodes initiate which traffic and how many simultaneous traffics a node will initiate or receive. This sort of load-balancing mechanism on the network is necessary to emulate real-life network usage.

8.9 Overview of the Proposed Scalable Cluster-Based Load-Balancing Scheme (SCLB)

The study proposes an improved energy-efficient cluster-based routing algorithm that is a hybrid of the existing intercluster and intracluster routing approach named SCLB. In SCLB, CHs are organized into several parallel chains where each chain contains CHs far away from the BS as well as those close to the BS, and uses a scheduling scheme to balance the traffic load among these chains as shown in Figure 8.9. Scalability is achieved by dividing the network into overlapping multihop clusters each with its own CH node. Each CH is responsible for building a local relative map corresponding to its cluster using intracluster node's range measurements.

The intrachain routing approach relays data through a multihop data-relay technique whereby data are relayed to the successor nodes until the last node forwards the data to the BS. However, the intrachain scheme schedules the data by occasionally skipping its successors and transmitting directly to the BS thereby avoiding load overhead on the node closest to the BS. Hence, the proposed model ensures balanced energy consumption and improves the network lifetime considerably.

8.10 Clusters Formation

Clustering offers a natural way of gathering information over smaller regions, and aggregating the data collected by the nodes in that cluster. In clustered sensor

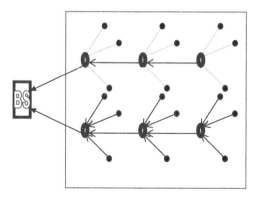

Figure 8.9   Proposed SCLB model.

networks, nodes are organized into smaller groups called clusters in a randomly deployed square unit. In order to reduce the energy consumption and traffic load of a sensor network, the amount of data transmitted in the network would have to be reduced. Data aggregation is a mechanism to reduce the size of packet transmissions. Instead of sending all data to the BS, data aggregation processes the data as intranetwork, and only sends processed data to the BS.

A CH is associated with each cluster for coordinating MAC and routing, as well as for aggregating the data collected by the nodes in the cluster. The CH selection process is performed according to some probability based on exponential-randomization. Each node decides whether or not to become a CH by choosing a random number between 0 and 1. If the number is less than a threshold, then the node becomes a CH. Node *n* calculates the threshold as follows:

$$
T(n) = \begin{cases} \dfrac{P}{1 - P \times (r \bmod \dfrac{1}{P})} & \text{if } n \in G, \\ 0 & \text{otherwise,} \end{cases}
\tag{8.1}
$$

where P is the percentage of nodes that are CHs, r is the current round, and G is the set of nodes that have not been selected as CHs in the last $1/P$ rounds.

Using this threshold function, all nodes take turns to be CHs in a random order. After all nodes have been CHs exactly once, that is, after $1/P$ rounds, all nodes start over to participate in the CH selection process again.

The node that elects itself as the CH broadcasts an advertisement message to notify other nodes. Nodes that are not CHs receive the CH advertisement messages. Since the nodes are deployed in a distributed system, sensor nodes join the cluster of the CH from where they receive the notification (node advertisement) with the highest radio signal strength. As shown in Figure 8.10, sensor nodes 1, 2, 3, 4, and 5 have overlapping signals and therefore they are within the same signal ranges. The sensor nodes send a joining message to the CH and consequently the CH assigns a time division multiple access (TDMA) time slot for each node to transmit in its cluster.

8.11 Intercluster Communication

After the CH formation, the nodes transmit their packets to the CH in their scheduled time slot. The CH receives all packets from its cluster members, compress the data into one packet, and send the packet to the closest CH in another cluster neighborhood in a P2P communication.

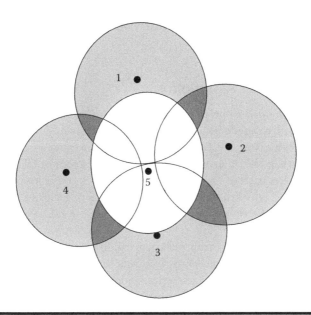

Figure 8.10   Sensor nodes within the same ranges of radio signal strength.

Based on the respective outputs of the sensor readings, the intercluster communication is formed where each CH has to produce an output to be transmitted over a wireless communication link. The packet outputs are finally received by the BS. Thus, the scheme is modeled as a distributed function of the observations since the sensors have a chance to cooperate to some extent.

8.12 Data Preparation

The simulation was carried out within a 200×200 m region. The network comprised 250 sensor nodes and these sensor nodes were randomly deployed with different levels of density. The communication range of each sensor was set to 20 m. Two nodes have a wireless link between them if they are within the communication range of each other. The study considers the amount of energy the radio dissipates to run the transmitter or receiver circuitry as $E_{elec} = 50$ nJ/bit and the amount of energy consumed by the transmit amplifier as $E_{amp} = 100$ pJ/bit/m^2.

8.13 Model Development

In this section, the model assumption, problem formulation, model description, and algorithm development would be considered. This aspect is very much relevant to the study because it underlines the basis upon which the simulation platform was developed.

8.13.1 WSN Model Assumption

The model was assumed to be of homogeneous sensor nodes where all the nodes are of equal sizes, having identical hardware capabilities, and uniform energy consumption, using the same IEEE 802.11 radio protocol. None of the sensor nodes is resource-rich than the others and they had equal chances of CH selection. The study also considers each of the CH to be stationary in order to provide a stable network topology. Each of the CH is dedicated to both intracluster and intercluster traffic and a way to distribute the radio resources for the two types of traffic was considered.

Three major issues within the intraclusters region were considered: cluster forming, CH re-election, and CH canceling. In cluster forming, when a node is powered on, it marks itself clusterless and sets up a random waiting timer T_t and starts to monitor the radio channel for signal request. If no request is heard within T_t, the node marks itself as a CH. Once a node is notified as CH, it sends a beacon immediately in form of a contention packet to the neighboring nodes within the cluster set up.

Every node has their budget variables for CH selection, which is usually in terms of residual battery power or elapsed time for acting as CH. When a node is set as CH, it starts a timer T_m for acting as CH. After the T_m has expired, it selects the node that has the highest budget variable as the next CH and sends it a packet for notification. CH canceling occurs when a CH hears a probe request from another CH and thereby sets itself as a slave. Subsequently, the node sends a beacon message to the CH for association.

8.13.2 Problem Formulation

In the proposed SCLB, WSN with n nodes and k number of nodes are considered which is denoted as n_1, n_2, n_3, n_4, ..., n_k and a BS b as shown in the Figure 8.11. A network of sensors are considered to be connected only if there is at least one path between each pair of nodes in the network. Connectivity, therefore, depends primarily on the existence of paths.

A SCLB model was developed as a directed graph approach. The nodes of the graph correspond to the transmitting and receiving units, and the edges of the graph correspond to the connections that link the nodes in the network and describe its topology. A directed graph is a graph $G = G(V, E)$ that consists of ordered pairs of distinct nodes with two sets V(G) and E(G) where, $V = V(G)$ is the set of $p > 1$ nodes or vertices of the graph, $E = E(G)$ is a set of $q > 0$ pairs of nodes or the edges or links of G, and edge $E = (u, v)$ is said to join nodes u and v in G.

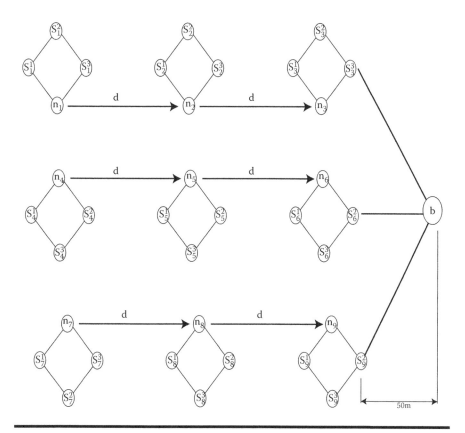

Figure 8.11 Annotated graph model of SCLB for WSNs. (From Oluwaranti AI, Ayanda DO, 2011. Performance analysis of an enhanced load balancing scheme for wireless sensor networks. *Wireless Sensor Network*, 3, 275–282, doi:10.4236/ wsn.2011.38028. Published Online August 2011.)

In this model, p is the number of vertices and q is the number of edges. This graph is said to be k-connected if there are at least k disjoint paths between every pair of nodes u, v ∈ V. If we set d (u, v) as the physical distance between nodes u and v, and R_c is the radius of communication, then E is defined as follows:

$$E = \{(u, v) \in V^2 | d(u, v) \le R_c\} \tag{8.2}$$

The CH nodes n_1, n_2, n_3, ..., n_k are equally spaced over a distance d and each of the nearest CHs to the BS are 50 m away from the BS b. Each of the sensed data from sensor nodes S_j^i within the cluster region of n_1, n_4, and n_7 are sent to their respective CHs. The aggregated data are compressed and the greedy search algorithm is used to select and transmit l_1, l_2, and l_3 bits of data from these nodes to the next CHs n_2, n_5, and n_8 respectively where they are compressed jointly with the data at that node.

Greedy search algorithm selects the closest node from the next tier of CH. When one node n_1 is closest to multiple nodes in the current tier, the node that is farthest to b wins and marks itself as selected. The other competing nodes have to select from the remaining unmarked nodes. In this case, we assume idealized entropy coding that achieves the maximum compression possible.

The l_2 bits of packet are transmitted from the second CHs to the third CHs and so on till the last node within the cluster. Then the jointly compressed l_k bits of data from each cluster are routed to the BS b in a P2P approach. SCLB incorporates a cluster-based joint routing and compression strategies that span from one extreme $(S_1^1, S_1^2, S_1^3; S_2^1, S_2^2, S_2^3; S_3^1, S_3^2, S_3^3)$ where no compression is performed and each node routes its information in a single-hop routing to the CH and to the other extreme (n_2 to n_3, n_5 to n_6, and n_8 to n_9) where data from every source is compressed sequentially before routing to the BS. The SCLB algorithm computes the following paths, one in each iteration, to route the packets l_1, l_2,...,l_k from source S_j^i through the nodes $n_1, n_2, ..., n_k$ to the BS b.

The research also considers the energy model of the SCLB where an l-bit packet is transmitted over a distance d. The corresponding transmission power E_{tx} and receiving power E_{rx} are:

$$E_{tx}(d) = E_{elec}l + E_{amp}ld^2 \qquad (8.3)$$

and

$$E_{rx} = E_{elec}l \qquad (8.4)$$

where E_{elec} is the amount of energy the radio dissipates to run the transmitter or receiver, circuitry = 50 nJ/bit, and E_{amp} is the amount of energy consumed by the transmit amplifier = 100 pJ/bit/m².

8.13.3 Algorithm Development

SCLB algorithm comprises three phases as depicted in Figure 8.12, namely: the setup phase, the steady phase, and the forwarding phase. In setup phase, the sensor nodes begin with the initialization state where the nodes are at sleep. At a particular threshold function say 5 s, the sensor nodes wake up and enter listening state where the nodes receive protocol data unit from other neighboring nodes to determine its role in the topology. After listening state, the sensor node enters learning state where the node can transmit and receive but not forwarding packets. This state compares the MAC address of the sensor nodes within the same radio signal to its own MAC address in its routing table to determine whether the MAC address of the incoming signal is higher than its own MAC address or not and subsequently

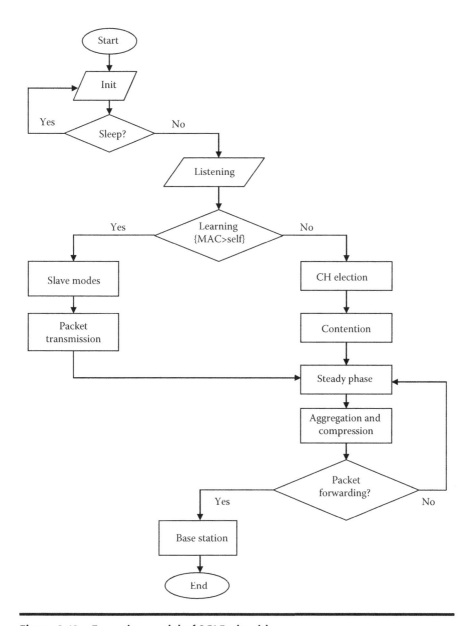

Figure 8.12   Execution model of SCLB algorithm.

drop its attempt to participate in CH election, otherwise the node drops the MAC address of the incoming signal and elects itself as the CH.

The sensor nodes enter a contention state where the node with the highest residual energy among the nodes within a particular cluster becomes the CH. The node broadcasts its status to the other sensor nodes in form of probe request and listens for beacons (probe responses) from the nodes. After the remaining nodes send the probe response, the CH initiates association and allocates a specified time slot for each of the sensor nodes to transmit to the CH.

The sensor nodes start transmitting in their allotted time slot using TDMA and this marks the beginning of the steady phase. When all nodes within an intra-cluster region finish the transmission of aggregated data to the CH, the forwarding phase begins. The forwarding phase marks the beginning of the intercluster communication where CH performs computation on the received data. CH also transmits the data using multiple-chain multihop approach to the neighboring CH until the data is received at the BS. Once all the CHs finish the transmission, the control returns to the steady phase again.

8.14 Benchmarking of the Proposed SCLB Model

In this research work, the proposed scheme is benchmarked against a HEED clustering model as discussed by Younis and Fahmy (2004). In HEED, all CHs send aggregated data to the BS via the shortest paths which form the shortest path trees. Theoretical analysis shows that HEED minimizes the total energy consumption; however, the energy consumption is still unbalanced due to the hot spot formed in the area close to the BS.

The proposed SCLB considers balanced intercluster sensor network routing where all nodes are location aware and have the same initial energy capabilities. The sensor network forms multiple chains of CHs and reliability of the packets transmitted is considered an improvement over HEED in order to ensure the validity of packets received at the other end of the cluster node head.

8.15 Overview of Network Simulators for WSNs

Research interest in the area of sensor networks has introduced a complete paradigm shift in networks from interpersonal communication to communication with the environment. For this reason, the performance of sensor-networking protocols cannot be evaluated without taking into account the characteristics of the environmental data and the sensor measurements which simulators represent.

According to Imran et al. (2010), analytical modeling, simulation, emulation, testbeds, and real deployment are the most commonly used techniques for performance analysis in WSNs. Both analytical modeling and real deployment may not be feasible or practical to test and evaluate the performance of WSNs since the two approaches are complex, costly, and time-consuming (Egea-López et al. 2005; Krop et al. 2007; Mekni and Moulin 2008) while sensor node deployments most often involve placement of hundreds or thousands of nodes in harsh or inaccessible terrains. On the other hand, simulators, emulators, and testbeds are effective tools to evaluate algorithms and protocols at design, development, and implementation stages. Quite a number of these tools are available each with different features, characteristics, models, and architectures for performance testing in WSNs. The range of simulators that have been targeted toward this field and the traditional wireless technology are highlighted in the subsections below.

8.15.1 Network Simulator-2

Network simulator-2 (NS-2) is a discrete event networks simulator. It is developed using the combination of two languages—C++ which is one of the programming languages for operating system and OTcl, an object-oriented scripting language. NS-2 as a very popular network simulator was designed to simulate IP networks due to low-level assumptions, but it was expanded based on research to wireless LAN protocols and MANETs. NS-2 supports only two wireless LAN protocols, 802.11 and a single-hop TDMA protocol (Carley 2005).

Cross-layer optimization can be implemented in NS-2 by adding interfaces to the various layers of the network stack. SensorSim developed at the University of California, Los Angeles (UCLA) extends NS-2 by adding modules for modeling sensing activity, battery models (to model power consumption in sensor nodes), a lightweight protocol stack specifically for sensor networks, and a module for the simulation to interact with actual sensors to form a hybrid simulation (Sridharan et al. 2004).

The major drawbacks of the simulator is the accuracy of results because of MAC protocols, packet formats, and energy models that are different from those of typical sensor network platforms (Bajaj et al. 1999; Levis et al. 2003).

8.15.2 Network Simulator-3

Network simulator-3 (NS-3) is an open-sourced discrete event network simulator designed primarily for research and educational use. It is written in C++ with python-scripting interface. It is a new simulator designed to replace NS-2 and is not backward-compatible with NS-2. Other features include: protocol entities designed to be closer to real computers; supporting the incorporation of more open-source networking software and reducing the need to rewrite models for simulation; the use of lightweight virtual machines; and allowing a tracing and statistics gathering framework to enable customization of the output without rebuilding the simulation core (Pan 2008). Since NS-3 is an Internet era simulator for layered network stacks, it has a strong bias for WSN.

8.15.3 GloMoSim

According to Zeng et al. (1998), GloMoSim began in 1998 as a simulator for mobile wireless networks. GloMoSim stands for GLobal MObile Information systems

SIMulator and is a scalable simulation environment for wireless and wired network systems (Bajaj et al. 1999). The simulator is written in a variant of C language called parsec with parallel programming extensions. GloMoSim was specifically designed to support wireless networks and thus has very good propagation models. Cross-layer optimization is also possible with GloMoSim.

The drawback of the simulator is its limitation to IP networks because of low-level design assumptions just as NS-2. Also GloMoSim does not provide support for sensors, actuator, and physical phenomenon (environmental conditions such as temperature, pressure, stress inclusive, and so on).

8.15.4 QualNet

QualNet (http://www.scalable-networks.com/) is a modeling and simulation tool developed to explore and analyze early-stage alternative device designs and application code in closed, synthetic networks at real-time speed, at a scale of up to thousands of network nodes. QualNet provides complementary tools such as EXata®, EXata/Cyber®, and VisNet™ that predict performance of network devices, transmitters, antennas, terrestrial characteristics, and human interactions at real-time speed. QualNet major drawback is that it is not freely available for research community.

8.15.5 J-Sim

J-sim is a generic component-based, compositional simulation framework written in Java and Jacl with a packet-switching network incorporated into its architecture. Unlike GloMoSim, J-sim provides considerable support for protocols, including a WSN simulation framework with a very detailed model of WSNs, and an implementation of localization, routing, and data diffusion WSN algorithms (Sobeih et al. 2005). J-sim models are easily reusable and interchangeable offering the maximum flexibility and provides a graphical user interface (GUI) library.

The drawback is that the only MAC protocol provided for wireless networks is 802.11 which makes the accuracy of the simulator to become a bottleneck. Also the execution time is 41% worse (Sobeih et al. 2005).

8.15.6 SENSE

SENSE is a component-based simulator developed recently (Chen et al. 2004). It is written in C++ and runs on top of COST, a component-based discrete event simulator. COST is written in CompC++ which is a component extension to C++. In the component repository of SENSE there are already different components available from the application to the physical layer including IEEE 802.11, AODV, DSR, SSR, SHR, as well as battery models and a power model (Kellner et al. 2010).

SENSE provides support for an energy model that is sufficient for WSNs but does not support sensor, actuator, and physical environmental phenomenon. Also, at the moment, there does not seem to be any further tools included in SENSE so that, for example, a visualization tool to analyze the network behavior graphically is missing.

8.15.7 VisualSense

VisualSense is a modeling and simulation framework for WSNs that builds on and leverages Ptolemy II (Baldwin et al. 2004; Baldwin et al. 2005). Models of VisualSense can be developed by subclassing base classes of the framework or by combining existing Ptolemy models.

Ptolemy II is a Java package that supports different models of simulation paradigms such as continuous time, data flow, and discrete event. It also addresses the modeling, simulation, and design of concurrent, real time, and embedded systems.

VisualSense provides an accurate and extensible radio model (Carley 2005) based on a general energy-propagation model that can be reused for physical phenomena. However, the simulator does not provide any protocols above the wireless medium, sensor and physical phenomena other than sound.

8.15.8 SENS

SENS stands for SEnsor Network Simulator. It is an application-oriented, high-level network simulator written in C++ and built with the GCC compiler (Sundresh et al. 2004). It has a modular, layered architecture so that components for applications, network communication, and the physical environment can be easily interchanged and extended.

SENS is platform independent so that new emerging sensor platforms can be added to the simulator. Also, it provides a platform for multiple sensor nodes to interact with an environment component. Unlike SENSE and VisualSense, SENS provides support for sensors, actuators, and physical phenomena only for sound. The major drawback is that the simulator does not accurately simulate a MAC protocol.

8.15.9 (J)Prowler

Prowler and (J)Prowler are probabilistic WSN simulators written in MatLab and Java respectively. (J)Prowler runs on berkeley MICA Mote hardware platform-running application built on TinyOS (Carley 2005). Its main feature is its ability to simulate transmission, propagation, and reception including collision in sensor networks (Simon et al. 2003). Although (J)Prowler provides an accurate radio model, the simulator does not provide support for sensors, actuators, or physical phenomenon.

8.15.10 TOSSIM and TOSSF

TOSSIM is a sensor network simulator that is targeted to simulate the Berkeley MICA Mote hardware platform-running applications built on TinyOS (Levis et al.

2003). It simulates the sensing channels on the motes and provides a mechanism to feed data into the motes from external sources.

TOSSIM provides a good empirical radio-propagation model and simulates the mote's devices including digital I/O and A/D which enables it to simulates sensors and actuators. However, TOSSIM does not simulate the physical phenomena and that each node must run the exact same code. Also the bit-level simulation makes the simulator to be very slow and not suited for large-scale simulation.

TOSSF is similar to TOSSIM in that it is specifically developed for a sensor network simulator (Perrone and Nicol 2002). The limitations with TOSSIM are addressed in TOSSF; however, the drawback is the accuracy of the simulated devices.

8.15.11 EmStar

EmStar is among the earliest simulators specifically developed for sensor networks. According to Girod et al. (2004), EmStar is a software framework for Linux-based systems that provides different levels of simulation environments starting from a pure simulation environment on a single PC to a real life deployment scenario. It also provides various services that are used and combined to provide network functionality for wireless embedded systems which link drivers for the lowest-layer interfaces to network resources, pass-through modules that implement various types of filter and passive processing, and routing modules such as flooding, DSR, Sink, StateSync, and Centroute (Elson et al. 2003).

Further study is required in order to ensure that the simulator covers all scenarios in term of deployment and scalability.

8.15.12 MANTIS

According to Abrach et al. (2003), MANTIS is an operating system built to run on the Nymph nodes as well as MICA2 motes from CrossBow. It is an open-source simulator developed in a C programming language specifically for sensor modules. The Mantis system comes with an emulator where each node is emulated as a separate process. Each of the emulated nodes can communicate with themselves as well as with the actual sensor nodes which is provided as a development tool, but could be thought of as a time-driven simulator. Its major drawback is the problem of scalability for different applications of WSN.

8.15.13 SIESTA

SIESTA is a sensor network simulator that is developed in Vanderbilt University and targeted toward testing and simulating middleware platforms in sensor networks (Simon et al. 2003). It has four components—application, middleware, hardware, and the sensing physical environment. The physical environment is the interface through which data could be fed into the simulator rather than a simulation of the sensed environment itself.

8.15.14 OMNET++

OMNET++ (Objective Modular Network Test-bed in C++) is a public-source, modular, discrete-event simulator implemented in C++. It provides a powerful GUI library for animation, tracing, and debugging support. Its primary application area is communication networks. It has generic and flexible architecture which makes it successful in other areas like queuing networks and hardware architectures (Pan 2008). OMNET++ represents a framework approach which provides an infrastructure for writing different simulations. Example of a WSN simulator built on top of OMNET++ is Castalia (Boulis 2009). It is a generic simulator intended for the first-order validation of high-level algorithms before moving to specific sensor platforms. Other simulation frameworks that enable OMNeT++ to be used for WSNs are mobility framework (Drytkiewicz et al. 2003), MiXiM (K"opke et al. 2008), Positif framework, INET framework (Ariza-Quintana et al. 2008), and NesCT.

OMNET++ has a major drawback which is the lack of available protocols in its library compared to other simulators, since it is not a network simulator itself. Instead, it is a network simulation platform that builds up a large user community. Further study from research contributions will have considerable improvements on the simulator.

8.15.15 OPNET

The OPNET modeler is a discrete event, object-oriented network simulator initially developed for military purposes. It is a large and powerful software that happens to be the first commercial network simulator that can be used as a research tool and also as a network design and analysis. Because of its industrial standard approach for commercial use, OPNET offers powerful visual or graphical support for users, with robustness and the ability to perform very fast large-scale simulations. It uses a hierarchical model to define each aspect of the system.

OPNET was originally built for the simulation of fixed network hardware and contains extensive libraries of accurate models from commercially available fixed network hardware and protocols. The simulator also supports recent wireless networks including Zigbee compatible 802.15.4 MAC (Korkalainen et al. 2009). Its major strength in wireless network simulation lies in the accurate modeling of radio transmissions and it can also be used to define custom packet formats. The drawback is that there exists only a few models ready for wireless systems (Prokkola 2006) and its bias toward traditional network design principles constrain its support for sensor networks directly.

8.15.16 ATEMU

ATEMU (ATmel EMUlator) is an open-source tool built as a software emulator for AVR processor-based systems such as MICA2 and its peripheral devices (Polley et al. 2004). It was developed at Maryland University and is currently available in version 0.4 which was released in March 2004.

ATEMU is built to simulate operations of various applications on the MICA2 platform with the capability to emulate various components such as processor, timer, and radio interface. This shows that, at the moment only the MICA2 hardware is supported, but ATEMU can be easily extended to support other sensor node platforms. Although ATEMU is the most accurate instruction-level emulator for WSN research, it lacks simulation speed, being 30 times slower than TOSSIM (Kellner et al. 2010).

8.15.17 Shawn

Shawn is a customizable sensor network simulator based on an algorithmic approach with the design goals of simulating the effect caused by a phenomenon: scalability and support for extremely large networks and free choice of the implementation model (Kroeller et al. 2005; Fekete et al. 2007). The major drawback is that there is no stable release of Shawn; hence, the same level of details when comparing other simulators cannot be reached.

8.15.18 Avrora

Avrora is a set of simulation and analysis tools for programs written for AVR micro-controllers with support for different sensor platforms (Titzer et al. 2005). Some of the sensor platforms include MICA2 and MicaZ that allow wireless network simulation, dynamic instrumentation, and static analysis. Avrora is implemented in Java as a more efficient emulator than ATEMU with accuracy closer to TOSSIM (Curren 2005). Each node has its own thread that runs the code instruction-by-instruction, but it improves scalability by not synchronizing all nodes after every instruction. This approach introduces some inaccuracy in results although with faster execution speed.

Avrora scales up to networks of 10,000 nodes and performs 20 times faster than previous simulators with the emulating approach and the same accuracy. Also, it contains further tools such as profiling utilities, an energy analysis tool, a stack checker, a control flow graph tool, and so on. The major drawback of Avrora is its lack of an integrated GUI so that everything has to be done manually on the command line.

8.15.19 COOJA/MSPSim Simulator

COOJA/MSPSim is a combination of two separate tools into one cross-level WSN simulator (Eriksson 2009). COOJA (Contiki Os JAva) is a simulator for the

Contiki sensor node operating system that allows simultaneous cross-level simulation at application, operating system, and machine code instruction set level. MSPSim, on the other hand, is an emulator for MSP430 embedded in COOJA. COOJA is able to execute the Contiki programs in two different ways by either compiling the program code directly on the host CPU or compiling it for the MSP430 hardware.

8.15.20 OLIMPO

OLIMPO is a discrete-event simulator designed to be easily reconfigured by the user, providing a way to design, develop, and test communications protocols (Barbancho et al. 2004). The simulator was developed with the capability to track specific traffic called supervisory control and data acquisition (SCADA) packets from one node to BS through multihops. OLIMPO as an open-souce simulator, was created with C++ Borland Builder 6, creating a friendly GUI and providing a useful tool to simulate new communication or application protocols for other researchers. The simulator also provides several simulation modes, for example, MAC step, logical link control step, network step, and application step which provide a way to debug protocols. Further study from research contributions will have considerable improvement on the simulator for new protocols and the implementation of SCADA applications over real sensor nodes.

8.16 Performance Evaluation

Performance metrics can be described as a measure to analyze or grade an activity or event. They act as yardsticks on which decisions about the network can be made. For this simulation, three metrics were used. Each of them is very important and constitutes a major trade-off for evaluating the actual behavior of the WSN modeling. The metrics are energy consumption, network lifetime, and reliability.

8.16.1 Energy Consumption

Schurgers et al. (2002) described radio communication function of sensor nodes as the most energy-intensive function in the network compared to the actual sensing operation, which consumes the least energy. The study also recognized two approaches to reducing energy consumption for sensor communications. The first approach is to design a communication scheme that conserves energy inherently (for instance, turning off the transceiver for a period of time). The second approach is to reduce the volume of communications through intranetwork processing which entail functions such as data aggregation and data compression at the CH node.

The energy model for communication considers energy for transmitting l-bit data over a distance d as E_{tx} (l, d) and the energy for receiving l-bit data over a distance d as E_{rx} (l, d). The mathematical relation is shown below.

$$E_{tx} (l,d) = lE_c + led^2 \tag{8.5}$$

$$E_{rx} (l,d) = lE_c \tag{8.6}$$

where E_c is the base energy required to run the transmitter or receiver circuitry with a typical value of 50 nJ/bit for a 1 Mbps transceiver, and e is the unit energy required for the transmitter amplifier with a typical value of 100 pJ/bit.m².

The total energy E_{ij} (l,d) for transmitting l-bit data from source node i to destination node j within a distance of d is

$$E_{i,j} (l,d) = E_{tx} (l,d) + E_{rx} (l,d) \tag{8.7}$$

In HEED, the energy consumed for the shortest path routing from CH n_1 till n_7, that is, the energy consumed to deliver a packet to the neighboring CH is given as

$$E^{HEED}(k) = \begin{cases} (n-1)E_{rx} + nE_{tx}(D) & : k = 1 \\ (n-k)E_{rx} + (n-k+1)E_{tx}\left(\dfrac{1}{n}L\right) & : k > 1 \end{cases} \tag{8.8}$$

Where the maximal single-node energy consumed for HEED is given as

$$E_{MAX.}^{HEED}(k) = (n-1)E_{rx} + nE_{tx}(D) \tag{8.9}$$

In SCLB, a greedy search algorithm is used to form a straight-chain model. The study adopted the algorithm of the research conducted by Mandala et al. (2007) where the average cost per round at node k is calculated thus

$$E(k) = \frac{1}{n}\sum_{r=1}^{n}\left[N_{rx}^{k,r}E_{rx} + N_{tx}^{k,r}E_{tx}\left(\frac{1}{n}L\right) + N_{tx'}^{k,r}E_{tx}\left(D + \frac{k-1}{n}L\right)\right] \tag{8.10}$$

where
$N_{rx}^{k,r}$ = number of packets received
$N_{tx}^{k,r}$ = number of packets transmitted to the successor
$N_{tx'}^{k,r}$ = number of packets transmitted to the BS

$$N_{rx}^{k,r} = \begin{cases} n - k & : r \le k \\ r - k - 1 & : r > k \end{cases} \tag{8.11}$$

$$N_{tx}^{k,r} = \begin{cases} 0, & : k = 1 \vee k = r \\ N_{rx}^{k,r} + 1, & : \text{otherwise} \end{cases} \tag{8.12}$$

$$N_{tx'}^{k,r} = \begin{cases} 0, & : k \ne 1 \wedge k \ne r \\ N_{rx}^{k,r} + 1, & : \text{otherwise} \end{cases} \tag{8.13}$$

Combining Equations 8.11, 8.12, and 8.13, it becomes

$$E_{SCLB} = \begin{cases} \dfrac{n-1}{2} E_{rx} + \dfrac{n+1}{2} E_{tx}(D) & : k = 1 \\[2ex] \dfrac{(n-k)(n+k-1)}{2n} E_{rx} & : \\[2ex] + \dfrac{(n-k+1)(n+k-2)}{2n} E_{tx}\left(\dfrac{1}{n}L\right) & : \\[2ex] + \dfrac{(n-k+1)}{n} E_{tx}\left(D + \dfrac{k-1}{n}L\right) & : k > 1 \end{cases} \tag{8.14}$$

8.16.2 Network Lifetime

Network lifetime is crucial to a large-scale sensor network since it is undesirable or infeasible to replace or recharge sensors once the network is deployed. The performance measure of network lifetime is, therefore, relevant to sensor networks where battery-powered, dispensable sensors are deployed to collectively perform a certain task.

According to Swami et al. (2007), network lifetime is the time span from the deployment to the instant when the network is considered nonfunctional. A network is considered nonfunctional based on the specificity of the application. For instance, the network becomes nonfunctional when a certain fraction of sensors die; loss of coverage occurs whereby a certain portion of the desired area can no longer be monitored by any sensor; loss of connectivity occurs whereby sensors can no longer communicate with the CH or BS; or the detection probability drops below a certain threshold value.

In modeling the sensor network lifetime, the study obtained formula applies to any definition of the network lifetime (Chen and Zhao 2005) that is derived based on the Strong law of large numbers (SLLN). This is given by

$$\epsilon[A] = \frac{\mathcal{E}_o - \epsilon[E_w]}{P_c + \lambda\epsilon[E_r]} \tag{8.15}$$

where

$\mathcal{E}_0$ = total nonrechargeable initial energy
$\epsilon[E_w]$ = expected wasted energy (i.e., total unused energy in the network when it dies)
P_c = constant continuous power consumption over the whole network
λ = average sensor reporting rate (defined as the number of data collections per unit time)
$\epsilon[E_r]$ = expected reporting energy consumed by all sensors in a randomly chosen data collection

The study considers a scalable sensor network with S homogeneous sensor nodes, each powered by a nonrechargeable battery with E_o initial energy. In SCLB model, applying Equation 8.15 to the current research study with an assumption that $\lambda = 1$, network lifetime becomes:

$$\epsilon[A]\frac{SE_o - \epsilon[E_w]}{\epsilon[E_r]} \tag{8.16}$$

where

E_o = initial energy of a nonrechargeable battery
S = size of homogeneous sensor network

8.16.3 Reliability

Reliability implies the extent to which a measurement instrument is said to yield a consistent result. In other words, reliability points to the degree in which a system measures the same way each time it is used under the same condition with the subjects. The reliability of a system is not measured rather it is estimated.

In WSNs, reliability is used as a measure to show how reliable the sensed event can be reported to the BS and is measured as the ratio of successfully received packets over the total number of packets transmitted. With an assumption of a secured environment, the study uses the metrics R_{SCLB} to measure the reliability of the system (El-Darymli et al. 2009) as follows:

$$R_{SCLB} = e^{-\lambda t} \tag{8.17}$$

where

$$\lambda = \text{system failure rate} = \frac{\text{no. of failures}}{\text{no. of component hours}} \qquad (8.18)$$

and t = operating time

8.17 Simulation Results and Discussion

The study assumes the radio communication to be IEEE 802.11b, network topology is stationary, and the node placement to be homogeneous. The results of the simulation based on the equations discussed earlier in the study are presented for the existing multihop model (HEED) and the proposed cluster-based model (SCLB).

8.17.1 Energy Consumption

The energy consumed by SCLB at maximal single node starting from two to eight CH nodes, that is, each of the closest CH nodes to the BS is 0.475, 0.650, 0.825, 1.000, 1.175, 1.350, and 1.525 mJ while HEED are 0.650, 1.000, 1.350, 1.700, 2.050, 2.400, and 2.750 mJ, respectively. This is depicted in Figure 8.13. Also, the SCLB values

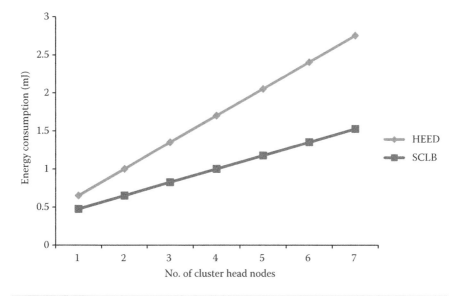

Figure 8.13   Graph of energy consumption for HEED and SCLB. (From Oluwaranti AI, Ayanda DO, 2011. Performance analysis of an enhanced load balancing scheme for wireless sensor networks. *Wireless Sensor Network*, 3, 275–282, doi:10.4236/wsn.2011.38028. Published Online August 2011.)

represent 26.92%, 35%, 38.89%, 41.18%, 42.68%, 43.75%, and 44.55% decrease in energy consumption when compared with HEED. This implies that SCLB achieves an average lower energy consumption of 39% as compared with HEED.

It can be observed that energy being consumed in SCLB increases from CH nodes n_2 to n_8. This is due to the increased number of CHs in the network topology which increase the energy consumed on nodes computation and communication. The distance of the sensing field also increases with attendant impact on overall energy consumption. In the same vein, HEED increases from a single chain two-CHs to eight-CHs which affect the energy consumed with increment from 0.650 to 2.750 mJ—more than three times the initial energy consumption. In essence, the study confirms the underline theory that the energy consumption is directly proportional to the square of the distance of sensor nodes from the BS.

8.17.2 Network Lifetime

The network lifetime is an important metric which the study considers for SCLB and HEED as shown in Figure 8.14. Starting from two single-node CHs to eight single-node CHs, SCLB decreases with values: 4.211, 3.077, 2.424, 2.000, 1.702,

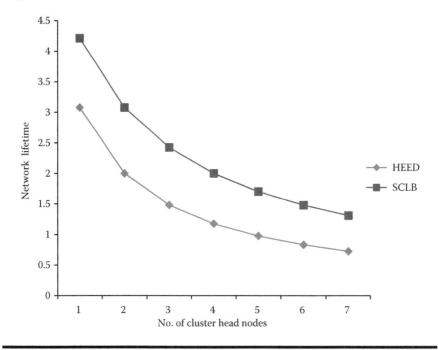

Figure 8.14 Graph of network lifetime for HEED and SCLB. (From Oluwaranti AI, Ayanda DO, 2011. Performance analysis of an enhanced load balancing scheme for wireless sensor networks. *Wireless Sensor Network*, **3, 275–282, doi:10.4236/ wsn.2011.38028. Published Online August 2011.)**

1.481, and 1.311. Also, HEED decreases with values: 3.076, 2.000, 1.481, 1.179, 0.976, 0.833, and 0.727. The percentage difference between the existing scheme (HEED) and the proposed scheme (SCLB) is 36.90%, 53.85%, 63.67%, 70.07%, 74.39%, 77.80%, and 80.33% respectively.

The study shows that network lifetime of SCLB and HEED decrease from CH nodes n_2 to n_8. The decrease is accounted for by the increased amount of energy being consumed since sensor nodes are battery powered and energy consumption is considered as a major performance metric that significantly improves the network lifetime. A sensor node with minimal energy being consumed can stay longer than a sensor node that consumes more energy in its circuitry. Hence, the lower the energy consumed the better the network lifetime.

The results generated further imply that SCLB has improved network lifetime with an average lifetime of 65.29%. The appreciable increases in network lifetime of SCLB is accounted for by the paradigm shift from the multihop load-balancing scheme to cluster-based load-balancing approach.

8.17.3 Reliability

The result from Figure 8.15 shows that R_{SCLB} drops gradually from 98% to 2% as the operation time of the system increases. This is due to the fact that at first transmission of the aggregated data from CH n_7, which is the farthest single-chain CH load balancing assumed for the research, all the data are transmitted to the next CH. The next CH node accepts the data and aggregates it with those sent to it from the

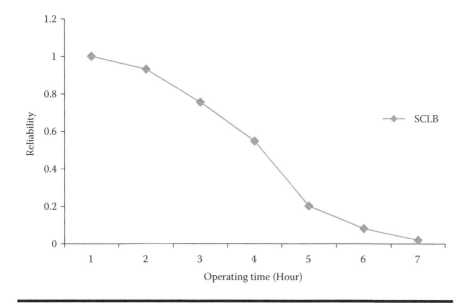

Figure 8.15   Reliability result of R_{SCLB}.

nodes within its cluster region. The total aggregated data are computed and sent to the next CH. CH n_5 also collects the data with the sensing data from the clustered nodes within its intracommunication domain. These processes continue until the CH node closest to the BS. Moreover, optimal reliability was attained at 75.6% for SCLB while no metric for reliability was provided for the existing HEED scheme.

8.18 Future Research Direction

In this study, data delivery was measured based on the RSSI to estimate packet reception ratio. Other metrics such as link quality indicators can be employed to estimate reliability. Since more than one retransmission is recommended for reliable data delivery, this would pose a challenge of delay in delivery data to the BS which can be considered for further study.

Furthermore, deploying SCLB with more real-world sensor nodes to discover how it behaves in the real-life setting is suggested as future work.

8.19 Conclusion

The field of WSNs offers a rich, multidisciplinary area of research, in which a variety of tools and concepts can be employed to address a diverse set of applications. As such, this research attempted to develop an enhanced load-balancing scheme that is an improvement over the existing multihop routing algorithm. The proposed scheme deployed greedy search algorithm to route packets across the sensing fields. This has a significant impact on minimizing the energy consumption and improving network lifetime during load-balancing process. Simulation results have shown that SCLB scheme demonstrates significant improvement over the existing HEED scheme.

In this study, efforts to achieve fault tolerance in communications were looked into. Moreover, due to random deployment of sensor nodes which most often involve placement of hundreds or thousands of nodes in harsh or inaccessible terrains, incorporating reliability into protocols for WSN enhances the extent to which the packets being transmitted is delivered to the BS.

Therefore, it is concluded that the proposed mechanism for load balancing is efficient in that it minimizes energy consumption of sensor nodes, which subsequently improves the overall network lifetime of the WSNs.

References

Abrach H, Bhatti S, Carlson J, Dai H, Rose J, Sheth A, Shucker B, Deng J, Han R, 2003. MANTIS: System Support for MultimodAl NeTworks of In-situ Sensors, *2nd ACM International Workshop on Wireless Sensor Networks and Applications (WSNA)*, San Diego, CA, pp. 50–59.

Akhtar A, Minhas AA, Jabbar S, 2010. Energy aware intra cluster routing for wireless sensor networks. *In International Journal of Hybrid Information Technology*, 3(1), 29–48.

Akyildiz IF, Su W, Sankarasubramanian Y, Cyirci E, 2002. Wireless sensor networks: A survey. *Computer Networks,* 38(4), 393–422.

Alrajeh NA, Khan S, Lloret J, Loo J, 2013. Secure routing protocol using cross-layer design and energy harvesting in wireless sensor networks. *International Journal of Distributed Sensor Networks*, 2013, Article ID 374796, 11, http://dx.doi.org/10.1155/2013/ 374796.

Ariza-Quintana A, Casilari E, Cabrera A, 2008. Implementation of MANET routing protocols on OMNeT++, in *Proceedings of the 1st ICST*, Brussels, Belgium.

Bajaj L, Takai M, Ahuja R, Tang K, Bagrodia R, Gerla M, GloMoSim: A Scalable Network Simulation Environment, UCLA Computer Science Department Technical Report 990027, May 1999.

Baldwin P, Kohli S, Lee EA, Liu X, Zhao Y, 2004. Modeling of sensor nets in Ptolemy II. In *Proceedings of Information Processing in Sensor Networks (IPSN)*, Berkeley, pp. 359–368, April 2004.

Baldwin P, Kohli S, Lee EA, Liu S, Zhao Y, 2005. VisualSense: Visual Modeling for Wireless and Sensor Network Systems. *Technical Memorandum UCB/ERL M05/25*, University of California, Berkeley, CA 94720, USA July 15, 2005.

Barbancho J, Molina FJ, León C, Ropero J, Barbancho A, 2004. OLIMPO: An ad-hoc wireless sensor network simulator for optimal SCADA-applications. In *Proceedings of the Communication Systems and Networks*, Spain.

Boulis A, 2009. Castalia, a simulator for wireless sensor networks and body area networks, version 2.2, User's manual, NICTA, Canberra, Australia. August 2009.

Carley TW, 2005. *Sidh: A Wireless Sensor Network Simulator*, University of Maryland at College Park, USA.

Chen G, Branch J, Pflug M, Zhu L, Szymanski B, 2004. Sense: A sensor network simulator. *Advances in Pervasive Computing and Networking*, pp. 249–267, 2004.

Chen Y, Zhao Q, 2005. On the lifetime of wireless sensor networks. *IEEE Communications Letters*, 9(11), 976–978.

Chong CY, Kumar SP, 2003. Sensor networks: Evolution, opportunities, and challenges. *Proceedings of the IEEE*, 91(8), 1247–1254.

Curren, D, 2005. A survey of simulation in sensor networks. Available online at: http://www.cs.binghamton.edu/kang/teaching/cs580s/david.pdf.CV. Accessed on July 10, 2013.

DARPA. 2010. Defence Advance Research Projects Agency—Self-healing Mines Available at (http://www.darpa.mil/ato/programs/SHM/.). (Accessed on July 5, 2010).

Ding N, Liu PS, 2004. Data gathering communication in wireless sensor networks using ant colony optimization. *Proceedings of the 2004 IEEE International Conference on Robotics and Biomimetics*, Shenyang, China, pp. 822–827.

Drytkiewicz W, Sroka S, Handziski V, Koepke A, Karl H, 2003. A mobility framework for OMNET++, in *3rd International OMNeT++ Workshop*, Budapest, Hungary.

Egea-López E, Vales-Alonso J, Martínez-Sala AS, Pavón-Mariño P, García-Haro J, 2005. *Simulation Tools for Wireless Sensor Networks, Summer Simulation Multiconference—SPECTS*, Philadelphia, USA.

El-Darymli K, Khan F, Ahmed MH, 2009. Reliability modeling of wireless sensor network for oil and gas pipelines monitoring. *Sensors & Transducers Journal*, 106(7), 6–26.

Elson J, Bien S, Busek N, Bychkovskiy V, Cerpa A, Ganesan D, Girod L et al., 2003. Emstar: An environment for developing wireless embedded systems software, Center for Embedded Networked Sensing (CENS) Technical Report, vol. 9, USA.

Eriksson J, 2009. Detailed Simulation of Heterogeneous Wireless Sensor Networks Published Dissertation for the degree of Licentiate of Philisophy in Computer Science at Uppsala University 2009.http://www.it.uu.se/ Accessed on July 5, 2010.

Fekete S, Kroller A, Fischer S, Pfisterer D, 2007. Shawn: The fast, highly customizable sensor network simulator. In *Networked Sensing Systems, 2007. INSS'07. Fourth International Conference on*, Braunschweig, Germany, pp. 299–299.

Ganesan D, Govindan R, Shenker S, Estrin D, 2001. Highly resilient, energy efficient multipath routing in wireless sensor networks. In *Proceedings of the 2nd ACM International Symposium on Mobile Ad Hoc Networking and Computing (MobiHoc'01)*, Long Beach, CA, pp. 251–254.

Girod L, Elson J, Cerpa A, Stathopoulos T, Ramanathan N, Estrin D, 2004. EmStar: A software environment for developing and deploying wireless sensor networks. In *the Proceedings of USENIX General Track* 2004. Boston, MA.

Goldsmith A, Wicker S, 2002. Design challenges for energy-constrained ad hoc wireless networks. *IEEE Wireless Communications Magazine*, 9, 8–27.

Haenggi M, 2003. Energy-balancing strategies for wireless sensor networks. In *IEEE International Symposium on Circuits and Systems (ISCAS'03)*, Bangkok, Thailand.

Haenggi M, 2004. Twelve reasons not to route over many short hops. In *IEEE Vehicular Technology Conference (VTC'04 Fall)*, Los Angeles, CA.

Hedetniemi S, Liestman A, 1988. A survey of gossiping and broadcasting in communication networks. *Networks,* 18(4), 319–349.

Heinzelmann H, Chandrakasan A, Balakrishnan H, 2000. Energy-efficient communication protocol for wireless microsensor networks. *In Proceedings of the 33rd Annual Hawaii International Conference on System Sciences*, Vol. 2, Hawaii, USA.

Imran M, Said A, Hasbullah H, 2010. A Survey of Simulators, Emulators and Testbeds for Wireless Sensor Networks 978-1-4244-6716-7/10, 2010 IEEE.

Israr N, Awan I, 2006a. Multihop routing Algorithm for Inter Cluster Head Communication. *22nd UK Performance Engineering Workshop*, Bournemouth, UK, pp. 24–31.

Israr N, Awan I, 2006b. Multihop clustering algorithm for load balancing in wireless sensor networks. *International Journal of Simulation,* 8(1), 13–25.

Kamal P, Zhang Y, Trappe W, Ozturk C, 2005. Enhancing source-location privacy in sensor network routing. *In Proceedings of the 25th IEEE Int. Conference on Distributed Computing Systems (ICDCS05)*, Washington DC, USA, pp. 599–608.

Kellner A, Behrends K, Hogrefe D, 2010. Simulation Environments for Wireless Sensor Networks, Technical Report, IFI-TB-2010-04, June 2010.

Köpke A, Swigulski M, Wessel K, Willkomm D, Haneveld P, Parker T, Visser O, Lichte H, Valentin S, 2008. Simulating wireless and mobile networks in OMNeT++ the MiXiM vision, in *Proceedings of the 1st ICST*, Brussels, Belgium.

Korkalainen M, Sallinen M, Karkkainen N, Tukeva P, 2009. Survey of wireless sensor networks simulation tools for demanding applications, In *Proceedings of the 2009 IEEE Fifth International Conference on Networking and Services*, Valencia, Spain.

Koubaa A, Alves M, 2005. A two-tiered architecture for real-time communications in large-scale wireless sensor networks: Research challenges. *Technical Report*, TR-050701, Version 1.0. Available at http://www.hurray.isep.ipp.pt, accessed on May 15, 2010.

Kroeller A, Pfisterer D, Buschmann C, Fekete S, Fischer S, 2005. Shawn: A new approach to simulating wireless sensor networks. In *Proceedings Design, Analysis, and Simulation of Distributed Systems (DASD05)*, San Diego, CA. pp. 117–124.

Krop T, Bredel M, Hollick M, Mogre PS, Steinmetz R, 2007. JiST/MobNet: Combined simulation, emulation, and real-world testbed for ad hoc networks. *Proceedings of the Second ACM International Workshop on Wireless Network Testbeds, Experimental Evaluation and Characterization*, Canada, pp. 27–34.

Levis P, Lee N, Welsh M, Culler D, 2003. TOSSIM: Accurate and scalable simulation of entire TinyOS applications. In *Proceedings of the ACM Conference on Embedded Networked Sensor Systems (SenSys)*, 2003.5, California, USA.

Lindsey S, Raghavendra C, 2002. PEGASIS: Power-efficient gathering in sensor information systems. In *Proceedings of the IEEE Aerospace Conference*, 3, 1125–1130.

Mandala D, Du X, Dai F, You C, 2007. Load balance and energy efficient data gathering in wireless sensor networks. *Wireless Communications and Mobile Computing*, 2008; 8, 645–659. DOI:10.1002/wcm.492.

Meguerdichian S, Koushanfar F, Potkonjak M, Srivastava M, 2001. Coverage problems in wireless ad-hoc sensor networks. In *Proceedings of the 20th Annual Joint Conference of the IEEE Computer and Communications Societies (INFOCOM'01)*, Vol. 3, Anchorage, AK, pp. 1380–1387.

Mekni M, Moulin B, 2008. A survey on sensor webs simulation tools. In *Proceedings of the 2nd International Conference on Sensor Technologies and Applications*, Agay, France.

Mozumdar MM, Goufang N, Gregoretti F, Lavagno L, 2009. An efficient data aggregation algorithm for cluster-based sensor network. *Journal of Networks*, 4(7), 598–606.

Oluwaranti AI, Ayanda DO, 2011. Performance analysis of an enhanced load balancing scheme for wireless sensor networks. *Wireless Sensor Network*, 3, 275–282, doi:10.4236/wsn.2011.38028 Published Online August 2011.

Pan J, 2008. A Survey of Network Simulation Tools: Current Status and Future Developments, Available at http://www.cse.wustl.edu/~jain/cse567-08/index.html, accessed on January 5, 2011.

Perrone LF, Nicol D, 2002. A scalable simulator for TinyOS applications. In *Proceedings of the Winter Simulation Conference*. San Diego, CA.

Polley J, Blazakis D, McGee J, Rusk D, Baras JS, Karir M, 2004. ATEMU: A fine-grained sensor network simulator, *Proceedings of First IEEE International Conference on Sensor and Ad Hoc Communication Networks (SECON'04)*, Santa Clara, CA.

Prokkola J, 2006. OPNET—Network simulator, VTT Technical Research Center of Finland, Finland.

Puccinelli D, Haenggi M, 2005. Wireless sensor networks: Applications and challenges of ubiquitous sensing. *IEEE Circuits and Systems Magazine*, Third Quarter 2005.

Puccinelli D, Haenggi M, 2009. Lifetime benefits through load balancing in homogenous sensor network. In *The WCNC Proceedings*, New Jersey, USA.

Reijers N, Loangendoen K, 2003. Efficient code distribution in wireless sensor networks. In *Second ACM International Workshop on Wireless Sensor Networks and Applications*, San Diego, CA.

Schurgers C, Aberthorne O, Srivastava M, 2001. Modulation scaling for energy aware communication systems. In *Proceedings of the 2001 International Symposium on Low Power Electronics and Design*, Huntington Beach, CA, pp. 96–99.

Schurgers C, Tsiatsis V, Ganerival S, Srivastava M, 2002. Optimizing sensor networks in energy–latency–density design space. *IEEE Transactions on Mobile Computing*, 1(1), 70–80.

Shah RC, Roy S, Jain S, Brunette W, 2003. Data MULEs: Modeling and analysis of a three-tier architecture for sparse sensor networks. In *Ad Hoc Networks Journal* (Elsevier), 1, 215–233.

Simon G, Volgyesi P, Maroti M, Ledeczi A, 2003. Simulation-based optimization of communication protocols for large-scale wireless sensor networks. *Proceedings of the IEEE Aerospace Conference*, Big Sky, MT, March 2003.

Sridharan A, Zuniga M, Krishnamachari B, 2004. Integrating Environment Simulators with Network Simulators, USC Department Computer Science Technical Report 04-386, 2004.

Sobeih A, Chen W, Hou JC, Kung L, Li N, Lim H, Tyan H, Zhang H, 2005. J-Sim: A simulation and emulation environment for wireless sensor networks. In *Proceedings of the Annual Simulation Symposium (ANSS 2005)*, San Diego, CA, pp. 175–187, April 2005.

Sundresh S, Kim W, Agha G, 2004. SENS: A sensor, environment and network simulator. In *Proceedings of the 37th Annual Symposium on Simulation*. IEEE Computer Society, Washington, DC, USA, 2004.

Swami A, Zhao Q, Hong Y, Tong L, 2007. *Wireless Sensor Networks: Signal Processing and Communication Perspectives*, John Wiley & Sons, Chichester, England, pp. 95.

Titzer B, Lee D, Palsberg J, 2005. Avrora: Scalable sensor network simulation with precise timing. In *Proceedings of the 4th International Symposium on Information Processing in Sensor Networks*. IEEE Press Piscataway, NJ.

Verdone R, Dardari D, Mazzini G, Conti A, 2009. *Wireless Sensor and Actuator Networks: Technologies, Analysis and Design*. Prepared Text at WiLAB, University of Bologna, Italy,

Woo A, Tong T, Culler D, 2003. Taming the underlying challenges of reliable multihop routing in sensor networks. *In 1ˢᵗ International Conference on Embedded Networked Sensor Systems (SenSys '03)*, Los Angeles, CA, USA.

Younis O, Fahmy S, 2004. HEED: A hybrid, energy-efficient, distributed clustering approach for ad-hoc sensor networks. In *IEEE Transactions on Mobile Computing*, 3, 366–379.

Yuan L, Qu, G, 2004. Energy-efficient design of distributed sensor networks. In *Handbook of Sensor Networks: Compact Wireless and Wired Sensing Systems*, Ilyas M and Mahgoub I, eds., CRC Press, Boca Raton, FL, pp. 38.1–38.19, 2004.

Zeng X, Bagrodia R, Gerla M, 1998, GloMoSim: A library for parallel simulation of large-scale wireless networks. In *Proceedings of the 12th Workshop on Parallel and Distributed Simulations (PADS)*, May 26–29, 1998 in Banff, Alberta, Canada.

Chapter 9

TinyKey

A Pragmatic and Energy Efficient Security Layer for Wireless Sensor Networks

Roberto Doriguzzi Corin, Giovanni Russello, and Elio Salvadori

Contents

9.1 Introduction

The use of wireless sensor networks (WSNs) in sensitive sectors (such as military, healthcare, and industrial fields) has increased the need of security mechanisms for WSNs. Although WSNs share some commonalities with computer networks, it is not always possible to migrate security mechanisms from computer networks to WSNs. Due to the limited computation and communication capabilities offered by sensors, introducing security often means to further reduce those resources for the application level. Therefore, introducing security mechanisms in WSNs requires the balancing of opposing requirement: on the one hand, the needs of security primitives to guarantee an adequate level of protection, on the other hand the need of limiting the impact on resource consumption.

In literature, there are several approaches to increase security in WSNs. Most of these approaches have been developed as stand-alone mechanisms without taking into account the stringent requirements of real-world deployments. Most security approaches have focused on the issue of message confidentiality using secure-group communication schemes where nodes share a given key used for encrypting the data. The main body of literature work has been addressing optimization problems related to how to distribute keys within a group using a low number of messages. However, some of these schemes have neglected other security aspects such as message authentication and message freshness [1]. Others, such as RiSeG [2], require complex synchronization schemes to be able to cater for such security aspects.

To address these limitations, we present our approach called *TinyKey*. TinyKey is a security architecture with an integrated key management solution. The key management in TinyKey is responsible for key generation, key distribution, and rekeying. Second, TinyKey is implemented for version 2.x of TinyOS making it promptly available for addressing security within current applications. Third,

TinyKey is a fully implemented prototype that has been tested in the real deployments of two projects for the local government of the province of Trento. Each project has specific security requirements, therefore, the architecture of TinyKey has to be flexible enough to accommodate both. This allowed us to perform some quantitative analysis regarding the performance of TinyKey in real-work deployments. The results of these analyses are presented in this chapter. Finally, the TinyKey prototype (written in nesc [3]) is available for the research community to download and test [4].

This chapter is organized as follows. We start by reviewing the related literature in Section 9.2. In Section 9.3, we provide an overview of the use cases where TinyKey has been deployed. Section 9.4 discusses in details the security requirements and adversary model. In Section 9.5, we describe the security properties supported by TinyKey. The details of the key management process are provided in Section 9.6. Performance measurements of TinyKey are presented in Section 9.7. We conclude this chapter in Section 9.8.

9.2 Related Work

The main motivation behind TinyKey is that most of the work related to security in WSNs is either an experimental exercise conducted either in small testbeds deployed in laboratories or in simulated environment. The realization of TinyKey was driven by our needs for a robust and practical security mechanism for WSNs to be deployed in real environments.

9.2.1 Security for WSNs

Very close to TinyKey is TinySec [5]. TinySec is one of the first works to address the security concern in WSNs. TinySec provides a fully implemented architecture for link-layer security in WSNs based on well-known cryptographic primitives. Another important aspect of TinySec is that its security properties exploit the limitations that are intrinsic of WSNs. For instance, WSNs have limited bandwidth, limiting the number of packets that an adversary can inject to or eavesdrop from the network. Therefore, it becomes possible to relax certain security properties and still guarantee adequate protection to specific attacks. Although TinySec provides a usable approach for WSN applications, pragmatic aspects that need to be addressed before a real deployment are left undefined. One of such aspects is key management. In TinyKey, we decide to provide within the security mechanism also an implementation of a key management system.

There are several research efforts targeting secure data communication in WSN. Most of these efforts have focused on realizing secure-group communication schemes where nodes are grouped together and able to securely communicate with each node member within the group. SLIMCAST [6] is one of the first works aimed at realizing secure group communication in WSNs. It implements a key infrastructure for secure multicasting to provide data confidentiality via a hop-to-hop re-encryption. SeGCom [7] proposes a scheme where nodes are clustered in tree-like structures. The nodes share, in a pair-wise manner, keys and require synchronization between nodes, which is very expensive for WSNs [8]. RiSeG [2] proposes a scheme where a logical ring is established among the nodes belonging to a group. The idea is that within each ring, a node is responsible for the key management, but at the price of a high latency due to the logical ring structure. In fact, rekeying messages have to be sent around in the ring until it reaches back the controlled node. This scheme does not scale well when the group size increases.

In recent years, a new type of WSNs is emerging: the wireless sensor and actuator networks (WSANs) [9]. In this type of WSNs, the data collected by the sensor nodes is re-used by the actuators that are specialized nodes in the network. This means that data does not need to be routed toward a base-station but is generated and consumed within the network. Since WSANs may operate in complete isolation, specialized secure communication schemes are required. The lightweight authenticated rekeying (LARK) scheme [10] was designed to cater for needs of WSANs. In LARK, nodes are logically grouped according to the application logic independently from network management purposes.

9.2.2 Key Management for WSN

In terms of key management, there are several works that provide schemes for optimizing key revocations and redistributions. To reduce the number of messages for rekeying, the logical key hierarchy (LKH) was proposed [1]. The main idea behind this scheme is to create several subgroups of nodes each with a specific key-encryption-key (KEK). These subgroups KEK are then used for securely delivering the group key to the nodes in each subgroup. Although it reduces the number of messages needed for rekeying, it also introduces higher computational and storage overhead. An optimization of this scheme is proposed in the one-key derivation (OKD) scheme [11]. In this scheme, the rekeying messages are further reduced by means of key local computations using a one-way hash function. To compute the new key k' each group member uses the hash function H with the current key k generating $k' = H(k)$.

The localized encryption and authentication protocol (LEAP) [12] provides a key management scheme for WSN that aims at reducing the security implications of a compromised node on the rest of the network. To achieve its goal, LEAP requires that each node has four different types of keys. This requires higher overhead for each node and, given the large number of keys, the generation and distribution of new keys is a very expensive procedure. Moreover, this scheme requires the μTESLA [13] scheme for broadcast authentication realized through node synchronization. In WSNs, node synchronization is very hard to achieve [8].

Another approach for key management is to deterministically predistribute the keys on each node to establish a pairwise session key with each neighbor. A naïve implementation of this scheme would require that each node carry a session key for each link established with a neighbor node. This approach is not very scalable especially in highly connected networks. To reduce the number of session keys, Choi et al. [14] propose to establish an ordered relation between the nodes in a network that would allow a node only to carry half of the required keys of the naïve scheme. Using the ordered relation, each node is able to compute the other half of the keys using a hash function. Although this approach reduces the number of keys, it is still not scalable.

In [15], the authors propose a new deterministic predistribution key scheme based on symmetric balanced incomplete block design (SBIBD). This scheme allows constructing key rings such that each two key rings share exactly one single key. The main advantage of this scheme is that each two nodes will be able to communicate because they share a common key. However, the scheme does not scale well in large networks of nodes. A variant of this scheme was proposed in [16] to support secure communication between nodes in the same region when the WSN is organized as a grid of nodes.

The approach proposed in [17] introduces trade-based key management where each two nodes in a deployment are able to establish a direct secure link by computing a session key based on a common secret that they might share. The sharing of the common secret is based on a Steiner trade scheme to guarantee that two nodes share a common secret to enable the session key derivation. However, this scheme has a low sharing probability of at most 0.25 and it is not very resilient. An attacker can construct part of the pairwise secret keys and compute the session keys with the rest of the noncompromised nodes.

Finally, the approach presented in [18] proposes the use of a unital design [19] to predistribute the key in the nodes. The main advantage of this scheme is that it allows the creation of session keys with a high probability. On the other hand, this scheme is quite expensive in terms of performance for each node to compute the secret key.

9.3 Use Cases

In this section, we present two real-world applications where TinyKey has been deployed and have driven the choices of the security design in TinyKey. In literature, the vast majority of the works on security for WSN focus on mechanisms that are tested in "protected" environments, such as simulators or dedicated WSN testbeds. To the best of our knowledge, there are no existing works describing the performance of security mechanisms in operational WSN dealing with real-world applications.

In fact, each application based on WSN-sensing technology has its own specific requirements which are derived from both the WSN setting specific to the application itself (nodes hardware, placement, access, just to mention a few) as well as from the overall requirements of the application where the information gathered through the nodes are collected. As we will see in the rest of this chapter, TinyKey is able to cater for each scenario thanks to its configurable architecture. Moreover, simplicity has been our first priority given that existing work presents overcomplicated security architectures.

9.3.1 WSNs for Adaptive Lighting Systems in Road Tunnels

The first application is related to a deployment of WSN nodes to provide adaptive control of the light intensity in road tunnels for the Trentino Research and Innovation for Tunnel Monitoring (TRITon) project.* The main goal of this project is to reduce the management costs of road tunnels and improving their safety. In this application, a set of light sensors is placed along the sides of the tunnel and provide information about measured light within a SCADA-based controller which tunes the lighting system to reduce energy consumption. The WSN-supporting adaptive light control is expected to become permanent in four existing tunnels, whose lengths range from 400 to 1400 m. The tunnels are situated on a major freeway nearby Trento and have an average traffic of 14,000 vehicles per day.

In this application scenario, the sensor nodes are deployed on the upper part of the tunnel walls to be serviced only by an aerial work platform. Moreover, the tunnels are mainly for vehicles and are video monitored 24/7.

9.3.2 WSN for Assisted Living Homes

The second application scenario is within the ambient aware assistance (ACube) project.† In this project, assisted living homes are equipped with a set of sensing technologies for monitoring the behavior of Alzheimer-diseased patients. The WSN application is deployed to help personnel in performing their duties and support the safety of users. A prototypical version of the whole system has been put in operation in a complex of several apartments located in Trento. Here, the rooms of the apartments are equipped with a set of WSN nodes anchored to walls and doors while hosted patients are expected to wear their own node. The information collected by the WSN-sensing subsystem (and handled by a central controller) allows the detection of not just patient presence in a specific room but also his/her proximity to exit doors and windows in order to prevent potential escapes or dangerous behavior. Furthermore, thanks to an accelerometer installed on the patient node, the WSN can also detect patients' falls and automatically trigger the intervention of medical personnel.

---

* http://triton.disi.unitn.it.
† http://acube.fbk.eu.

9.4 Security Requirements

In this section, we review the adversarial model together with the security requirements of TinyKey.

First of all, given the deployments of our scenario, an attacker cannot easily gain physical access to the sensor nodes. In fact, for the tunnel deployment the nodes are not easy to reach. Moreover, the tunnels where the nodes are deployed are only for vehicles and not for pedestrians. Therefore, an attacker that wants to access the nodes has to stop her vehicle in the tunnel and use a ladder. Such behavior can be easily spotted by the video surveillance system deployed in the tunnels. Similarly, access to the assisted living facilities is granted only to authorized personnel and patients' relatives which we assume are trusted entities. However, also in this case there is a video surveillance system deployed in the houses that can spot any suspicious activities. With this in mind, we can assume that secret information stored in a node is not easily reachable for an attacker. However, in the event the attacker is able to gain access to such information, we can safely assume that we can spot such an attack and take some countermeasures (such as revoking the existing secret material and deploy new secret information). This might introduce some interruption of the service but at least we can avoid that an adversary is able to inject rogue information in the WSN.

Given that the sensor nodes communicate via a wireless connection, there are still the following issues to be considered. In particular, for the tunnel scenario we can envisage the following attacks can be still mounted:

- Impersonation attack: An attacker tries to impersonate a valid node to inject false and/or erroneous information. A solution against this attack can be message authentication. Also to protect the network from distributing erroneous messages injected by the attacker, an integrity solution must be provided.
- Replay attack: An attacker tries to replay an old message to disturb the operations of the WSN. The needs of protection mechanisms against this threat have been reduced due to the fact that it is supposed to have a monitoring facility able to detect anomalous node consumption of battery. However, a simple protection scheme based on sequence numbers can further reduce the impact of this attack. Yet it is up to the network deployment phase to decide the exact level of protection with respect to this kind of attacks.
- Node integrity: The tunnels are still accessible to the public. Thus, for an attacker it is still possible to damage or destroy the nodes just to disrupt the functionality of the WSN. Against this attack we can employ replication of resources meaning that we can deploy a larger number of nodes that is actually required so that even if some nodes are down we can still guarantee the functionality of the WSN. Moreover, the attacker would require more time to destroy a larger number of nodes to make the attack successful.

In the second-application scenario, together with the above security threats there is also the threat of guaranteeing the confidentiality of the data exchanged through the WSN. Here, an adversary eavesdropping messages represents the main threat. Because the message exchanged contains sensitive information about patients, the adversary could potentially compromise the patients' privacy. For example, by using data-mining techniques an adversary could observe patterns in the patients' behavior within the assisted living facility or reconstruct biometric information related to the patient. To protect against this attack, we can use encryption to guarantee confidentiality of the information collected through the WSN.

9.5 Architecture of TinyKey

The TinyKey architecture covers the three main security properties for messages: *authentication, integrity,* and *confidentiality.* Additionally, TinyKey provides protection against *replay attacks* and has a built-in *key management* mechanism. This last feature sets TinyKey apart from other similar approaches.

9.5.1 Authentication and integrity

Authentication is the capability of the system to allow only legitimate nodes to participate in the network. At the same time, message authentication can be used for checking the message integrity. If a corrupted message arrives to a node, the node should be able to assess that the message is corrupted (or has been tampered with). Entity authentication and message integrity are always enabled in TinyKey and are supported by means of the message authentication code (MAC) provided in each packet. TinyKey uses the TinySec implementation of CBC–MAC [20] algorithm for computing and verifying MAC codes.

9.5.2 Message Confidentiality

Confidentiality is a property applied to the message content to keep its information undisclosed to unauthorized parties. Message confidentiality in TinyKey is provided by means of symmetric encryption using TinySec implementation of the Skipjack [21] algorithm. To make matters more difficult to a hacker using a dictionary attack to find patterns in the encrypted messages, TinyKey supports initialization vectors (IV) for the message encryption. The use of an IV prevents repetition in data encryption, that is, the same message will be encrypted as different ciphertext. An IV is an arbitrary number that can be used along with a secret key for data encryption. Like TinySec, TinyKey uses IVs of 8 bytes based on packet header format represented in Table 9.1, where *dst* is the destination address, *AMType* is the active message type, *len* is the length of the data payload, *src* is the source address, and *counter* is a 16-bit

Table 9.1 IV Format

dst	AMType	len	src	counter
(2)	(1)	(1)	(2)	(2)

Note: Size in bytes enclosed within parentheses.

counter that starts from 0 when the node boots and is increased by 1 after each message is sent.

9.5.3 Replay-Attack Protection

A replay attack consists in an adversary capturing a valid message and sending it again to the network. Since the message is generated by a legitimate node, the receiving nodes cannot have any clue the message was resent by an adversary. To distinguish between valid messages and replayed messages a node must keep track of the received messages; therefore, each message must be identified by a unique id.

In TinyKey, each packet sent on the network includes a 16-bit increasing sequence number (the same used to compute the IV) that is tracked by the recipients that reject packets with old values.

On each node of the network, which must maintain the history of all the other nodes in the network, this mechanism is quite expensive in terms of RAM memory occupation. The main sources of RAM memory usage are the arrays storing the addresses and the sequence numbers of the other nodes. The total size of the two arrays (in bytes) is computed as follows:

$$N*sizeof(am\_addr\_t) + N*sizeof(seqnum\_t)$$

where N is the number of nodes, am_addr_t is the variable type for node addresses (16 bits), and seqnum_t is the variable type for sequence numbers (16 bits). Since in *nesc* the dynamic allocation of memory is not allowed, the value of N is set at compilation time and must be equal or greater than the number of nodes in the network. The choice of setting N greater than the network size allows the addition of new nodes after the initial deployment without the need of reprogramming the whole network.

9.5.4 Key Management

Key management refers to the activities related to generating new keys and distributing the keys to the valid nodes in a reliable manner. Additionally, key management must support node addition in the network. The key management requires not only cryptographic support but also access to the network layer API and to nonvolatile storage API. TinyKey provides a key management submodule (KMS box in Figure 9.1) which includes key generation, distribution, and storage. We will

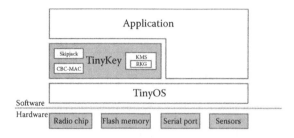

Figure 9.1 WSN node architecture with TinyKey enabled. KMS functionalities and hardware access depend on the role of the node in the network.

discuss in more detail KMS and its random key generator (RKG) submodule in the next section.

9.6 Key Management

Key management, including key generation and distribution, is handled internally to TinyKey by the KMS. KMS is based on centralized and symmetric key management schemes where only one entity (the *sink* node) controls the generation, regeneration, and distribution of the keys over the network.

9.6.1 Generation and Distribution of Keys

KMS operations, depicted in Figure 9.2, begin on the sink node with the generation of authentication and encryption keys performed by the RKG submodule

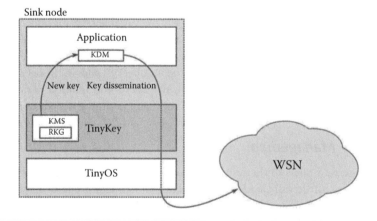

Figure 9.2 The key generation and distribution process work-flow on the sink-node.

(presented at the end of this section). Afterwards, the keys are passed to the application-level key distribution manager (KDM) which sends the keys to the other nodes of the network through dedicated interfaces of TinyKey.

9.6.2 Key Renewal

The renewal of the keys is performed periodically and the period is set at the compilation time. The new generated keys are sent in broadcast and are encrypted with a special set of predistributed keys shared by all the nodes in the network. In this way, only legitimate nodes (i.e., the nodes with the correct preinstalled key set) are capable of authenticating and decrypting the packet with the new keys. These preinstalled keys are never changed unless the nodes are redeployed.

9.6.3 Establishing Node Keys

When TinyKey receives a message containing new keys, it signals the event to the application which eventually calls the KMS to validate and save the new keys on the nonvolatile memory.

Key validation is performed by comparing the version number included in the message with the version of the current keys. If the new version number is higher, the current keys are replaced by the new keys and stored to the nonvolatile memory. After this step, KMS returns a value to the application (SUCCESS if the new keys are valid, FAIL otherwise) which can redistribute the keys to the nodes more than one hop away from the sink. The process just described makes TinyKey compatible with multihop WSN topologies.

All the nodes store the last two versions of the keys. If a node does not receive the new keys, it continues to operate with its current keys (that are the old keys for all the other nodes on the network) and its messages are still correctly received by the other nodes. However, these keys do not allow the node to receive messages from the rest of the network. When TinyKey receives a message secured with the old keys, it signals an event to the application that can decide whether to resend the new keys in unicast to the node which sent the message.

9.6.4 Addition of Nodes

When a new node is added to a WSN already in operation, it cannot communicate with the other nodes since it can only use the preinstalled key material. Its integration with the WSN is complete only after a key update. Since this could happen after a long time (depending on the configuration of the sink), the generation of the new keys can be forced from remote through the *WSN gateway*.[*]

---

[*] The WSN gateway is the device, often an embedded PC, which provides interconnection between the WSN and remote servers.

9.6.4.1 RKG Submodule

In the early releases of TinyKey, the Key Generator sub-module was built around the TinyOS *Random* function. This function is a pseudorandom generator and is seeded with the ID of the node. Therefore, every time the node boots, this function always outputs the same sequence of numbers. This predictable behavior can be easily understood by an attacker and consequently compromise the security of the whole network.

On the basis of these considerations, we tried to find a simple and efficient pseudorandom number generator with the following design constraints imposed by the two use cases presented in Section 9.3:

- The sink node does not mount any sensor, therefore, the generator cannot use any sensor-specific features as source of randomness.
- The ROM memory overhead of the generator must be less than 300 bytes, since the remaining portion of memory is assigned to upper-layer code (middleware and applications) and the other modules of TinyKey.

For our implementation we started considering TinyRNG [22], a pseudonumber generator that uses the transmission bit errors on the wireless channel as the main source of entropy. Unfortunately, we could not use TinyRNG because, as documented in [22], it requires at least 1392 bytes of additional ROM memory, therefore, it did not meet our constraints.

In the end, we decided to implement a pseudorandom key generator based on the traffic flowing from the sink to the WSN gateway through the serial connection. This traffic includes all the data collected by the WSN (mostly sensors values) and the messages of the protocol governing the communication between sink and gateway. Since in our deployments the sink node does not mount any sensor, the idea behind our key generator is to leverage on the sensor values coming from the WSN (sent to the sink and then to the gateway through the serial connection) as source of randomness.

We designed the generator in the following way: a recursive XOR bitwise operation is performed between a 16-bit value and the bits of the sink-gateway communication. More precisely, let $a \oplus b$ denote the bitwise exclusive-or of a and b and let $\gg_L$ denote the L-bit right circular bit shift operation, we have that

$$seed_i = seed_{i-1} \oplus (w \gg_{16} (i \bmod p)), \quad seed_0 = 0 \qquad (9.1)$$

where w represents a 16-bit word of the sink-gateway communication and p the largest prime number below 16 and different from the message length. The computation of Equation 9.1 is performed for each 16-bit word going from the sink to the gateway and updates the 16-bit value *seed*. The current value of *seed* and its complement are used to seed the TinyOS pseudonumber generator *RandomMlcgP* during the periodic generation of authentication and encryption keys.

9.7 Evaluation and Results

To measure the performance of TinyKey, we focus on three aspects: energy, packet loss, and memory. The analysis is mostly based on the comparison between TinyKey and the standard TinyOS network stack and is organized in two parts: the first part reports specific results related to the use cases presented in Section 9.3, the second part analyses the common features of any TinyKey deployment, that is, the performance of the RKG introduced in Section 9.6 and memory occupation of TinyKey submodules.

9.7.1 Application-Related Results: Packet Loss and Energy Consumption

In this section, we describe the performance of TinyKey in two different configurations and environments (road tunnel and assisted living home). The evaluations are presented in terms of the number of packet loss and the power consumption overhead. For each scenario, we considered only the metric that would perform in the worst-case conditions due to the specific requirements of the scenario. Therefore, for the tunnel where a large number of nodes are deployed we will present the results on the packet-loss rate. In fact, in this condition the number of packet losses due to communication conflicts should be the highest. For the assisted living home, we will provide the measurements of the energy consumption. The rationale is that in this scenario TinyKey has to encrypt each packet (while in the tunnel this is not required). Therefore, more computations have to be performed by the nodes requiring a larger amount of energy.

9.7.1.1 Packet Loss: The Road Tunnel Scenario

Security requirements for this scenario are message authentication, message integrity, and key management. Confidentiality is not required in this scenario as we have explained in Section 9.4. Therefore, possible causes of the packet loss overhead are the CPU overload due to the computation of the CBC–MAC algorithm and failures in the key-update process.

The hardware we used in this scenario consisted of WSN nodes manufactured by a local company and equivalent to the TMote Sky motes [23]. These nodes were equipped with the MSP430 processor and the Chipcon2420 2.4 GHz RF transceiver, both by Texas Instruments, and mounting an expansion board containing 4 ISL29004 luminance sensors. All the nodes were powered by 4 Duracell Procell D-size batteries except two nodes acting as sinks that were powered by the WSN gateway through the USB connection.

The experimentation was performed in a two-lane road tunnel where we deployed two WSNs of 16 nodes transmitting on noninterfering channels. As depicted in Figure 9.3, for each of the two walls of the tunnel we deployed an

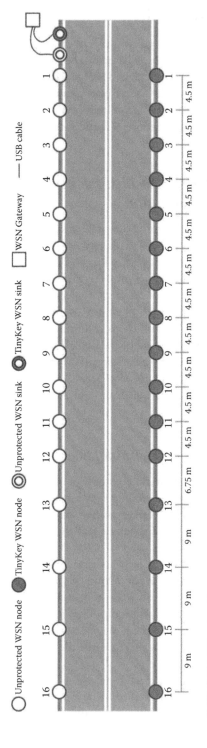

Figure 9.3 The deployment of the two arrays of WSN nodes in the testbed tunnel (view from above).

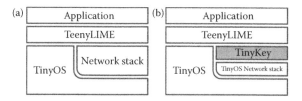

Figure 9.4 **WSN node software architecture for the road tunnel scenario. (a) Generic node software architecture. (b) Node software architecture including TinyKey.**

array of nodes, each array representing a different WSN with a dedicated sink node. In order to evaluate the packet loss overhead incurred by the use of TinyKey, we deployed an array of nodes without TinyKey (represented as full-white nodes in Figure 9.3) that we will refer to as *unprotected WSN* and a second array using TinyKey for protecting the data transmitted by the nodes (shown as full-grey nodes in Figure 9.3), referred to as *TinyKey WSN*. The software deployed on the nodes of the unprotected WSN is depicted in Figure 9.4a. Figure 9.4b shows the software deployed on nodes forming the TinyKey WSN where TinyKey was enabled.

In accordance with TRITon project requirements, on both networks the application ran on top of TeenyLIME middleware [24] which was responsible of packet routing and forwarding operations and enabled the nodes with multihop capabilities. On nodes of the TinyKey WSN, TinyKey was placed between TeenyLIME and the network stack providing message authentication, message integrity, and key management. Finally, to obtain energy savings, a duty-cycling scheme for the radio known as low-power listening (LPL [25]) was employed with sleep interval set to value 250 ms.

In both WSNs, the main task of the application was reading the luminance value from all the four sensors of the expansion board at a fixed rate of 5 s and reporting the readings to the sink every 30 s in packets with fixed payload size of 105 bytes. Beside the light samples, the application reported other values including the battery level, the temperature, and system information once a minute. On the sink of the TinyKey WSN, a new authentication key was generated every hour in order to evaluate the reliability of the key management mechanism.

Statistics over 7 days of experimentation recorded on the nonvolatile memory of the gateway are summarized in Table 9.2. As can be noticed from Table 9.2, the two loss rate values are comparable; therefore, the additional CPU load introduced by TinyKey algorithms does not seem to degrade the WSN operations in terms of packet loss.

9.7.1.2 Energy Consumption: The Assisted Living Home Scenario

In this scenario, WSN technology is adopted for the development of a localization and proximity system that detects when actors (patients/caregivers) are in the range

Table 9.2 Light Sample Statistics

	Unprotected WSN	*TinyKey WSN*
Collected samples	424134	437469
Lost samples	275	224
Loss rate	0.064%	0.051%

of a hazard area or when actors pass from one room to another. To protect actors' privacy, it is required to transmit data captured by the sensors to the base station on a secure channel providing data confidentiality.

The ACUBE implementation rely on TeenyLIME and TinyKey but, differently from tunnel scenario, the experimentation of TinyKey was performed before the final integration with the project-specific application, although packet rate (one packet sent every 3 s), packet routing (single hop), packet size (100 bytes), and LPL sleep interval (50 ms) was compliant to the project specifications. Also in this scenario, we deployed two WSNs, one unprotected and one with security enabled. As shown in Figure 9.5, each WSN consisted of four nodes and one sink installed in a room at approximately 1 m from the floor. We used TMote Sky motes [23] powered by four Sony Alkaline StaminaPlus D-size batteries and transmitting on two noninterfering channels (one for each WSN).

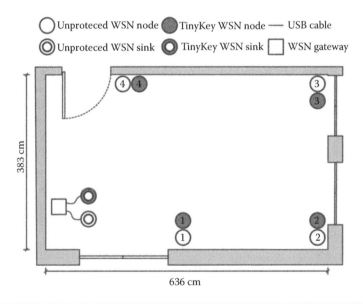

Figure 9.5 The assisted living home test deployment.

TinyKey configuration for this test included message authentication, message integrity, message confidentiality, and replay-attack protection. KMS was also enabled and new keys were generated and disseminated in broadcast every 60 min.

Evaluation of energy consumption overhead introduced by this configuration of TinyKey was performed by connecting an oscilloscope to the node #1 of the TinyKey WSN and to the node #1 of the unprotected WSN. We indirectly measured the electrical current with the oscilloscope by attaching a resistor of known value (100 Ω) to the oscilloscope's probes, measuring the voltage across the resistor, and then using Ohm's Law to calculate the electrical current. Plots presented in Figure 9.6 report samples taken on the nodes while sending packets to the sink during a 500 ms interval.

Two considerations can be made by comparing the plots in Figure 9.6: first, on the TinyKey node the microcontroller is awakened before the packet transmission for computing the security algorithms (for about 32 ms). Second, the total energy consumed for sending a packet with TinyKey (obtained by calculating the area under the curves in Figure 9.6 during the sending periods) is 0.000617 mAH, while sending a packet with the TinyOS stack consumes 0.000589 mAH. Therefore, we can conclude that the energy consumption overhead for sending a 100-bytes packet with TinyKey in this configuration is about 4.5%.

In this scenario, we also measured the energy consumption on a TinyKey WSN node when processing new security keys coming from the sink node. Keys were sent in broadcast in packets with 22 bytes payload: 10 bytes for the authentication

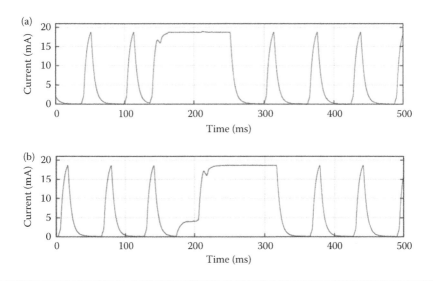

Figure 9.6 Energy consumption as a function of time. The periodic spikes are the consequence of the radio duty cycle while the longer power draws are caused by the sending process. (a) Unprotected node #1. (b) TinyKey node #1.

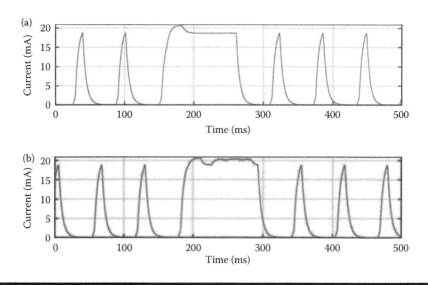

Figure 9.7 Energy consumption for processing the incoming keys. The periodic spikes are the consequence of the radio duty cycle while the longer power draws are caused by the receiving and storing processes. (a) New keys reception. (b) New keys reception and storage.

key, 10 bytes for the encryption key, and 2 bytes for the version number. As a result, we obtained that processing the incoming security keys causes a consumption of about 0.000623 mAH and, as can be noticed by comparing plots in Figure 9.7a and b, part of the power draw (about 7.3%) is due to the process of storing the keys to the program memory.

9.7.2 Further Considerations on TinyKey: Evaluation of the RKG and Memory Occupation

9.7.2.1 Evaluation of RKG

TinyKey relies on the unpredictability of the keys it uses; therefore, the random number generator which seeds RKG must meet two main requirements:

- ◼ The generated numbers must be uniformly distributed within the interval [0,65535].
- ◼ Absence of repeating patterns in the number-generation process.

As described in Section 9.6, RKG relies on the traffic transmitted from the sink node to the gateway through the serial connection to generate 16-bit random numbers. In the two use cases introduced in Section 9.3, this traffic is mainly composed of messages sent from the WSN to the sink with a payload size of around 100 bytes.

These messages carry sensors values (light, temperature, acceleration, etc.), system information (battery level, status of sensors, etc.), and routing statistics. Although these messages are a good source of randomness, to verify whether RKG meets the above requirements we performed the evaluation as close to the worst case condition as possible (i.e., the case where the same byte is repeated overtime), that is, we decided to send 8-bytes messages containing an incremental sequence number and the charge level of the batteries (both used in the implementation of the two use cases). We setup the test environment with eight battery-powered nodes sending one message per second to the sink connected to a PC through a serial connection. The analysis of the 200,000 random numbers generated by the sink in about 7 h is summarized in Table 9.3.

The entropy value, introduced by Shannon in [26], is determined as defined in Equation 9.2.

$$H = -\sum_{i=0}^{M} p_i \log_2(p_i) \qquad (9.2)$$

where $M = 2^{16}$ is the number of possible outputs of the generator with probability $p1,... pM$.

The resulting value should be as close as possible to 16 bits (the output bit length of the generator) and measures the degree to which the bits in the RNG are uniformly distributed.

The Monte Carlo value for π is computed by using each successive pair of randomly generated numbers as X and Y coordinates of the Cartesian plane.

$$\pi = 4 * \frac{inPoints}{totalPoints} \qquad (9.3)$$

In Equation 9.3, *inPoints* is the number of (X,Y) points lying within the circle of radius 65,535 while *totalPoints* is the total number of (X,Y) pairs used for the test (i.e., the number of samples divided by two).

Table 9.3 RNG Analysis

	Result	*Ideal Result*
Entropy	15.7423 bits per word	16 bits per word
Monte Carlo value for π	3.14708584 (error 0.17%)	Value of π
Autocorrelation	−0.01165	0
Arithmetic mean	32773.15	32768

The result of Equation 9.3 should be as close as possible to the value of π and measures the distribution of the generated numbers within the interval [0,65535].

The autocorrelation function, used to detect repeating patterns in a sequence of numbers, returns a value ranging from −1 to +1 and is defined as

$$r = \frac{\sum_{i=1}^{N}(X_{i-1} - \bar{X})(X_i - \bar{X})}{\sum_{i=0}^{N}(X_i - \bar{X})^2} \tag{9.4}$$

where N is the amount of numbers X_i generated during the experiment and $\bar{X}$ is the arithmetic mean of the numbers.

In conclusion, the results of the test reported in Table 9.3 are all in reasonable ranges of random numbers, therefore, RKG represents a secure alternative to the TinyOS *rand()* function used in the first releases of TinyKey.

9.7.2.2 Memory Occupation Overhead

In this section, we compare the memory usage overhead of various configurations of TinyKey. Values reported in Table 9.4 are the result of the comparison between TinyKey and the communication stack of TinyOS release 2.0.2.

In Table 9.4, *KMS* is the key management submodule, *Auth* is the CBC–MAC algorithm implementation providing message authentication and integrity, *RKG* is the random key generator, *ENC* is the Skipjack algorithm implementation providing message confidentiality, and *RP* is the replay attacks protection module configured for a 100-nodes WSN.

Table 9.4 contains the values of additional ROM and RAM memory required by an application using TinyKey for protecting the communication among the WSN nodes. For instance, the sample application included in [4] with a basic configuration of TinyKey (key management and authentication) requires 16732 bytes

Table 9.4 Memory Occupation Overhead of TinyKey in Various Configurations

TinyKey Configuration	Additional ROM Memory (Bytes)	Additional RAM Memory (Bytes)
KMS + Auth	3706	384
KMS + Auth + RKG	3960	386
KMS + Auth + RKG + ENC	5746	386
KMS + Auth + RKG + ENC + RP	5990	790

of ROM and 1227 bytes of RAM compared to the 13026 bytes of ROM and 843 bytes of RAM required by the same application without TinyKey.

We can notice that significant sources of memory usage are the Skipjack implementation (1786 bytes of ROM used) and the replay attacks protection module (400 bytes of RAM, 4 bytes for each node in the network).

9.8 Future Work

As future directions of our research, we are currently focusing on the following aspects.

TinyKey currently relies on the distribution abstractions provided by TeenyLIME. Although TeenyLIME provides the advantages typical of shared tuple spaces, is also limited in terms of flexibility to be employed in WSNs where actuator nodes play an important role. In this respect, we are working in integrating TinyKey with ESCAPE [27], a policy-based pub/sub middleware for sense-and-react applications based on WSNs. The main advantage of integrating TinyKey with ESCAPE is that we can widen the type of applications we can support with TinyKey given that ESCAPE is highly configurable and flexible as communication middleware.

Another area in our future work is to expand the capability of TinyKey to be able to detect when a node has been compromised. This capability will allow TinyKey to exclude the node from the computation environment and to report an alarm to the network manager for further countermeasures.

Finally, we are exploring the possibility of introducing more advanced encryption techniques. In particular, we are investigating the integration in TinyKey of encryption schemes to support encrypted data on which it is possible to perform some computation as done in [28]. In this way, we could deploy applications in outsourced WSNs managed by an untrusted third-party. However, because the data is distributed through the WSN in an encrypted format we can guarantee its confidentiality. This could open very interesting scenarios where two or more parties might share the same WSN, but without having to share the data that has been collected.

9.9 Conclusions

This chapter presents TinyKey, a security architecture for WSNs that takes into account pragmatic concerns of a real-work deployment. The deployment and security of real applications have been the main drive behind the developing of TinyKey. In particular, TinyKey has been used in the deployment of two sensitive applications. Although both applications are related to sensitive data, they have different security requirements. One application is concerned with the safety of the road tunnels around the city of Trento while the second application focuses on improving the quality of life of elderly people in a shelter home. As a result, we have been able to measure the performances of TinyKey in real deployments. The main contribution of this chapter is to present an evaluation of a security architecture that has been tested in real-world deployments. Related approaches in the literature only present evaluations carried out in "protected" environment like simulators or dedicated WSN testbeds. To the best of our knowledge this is the first work to present performance results conducted on full-scale deployment. Another major contribution of our work is to provide a complete key management mechanism within the architecture of TinyKey that takes into account of not only key generation and distribution but also message authentication and message freshness. Existing security approaches for WSNs have either neglected mechanisms related to key management or only focused on the specific distribution of the keys according to some particular scheme.

References

1. C. K. Wong, M. Gouda, and S. S. Lam, Secure group communications using key graphs, in *IEEE/ACM Transactions on Networking*, 1998, pp. 68–79.
2. O. Cheikhrouhou, A. Koubaa, O. Gaddour, G. Dini, and M. Abid, Riseg: A logical ring based secure group communication protocol for wireless sensor networks, in *Communication in Wireless Environments and Ubiquitous Systems (ICWUS,) International Conference on Wireless and Ubiquitous Systems*, Sousse, Tunisia, October 2010, pp. 1–5.
3. D. Gay, P. Levis, R. von Behren, M. Welsh, E. Brewer, and D. Culler, The nesc language: A holistic approach to networked embedded systems. In *Proc. of ACM SIGPLAN 2003 Conference on Programming Language Design and Implementation*, San Diego, California, USA, 2003.
4. Tinykey project web-page. [Online]. Available: http://gforge.create-net.org/gf/project/tinykey (accessed November 2012).
5. C. Karlof, N. Sastry, and D. Wagner, TinySec: A link layer security architecture for wireless sensor networks. in *Proc. of the Second ACM Conference on Embedded Networked Sensor Systems (SenSys 2004)*, Baltimore, Maryland, 2004.
6. J.-H. Huang, J. Buckingham, and R. Han, A level key infrastructure for secure and efficient group communication in wireless sensor network, in *Security and Privacy for Emerging Areas in Communications Networks, 2005. SecureComm 2005. First International Conference*, Athens, Greece, September 2005, pp. 249–260.
7. O. Gaddour, A. Koubaa, and M. Abid, Segcom: A secure group communication mechanism in cluster-tree wireless sensor networks, in *Communications and Networking, 2009. ComNet 2009. First International Conference*, Hammamet, Tunisia, November 2009, pp. 1–7.
8. S. Lasassmeh and J. Conrad, Time synchronization in wireless sensor networks: A survey, in *IEEE SoutheastCon 2010 (SoutheastCon), Proceedings of the*, Concord, North Carolina, March 2010, pp. 242–245.
9. I. F. Akyildiz and I. H. Kasimoglu, Wireless sensor and actor networks: Research challenges, *Ad Hoc Networks*, 2(4), 351–367, 2004. [Online]. Available: http://www.sciencedirect.com/science/article/pii/S1570870504000319 (accessed December 2012).
10. G. Dini and I. M. Savino, Lark: A lightweight authenticated rekeying scheme for clustered wireless sensor networks, *ACM Transactions in Embedded Computing*

Systems, 10(4), 41:1–41:35, 2011. [Online]. Available: http://doi.acm.org/10.1145/2043662.2043665 (accessed December 2012).

11. J.-C. Lin, F. Lai, and H.-C. Lee, Efficient group key management protocol with one-way key derivation, *Local Computer Networks, Annual IEEE Conference on*, 0, 336–343, 2005.

12. S. Zhu, S. Setia, and S. Jajodia, Leap+: Efficient security mechanisms for large-scale distributed sensor networks, *ACM Transactions on Sensor Networks*, 2(4), 500–528, 2006. [Online]. Available: http://doi.acm.org/10.1145/1218556.1218559 (accessed December 2012).

13. A. Perrig, R. Canetti, D. Tygar, and D. Song, The TESLA Broadcast Authentication Protocol, 2002. [Online]. Available: http://www.cs.berkeley.edu/~tygar/papers/TESLA_broadcast_authentication_protocol.pdf (accessed March 2013).

14. T. Choi, H. B. Acharya, and M. G. Gouda, The best keying protocol for sensor networks. In *IEEE WoWMoM*, 1–6, 2011.

15. S. A. Camtepe and B. Yener, Combinatorial design of key distribution mechanisms for wireless sensor networks, *IEEE/ACM Transaction on Networking*, 15, 346–358, 2007.

16. S. Ruj and B. Roy, Key predistribution using combinatorial designs for grid-group deployment scheme in wireless sensor networks, *ACM Transactions on Sensor Networks*, 6, 4:1–4:28, 2010.

17. S. Ruj, A. Nayak, and I. Stojmenovic, Fully secure pairwise and triple key distribution in wireless sensor networks using combinatorial designs. In *IEEE INFOCOM*, pp. 326–330, 2011.

18. W. Bechkit, Y. Challal, A. Bouabdallah, A new scalable key pre-distribution scheme for WSN, In *Proceedings of 21st International Conference on Computer Communications and Networks (ICCCN 2012)*, Munich, Germany, 1–7, August 2012.

19. E. F. Assmus and J. D. Key, Designs and their codes. *Cambridge Tracts in Mathematics*. Cambridge University Press, 1992.

20. M. Bellare, J. Kilian, and P. Rogaway, The security of the cipher block chaining message authentication code, *Journal of Computer and System Sciences*, 61(3), 362–399, 2000.

21. E. Brickell, D. Denning, S. Kent, D. Mahler, and W. Tuchman, Skipjack review. *Interim report*, 1993.

22. A. Francillon and C. Castelluccia, TinyRNG: A cryptographic random number generator for wireless sensors network nodes. in *WiOpt. 5th International Symposium on Modeling and Optimization in Mobile, Ad Hoc, and Wireless Networks*, Limassol, Cyprus, 2007.

23. J. Polastre, R. Szewczyk, and D. Culler, Telos: Enabling ultra-low power wireless research. In *Proceedings of 4th International Symposium on Information Processing in Sensor Networks (IPSN 2005)*, UCLA, Los Angeles, California, USA, April 2005.

24. P. Costa, L. Mottola, A. Murphy, and G. Picco, TeenyLIME: Transiently shared tuple space middleware for wireless sensor networks. In *Proceedings of 1st International Workshop on Middleware for Sensor Networks (MidSens)*, Melbourne, Australia, 2006.

25. TinyOS. TEP 105 - Low Power Listening. [Online]. Available: http://www.tinyos.net (accessed November 2012).

26. C. E. Shannon, A mathematical theory of communication. *The Bell System Technical Journal*, 27, 379–423, 623–656, 1948.

27. G. Russello, L. Mostarda, N. Dulay, A policy-based publish/subscribe middleware for sense-and-react applications. *Journal of Systems and Software*, 84(4), 638–654, 2011.

28. M. Ion, G. Russello, B. Crispo, Design and implementation of a confidentiality and access control solution for publish/subscribe systems, *Computer Networks*, 56(7), 2014–2037, 2012.

Chapter 10

Secure Multipurpose Wireless Sensor Networks

Daniel Jacobi, Marc Fischlin,
and Alejandro Buchmann

Contents

10.1 Introduction

Recent advances in microcontroller technology and miniaturization result in the development of computation capabilities in low-power environments that were seen before only in the area of mobile phone-sized devices. Many approaches that were only available to full-blown computers can now be adapted to these low energy, but capable devices.

Back in 2004, Culler et al. described a container–harbor scenario [4] that is still challenging today. In this scenario, every container has sensor nodes both inside and out. The ones inside are able to detect the cargo via radio-frequency identification (RFID) and can monitor various states of the container itself. The nodes on the outside communicate with the nodes from other containers and build a harbor-wide wireless sensor network (WSN).

Parties like customs officials or shipping agents can use the sensor network to retrieve different kinds of data. For the customs official, it may be interesting to know the origin of the containers that will be unloaded or if some of them got opened en route to plan further in-depth inspections of suspect containers. The shipping agents may be more interested to know if the containers of their company were handled with care or if the refrigeration worked without interruption. In this scenario, it is easy to see the two most important issues, the efficient management of this huge WSN of several hundred or thousands of nodes and security of management operations and node data.

In this work, we show the integration of an easy grouping scheme for WSNs and security measures that support this kind of grouping. The *Scopes Framework* [9] was developed to structure a WSN by defining groups, so called *scopes*, with a declarative language at runtime and to be flexible enough to be adapted to multiple scenarios. Securing structured WSNs includes the secure setup of the scope structures, and the maintenance operations including key management and rekeying. To integrate security in a transparent way for applications, we require an algorithm that can be applied in a distributed setting, is able to cope with the declarative scope definition language, and allows a secure group or session key exchange. We found these features in *ciphertext policy–attribute-based encryption (CP–ABE)* [1], however this algorithm was constructed for desktop computers and not sensor nodes. We adapted CP–ABE to the scopes environment which we call *Secure Scopes (SecScopes)*.

For the work described here we made some assumptions. First, we assume that all measures are taken to secure a deployed sensor node from physical attacks. Second, we assume one-key authority. For the selected scenario, multiple key authorities or at least multiple property issuing authorities make sense, however we defer this aspect to future work.

The contributions of this work are the following: We combined the scopes framework, a structuring mechanism for WSNs, with a suitable security algorithm, CP–ABE, to enable secure structuring and communication in a WSN. With the declarative definition of a scope, it is easy to restructure the network in a secure way. The secured scope still provides the ability of new group memberships, allows rekeying and supports forward secrecy. An additional question to answer is what the minimum resource requirements of low-power WSNs are in order to run the proposed system.

The remainder of the work is organized as follows. In Section 10.2, we show the design of the scopes framework followed by the security models (Section 10.3) and description of CP–ABE and its use on a single node in Section 10.4. In Section 10.5, we describe the network wide integration of the scopes framework and CP–ABE. We then present details of our CP–ABE implementation and its evaluation in Section 10.6. Related work is presented in Section 10.7. Section 10.8 concludes the chapter.

10.2 Structuring WSNs

The scopes framework is a modular system that enables multitasking by introducing a logical structure to the network. Nodes are organized into groups, which are called *scopes*. A scope is the central abstraction provided by the framework to applications. A scope can be defined by means of a logical expression, which must be satisfied by a node to become its member. Once a scope is created, it continues to exist until it is explicitly removed, even as nodes fail or if they temporarily leave and rejoin. The framework takes care of reliably maintaining the scope membership.

The scopes framework is composed of two main modules, ScopeMgr and Scope-Membership. They perform all the management operations and provide APIs to the upper application layer and lower routing layer, see Figure 10.1. For each scope, the ScopeMgr module maintains information like timestamps, scope properties, or ID of the root node. The ScopeMembership module evaluates

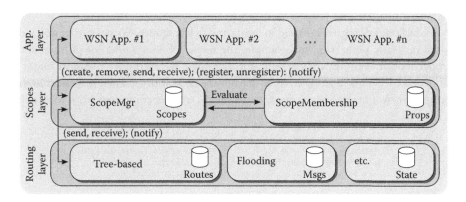

Figure 10.1   Architecture of scopes framework.

the local membership condition, given a scope definition and the node's properties. When a property changes, a scope reevaluation is triggered locally. This yields minimal energy consumption overhead as no communication is needed, only the CPU cost. To further reduce energy consumption for a small loss in responsiveness, this is done in configurable intervals, if needed.

The `ScopeMgr` module interacts with the *routing layer* to send and receive data, and to signal scope membership updates, which can be used to optimize the memory management of routing tables.

To create a scope, applications call an *open* method. The `ScopeMgr` (scope manager) module sends a scope creation request (SCR) to all potential scope members: if it is a new top-level scope, it is sent to all nodes; otherwise it is only sent to members of the parent scope. Whenever a node becomes a member of a scope, this is signaled to applications. Once a scope has been created, a bidirectional data channel between the scope root and its members is established. After finishing its task, an application can delete its scope by *closing* it. A removal message is then sent from the root node to the member nodes. Applications also get notifications about the dynamic scope membership changes. These are asynchronously delivered to them to avoid polling.

To address communication unreliability, topology changes and the update of dynamic properties, the *scope refresh* mechanism is implemented. A refresh message triggers a scope reevaluation, and contains the scope definition. Once a scope is removed, no further refreshes are sent. In case some nodes miss the removal message, configurable local lease will timeout. The scope is then invalidated and marked for deletion and associated resources are released. If the root node of a scope fails, scopes created at other nodes are not affected. Also, if the network gets physically partitioned, the partition that includes the scope root remains unaffected. Due to the missing connectivity to the root, nodes in the other partitions assume a failure of the root node and automatically invalidate the scope to save resources. After network reconnection and upon a refresh, the nodes evaluate their membership again and resume their tasks.

A declarative language is used to create and delete scopes. Scopes can be defined based on static properties, for example, the owner of a container, or on dynamic properties, for example, the content of a container. We exemplify the most important constructs within the container–harbor scenario above, shown in Figure 10.2. Imagine we want to define a scope `ferdexMotion` over containers from company

```
CREATE SCOPE ferdexMotion AS (        CREATE SCOPE dhllCooling (
    company = 'Ferdex' AND                EXISTS SENSOR Temperature AND
    EXISTS SENSOR Acceleration3D AND      containerType = cooling
    (accel_x > 3.0 OR                 ) AS SUBSCOPE OF dhll;
     accel_y > 3.0 OR
     accel_z > 3.0) );
```

Figure 10.2   Declarative scope definitions.

Ferdex that have been dropped or otherwise shaken above a defined threshold, here 3 g. `ferdexMotion` is able to be satisfied by nodes with a 3D-acceleration sensor, whose values are greater than the given threshold, and the user-defined variable `company` equals `ferdex`. Properties included in an expression may have a more static (e.g., `company`) or dynamic (e.g., `accel _ {x,y,z}`) nature. Dynamic properties are very powerful, but their use is expensive since every change results in a reevaluation of the scope membership expression. Such expressions are parsed and flattened into a preorder network format specifically designed for sensor networks.

Finally, scopes can be nested and therefore, form a hierarchy. Nested scopes specialize a scope definition by implicitly restricting the membership condition of their parent-scope. For example, if a scope `dhll` is defined over all nodes belonging to company DHLL, the scope `dhllCooling` could specialize it by selecting those nodes attached to a cooling container. A node that does not belong to any user-defined scope is a member of the artificial *world* scope each node belongs to. Conceptually, nesting scopes contribute clearer definitions and a better organization. Technically, they reduce the communication overhead thus, improving the performance and the energy efficiency.

10.3 Models and Assumptions

Analyzing the security needs of our approach, we identified the following models and requirements to give us a set of basic security parameters.

10.3.1 Network and Trust Model

In this work, a WSN is considered that executes the scopes framework with the further proposed security enhancements. A *trusted authority* is part of the network, but it will only be online when necessary, like in the event of a revocation of a sensor node, or in the bootstrapping phase, to issue the secret keys (SKs) for the sensor nodes. This trusted authority is also holder of the network's master key (MK). To ensure security of cryptographic keys on sensor nodes, measures must be taken to prevent physical compromise of sensor nodes.

10.3.2 Adversary Model

The adversaries are considered to attempt to access a scope they are not authorized to access. Attackers may be external intruders or users of the network. Internal users have access to their nodes, but none of them has a secure key that complies with the intended scope. Attackers can eavesdrop all message traffic in the WSN, they can collude with each other and, for example, exchange keys of their sensor nodes. In general, it is assumed that attackers have an interest in keeping the network up and running.

10.3.3 Security Requirements

To secure a distributed scope, common security aspects, such as data confidentiality and integrity that are also desirable in other sensor network security schemes, should be provided. The following requirements were recognized in conjunction with access control in the scopes framework.

- *Secure group key exchange.* The basis for scope-level access control is to be able to securely exchange scope keys to establish a trusted group communication. This scheme should provide a secure exchange of symmetric scope keys to enable efficient cryptography between scope members.
- *Group-level access control.* The scope-level access control extends the secure key exchange and adds access control. This ensures that only authorized sensor nodes join a scope and exchange data. This supposes that the access control scheme provides a system to precisely describe the properties of potential scope members and define expressive attributes.
- *Collusion resilience.* As described in the adversary model, users of the sensor network may cooperate to access an unauthorized scope. Therefore, it is vital to include in the access control scheme mechanisms so that colluding unauthorized users do not gain an advantage over what they can get from individual attacks.
- *Forward secrecy.* Managing the sensor nodes and to be able to exclude malicious or malfunctioning nodes is an important functionality in most scenarios. To support this, the access control scheme should ensure that after revoking a malicious node, it may not be able to join a secured scope again, even though the properties of the access structure are satisfied by the node's properties. Also, the access to other scopes that the node is already a member of should be prohibited.

10.4 Securing Sensor Nodes

As mentioned above, we were looking for a security algorithm that works in a distributed environment without additional infrastructure and nicely integrates with our scopes property structure. The elliptic curve-based *CP–ABE* proposed in [1] provides SKs with associated attributes that are used for the sensor nodes, and ciphertexts associated with a policy, which are used to transfer the group keys for a scope. These policies reflect only the static part of the scope properties in the process of creating a new scope. This is due to the fact that the properties in the SKs cannot be changed. A dynamic property is replaced by a check for a matching sensor. The check for the property itself is done outside CP–ABE.

Bethencourt et al. defined five basic operations in CP–ABE of which we use four: *setup, key generation, encryption,* and *decryption.* Delegation is not used here.

- *Setup.* The setup does not take any input. With the given elliptic curve parameters it generates the public parameters public key (PK) and an MK.
- *Key Generation.* The key generation takes the MK and a set of attributes *S* and computes a matching SK. The SK includes each attribute from *S* with a corresponding component needed for the decryption process.
- *Encryption.* The encryption takes as parameters the PK, a message *M* and an access policy A with a set of attributes. The message *M* will be encrypted to the ciphertext CT in a way that only with a matching SR which satisfies the policy A can it be decrypted. The CT contains implicitly the specified access policy.
- *Decryption.* The decryption takes a CT including an access policy A and an SK which represents a set *S* of attributes. If the attributes of SR satisfy the access policy A, the CT will be decrypted and the message *M* will be returned.

The above-mentioned access policy in the CT is created using a tree structure. The tree is built by so-called threshold gates, where each interior node is a threshold gate and each leaf in the tree is associated to an attribute of the policy. With threshold gates it is quite easy to model "AND" or "OR", for example, 2of2 or 1of2. Now, to be able to decrypt a ciphertext, there has to be an assignment of attributes from the SK to nodes of the tree such that the threshold gates are satisfied. This can be efficiently evaluated and if the tree is satisfied, the computation to retrieve the message M is executed. Together with the threshold gate approach, comparisons like <, >, =, ≤, and ≥ can be modeled. CP–ABE is collusion resistant; this is typical in attribute-based encryption [8,16]. A general problem that comes with attribute-based encryption is key revocation [2,18], since the encrypting sender cannot verify that an encrypted message can only be decrypted by a nonrevoked receiver and with CP–ABE there may be many SKs that match a given policy.

To be able to support forward secrecy, it is important to be able to revoke specific SKs. To do so, we introduced a revocation mechanism proposed in [21] to CP–ABE. With this mechanism we are able to update parameters of the MK, PK, and SK and so exclude all SKs that were not updated from the ability of decrypting CTs generated after the parameter update. The distribution can be done by the scopes framework.

To save ROM space on an individual node, we can limit the functionality in the following way: decryption has to be available on all sensor nodes in the network, encryption has to be available on all nodes that may create a new scope to be able to encrypt a new SCR. Setup and key generation can take place outside the sensor network. The MK can be kept outside the network, which is important to security, as no one can retrieve the MK to generate arbitrary keys.

10.5 Securing Structured Sensor Networks

The CP–ABE algorithm is based on elliptic curve cryptography. Although elliptic curve cryptography was shown to be feasible in sensor networks [11], its application here is more expensive because we also use pairings (see Section 10.6.3). This means, it is too costly to encrypt data traffic with it. For this purpose, symmetric algorithms are better suited. Additionally, the advanced encryption standard (AES) algorithm is available as a hardware module on nearly all sensor nodes that support IEEE 802.15.4 radio communication, for example, Tmote Sky or Z1. We use CP–ABE for secure key exchange and access control to scopes, and the scope key is used for AES encryption to enable secure communication inside the scope.

The integration of CP–ABE and the scopes framework into SecScopes introduces changes in three phases. First, in the bootstrapping phase all the necessary keys are generated and distributed. The second phase is scope creation, where the properties are divided into static and dynamic ones and the static attributes are used to encrypt the scope key with CP–ABE. The dynamic properties are encrypted using the scope key, adding a static placeholder for the type of property. For our previous example this would be a HAS _ ACCEL3D. The scope properties for scope ferdexMotion from Table 10.2 are separated in the dynamic part of the acceleration sensor and the others, where the others are used to encrypt the scope key. The last phase is the maintenance phase, where new group keys are distributed with the regular scope refreshes and data is encrypted.

Bootstrapping is performed before the sensor node is deployed. In our scenario this would be before the container is shipped, for example, while it is loaded with goods, information is updated on the node. First, the MK and the public parameters PK are generated with the setup routine, if not already present. Then, an SK is generated for the container node with suitable properties. Here we would have properties like company or containerType or types of sensors that are available. The SK and PK are then stored on the node, ready for deployment.

The scope creation procedure remains unchanged, but the SCR message is extended. Before it was mainly the parsed scope property, with some additional fields. Now there are two parts. First, there is the CP–ABE encrypted scope key for the AES encryption of data, and second the dynamic property portion of the scope properties and the additional fields encrypted with the scope key. This ensures that only nodes that can extract the scope key can access the full scope property string and so join the scope.

Scope maintenance describes the time span between the scope creation and deletion. Here, keeping the scope up and running, and handling leaving and joining nodes are the major tasks of the framework. The scope refreshes, used to keep a scope alive, are similarly structured as the SCR, so they undergo the same changes.

An important feature for lasting security is the ability to exchange the encryption keys used. Every message encrypted using the scope key has an ID attached that identifies the used encryption key. With the scope refreshes we can distribute a new encryption key just by replacing the current key with a new one. For each scope two keys can be stored, this enables a slow roll out of a new key and after some time switching the used key. This mechanism can be done automatically in predefined intervals. If a scope key is compromised, rekeying this way is still a secure way to exclude unauthorized nodes. To achieve forward secrecy, this mechanism is combined with CP–ABEs revocation, this way an attacker can be excluded from the scope's framework by revoking its CP–ABE key and renew the used scope key to exclude the node from previously joined scopes.

Deletion of a scope is only allowed at its root node. The root node sends a deletion message to all nodes of a scope. After this message is sent, nodes delete all information of the scope. If a scope times out because it did not receive a refresh for a period of time, the scope is assumed dead and all the data is deleted to free the occupied resources.

Our approach allows controlling access to a scope by setting a policy that a node has to fulfill to retrieve the scope key from the SCR. If the node can decrypt the scope key, it can further check if its dynamic properties match the ones in the scope properties and so join the scope. With scope refreshes, new scope keys can be introduced in predefined intervals and so a rekeying is performed. With the secure key exchange and rekeying of symmetric keys the data traffic inside a scope can be efficiently encrypted.

10.6 Implementation and Evaluation

We perform a series of experiments to evaluate our CP–ABE and SecScopes in WSNs. Therefore, we run tests on a range of platforms available in sensor networks, including Tmote Sky, Z1, Stargate and an Intel-based PC and in a live WSN testbed.

We will first have a short look at the implementation and afterwards compare the results of our underlying libraries (TinyECC) and algorithms (Tate pairing) with values from their respective publications [10,11] to ensure proper functionality and performance. Getting to CP–ABE we will evaluate the algorithm on a range of common platforms in sensor networks and take a closer look at the overall performance. We try to give an impression of the energy consumption, as this is important in WSNs. For the SecScopes framework we study the influence of security on stability and reliability of the scope creation process.

10.6.1 Implementation

Our implementation of the CP–ABE algorithm uses the Contiki operating system, but we have also tested the code on an embedded Linux and a desktop PC. The code is written in ANSI C for portability reasons and it does not depend on specific functions provided by Contiki, so it can be easily integrated in other projects based on C and its standard library. It implements all the basic features of CP–ABE: setup, key generation, encryption, and decryption.

To save implementation efforts, we ported the well-known TinyECC [11] library to C, so it can be used with Contiki. This library includes standard mathematical operations with large integers, on elliptic curves and was extended by Kampanikis with a Tate pairing algorithm, which is described in [10]. We also had to extend the library with some smaller functions, like generation of random elliptic curve points, and the Tate pairing to compute an uncompressed pairing [17]. In our port of TinyECC we still provide the ability to choose from the different optimization switches provided by the library. Although the assembler code used for optimizing the supported platforms is available, we could use it only on the Tmote Sky nodes in our experiments. Finally, CP–ABE's policy syntax is slightly changed to make it more lightweight for sensor networks and revocation was added.

The SecScopes framework is also based on Contiki and implements the aforementioned features. It comes with two routing algorithms implemented for it. First, a simple flooding as a baseline for our evaluation and a tree-based algorithm that uses a routing tree to send messages from the scope members to the root node and a selective flooding that is restricted to the parent scope to send data to the member nodes.

10.6.2 Methodology and Setup

For the evaluation of TinyECC on Contiki we reconstructed most of the parameters of the original tests to only have changed the underlying operating system and programming language. Therefore, we used the Tmote Sky as test platform and ran the test cases that came with the TinyECC distribution with the same elliptic curve. To save effort we selected the best and the worst case in the sense of runtime. So all optimizations of TinyECC are enabled or all optimizations are disabled. This will give us an overview of the performance which we will then compare to the original results.

The original results for the Tate pairing algorithm are only given for Imote2 and Micaz nodes, so we cannot directly compare these. We decided to use the Tmote Sky as our smallest platform as a reference to estimate the performance. We again used the test case and elliptic curves that came with TinyECC to evaluate the performance. This time with all optimizations enabled.

For the results of CP–ABE we evaluated it on four platforms, namely Tmote Sky, Z1, Stargate, and a PC. On the Tmote Sky we ran Contiki 2.4. We used the Tmotes at 4 MHz and at the Tmotes maximum frequency of 8 MHz. Similarly to the Tmote Sky nodes, with the Z1 we did tests at 8 and 16 MHz to compare the results. On the Z1's we used Contiki 2.5 RC1, as the Z1's are supported since this version. The Stargate was chosen due to its similarity with the Imote2 sensor nodes. The experiments were executed on an embedded Linux at 200 and 400 MHz. As a desktop PC-sized device we did tests on a 2200 MHz Core2Duo, where only one core was used for our experiments.

Because of the diversity of platforms, operating systems, and build environments, not all experiments and analysis could be performed on all platforms.

The most important part of our results for CP–ABE is the runtime evaluation. It was done on all platforms and with three different sets of properties shown in Table 10.1. The *small set* has a property `attrib1` and the policy uses the same. The

Table 10.1 Property and Policy Sets

Sizes	Small	Medium	Large
Properties	1	5	8
Policy	2	5	7

medium set has five properties and its policy is `attr1 attr3 attr9 attr4 2of4`. The *large set* has seven properties and its policy is the exploded version of `attrib4 > 6`. This settles in eight terms for an 8-bit unsigned integer. We did also experiments on ROM size and static and dynamic RAM usage and used a script from TinyOS to retrieve this data. This was done on all platforms except the PC. The energy consumption on the Z1's is measured by Contiki's ENERGEST framework [5] and on the Stargate it is calculated via the runtime and the specified supply voltage. To measure energy consumption on the Z1 and Stargate platforms, we calculated $U \times I \times t$ (or $P \times t$ for the Stargate platform). Both equations return milli-joules and are based on voltage (U), current (I), and execution time (t). The power (P) equals $P = U \times I$.

The Tmotes are the only platform available in this evaluation where TinyECC provides assembler code to speed up computation, so we show a performance comparison for this platform. The Z1's MSP430 uses an advanced assembler instruction set that seems to be incompatible, as we could not get the code to run on them. Also the X-scale processor on the Stargate seems to have little differences from the Imote2's.

For our experiments we obtained a 192 bit and a 512-bit curve that are both suitable for pairing operations on elliptic curves. The 512-bit curve is rated secure for pairings, but puts too much load on low-power sensor nodes, so that we also did the experiments with the 192-bit curve.

All our CP–ABE experiments were repeated 10 times and the results were averaged. The code used was the same for all platforms, except for methods provided by the operating systems and in the case of TmoteSky the assembler optimizations.

Tests for the SecScope framework are run on a live 32 Tmote Sky office testbed that is evenly distributed over 10 office rooms, and were repeated 5 times. The results shown are averages of these tests, where the 192-bit curve was used. Due to space limitations of the Tmotes flash the framework on the nodes uses a dummy CP–ABE, which excludes the code for the actual de-/encryption, but leaves everything else intact, like the memory footprint, network message size, and matching of attributes and policies. Also the AES encryption is unchanged.

10.6.3 TinyECC on Contiki

The evaluation of TinyECC on Contiki was done on a Tmote Sky with 160-bit elliptic curve parameters. We will first show our results in Figure 10.3b, and then compare them to the reported values in [11] in Figure 10.3d.

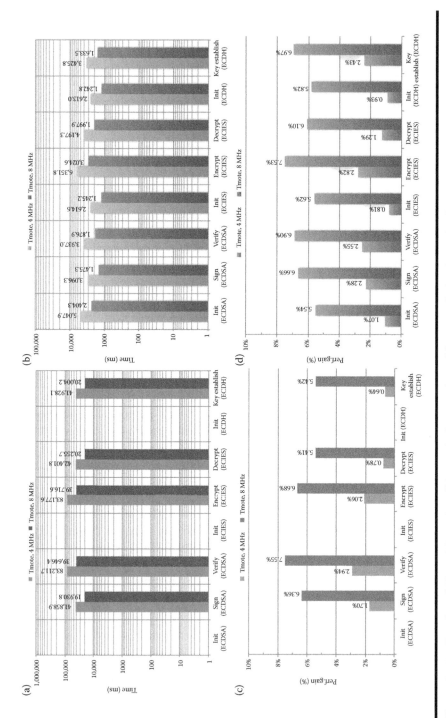

Figure 10.3 Results of TinyECC on Contiki. (a) Timing results, without optimizations; (b) timing results, with optimizations; (c) performance gain, without optimizations; and (d) performance gain, with optimizations.

Figure 10.3b shows the timing results of all three Tiny ECC algorithms. These are ECDSA, a signature scheme; ECIES, an encryption/decryption algorithm; and ECDH for key agreement using elliptic curves. In Figure 10.3a, all optimization switches of TinyECC were turned off. This results in zero runtime, as no pre-computations are required. In this experiment encryption/verification, and signing/decryption take almost the same amount of time. Comparing this to Figure 10.3b where all the optimizations are enabled revealed that the encryption is here the most expensive operation. Also the performance increased for all operations more than 10 times. In Figure 10.3b with all optimizations the initializations also need a considerable amount of time. For example, the initialization of ECDSA is the operation with the second highest costs overall. The advantage here is that the initialization operation only has to be performed once. Having a look at all values the results for 8 MHz are constantly 50% better as the 4 MHz results.

Comparing the results with the originally reported ones can be seen in Figure 10.3d. First, we note that in all cases and operations we have at least a slight gain in performance. For the 4 MHz results the gain varies between 0.93% and 2.82%, which is not very significant and at least confirmed the results from the original publication. However, the 8 MHz results show a gain of 5.41% to 7.55%. This in contrast is a significant gain achieved by porting the code to C and Contiki.

Here we were able to show that TinyECC is performing slightly better on Contiki than on TinyOS and we made sure that our base library is fully functional.

10.6.3.1 Tate Pairings in TinyECC

For the Tate pairing evaluation we had no numbers to directly relate to. But still the results in Figure 10.4 show for a 192-bit Tate pairing computation that it is

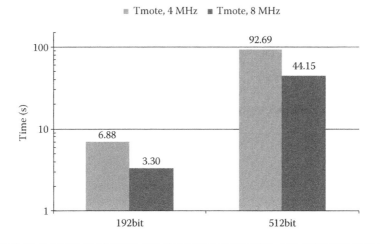

Figure 10.4   Timing of Tate pairing with all optimizations.

roughly as expensive as the encryption operation from basic TinyECC. The time needed for the computation of a pairing in a 512-bit curve is 92.7 s and 44.2 s, for 4 and 8 MHz respectively, more than 13 times longer than for the small curve. Depending on the application these large values may be acceptable in a WSN. For example, for a monitoring application that runs for a long time it may not matter if from time to time such a long operation takes place.

10.6.4 CP–ABE in WSNs

We started our evaluation of the CP–ABE algorithm with detailed measurements of execution time on all platforms and, as far as possible, with all property sets and curves we presented earlier. After that we give numbers for the energy consumption on the Z1 and Stargate platforms.

Table 10.2 shows the computations used over all CP–ABE operations. We defined three classes of operations: multiplications, additions, and pairings. As we have seen in the last section, pairings are very expensive, and so they are in a separate class. The setup operation is the only constant one. It uses 1 pairing and 4 multiplications. Key generation depends on the number of properties to be included in the SK and it uses 2 multiplications and 1 addition, plus 3 multiplications and 1 addition per property. Encryption and decryption highly depend on trees, the policy tree for encryption and the minimal satisfied policy tree for decryption. The tree used for decryption is smaller or has the same size, this depends on the actual policy and property set used to satisfy the policy. The base costs for encryption are 2 multiplications and 1 addition. For each node except the tree root we add 3 multiplications and 1 addition, plus 2 multiplications and 1 addition for every leaf node. Leaf nodes represent properties and inner nodes threshold gates. For decryption the costs are a bit higher, but in general this tree is smaller than for encryption. The base cost is 1 pairing and 2 additions, plus 6 multiplications and 2 additions

Table 10.2 CP–ABE Theoretic Cost Estimation

Operations	Setup	Keygen	Enc	Dec
Pairings	1	0	0	1
				leaf: 2
Mult	4	2	2	0
		prop: 3	(n − 1): 3	(n − 1): 6
			leaf: 2	leaf: 1
Add	0	1	1	2
		prop: 1	(n − 1): 1	(n − 1): 2
			leaf: 1	leaf: 1

for all nodes except the tree root, and 2 pairings, 1 multiplication and 1 addition for all leaf nodes.

Next we will have a closer look at Figures 10.5 and 10.6, which show the execution times on all platforms with a 192-bit elliptic curve. In our experiments we used all optimizations of TinyECC, except the assembler optimization, as this is only available on Tmotes and except the curve optimizations, as these are not available for the curves we use for pairing computations. Figures 10.5 and 10.6 show results of the small and medium property sets.

The execution time of setup was constant over all sets. This was to be expected, as the same keys with the same sizes had to be generated. Looking at the key generation we can see, that this was by far the most expensive operation in our setup, even though it is not using pairings. For a Tmote with 4 MHz and the medium set, it took more than 2.5 min to compute the SK. Comparing the three sets for key generation and encryption we see a steady increase of execution time, as these two functions compute values on properties and the complete policy tree.

Comparing the different platforms, we recognized the different execution times. Of course for a desktop PC this computation is no big deal, so we got 21 and 47 ms for the key generation, as the slowest, and 15 ms for setup, as the fastest operation. The Stargate also achieved very good performance values with times between 0.4 and 1.3 s. For Tmotes and Z1s we see a constant improvement; doubling the frequency halves the execution time. At 8 MHz, where we had data for both platforms, timings were almost equal. The small difference arises from the different compilers used, as the compiler for the Z1s produced smaller code than the one of the Tmotes.

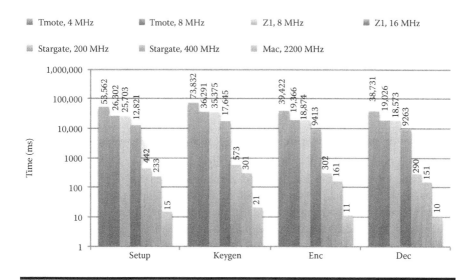

Figure 10.5 CP–ABE execution time for 192-bit curves, small set.

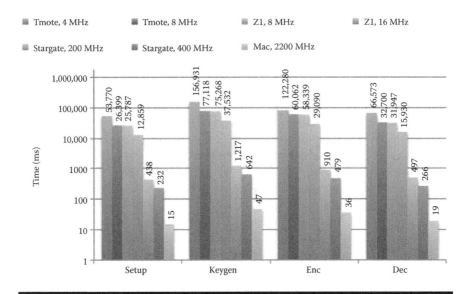

Figure 10.6 CP–ABE execution time for 192-bit curves, medium set.

Besides this, the execution times for these small nodes were very high. Keep in mind that we only have to decrypt a new scope key once in awhile for rekeying or, when a new scope is created these may still be acceptable values. Of course nodes should run at full power for this kind of operation.

In getting these numbers, we wanted to check if the provided assembler code for several methods inside the TinyECC library could speed up execution times. As we only had a working assembler code for the Tmotes, we compared execution times with and without the assembler code for the small property set (see Figures 10.7 and 10.8). In general, using the assembler optimization speeds up the execution, but the gain we achieve is small. On average we gain 1 s. With execution times between 40 and 150 s this is no significant improvement.

For execution times of 512-bit curves, Figures 10.9 and 10.10 show the results for the medium and large sets. The CP–ABE operations show a similar footprint in this experiment as for the 192-bit case, just the absolute timings are higher. Looking at them reveals for the Stargate, values of 1848, 5360, 4801, and 2266 ms for the four operations of CP–ABE with the large property set. At decryption, execution time increased from small to medium, but for the large set it took the same time as for the medium. This was because, for decryption, only properties that were needed to satisfy the policy are included in computation. These trees were of the same size for the medium and large sets. This shows that the algorithm is definitely feasible on more capable platforms, and can handle even larger property sets with tolerable execution times.

An important issue in WSNs is energy consumption. Therefore, Figures 10.11 and 10.12 show energy consumption measurements for Z1 and Stargate. For Z1 we

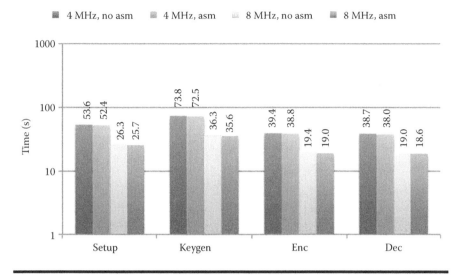

Figure 10.7   Influence of assembler optimization on execution time, small set.

used the ENERGEST framework [5] to retrieve the data for our calculations. For the Stargate we used the timings we got from the previous experiments. We omit the results for Tmotes here as they use the same microcontroller family as the Z1s and their current draw by frequency scales linear, so the results for 8 MHz are the same for both platforms. As mentioned before, the datasheet for Z1s microcontroller specifies a current draw of 0.41 mA for 8 MHz and 0.82 mA for 16 MHz

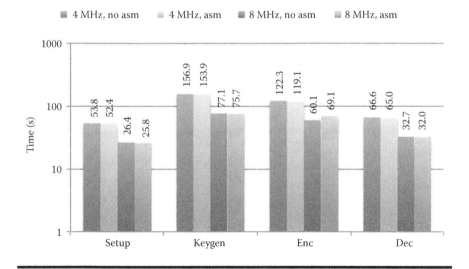

Figure 10.8   Influence of assembler optimization on execution time, medium set.

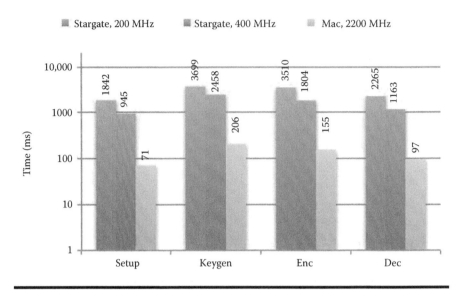

Figure 10.9 CP–ABE execution time for 512-bit curves, medium set.

what results, with the previously observed doubling of performance with a doubling of frequency, in the same energy consumption for 8 and 16 MHz, shown in Figure 10.11. For the Stargate we got 178 mW for 200 MHz and 411 mW for 400 MHz from the datasheet. The results in Figure 10.12 are all much smaller than the ones estimated for Z1s; even the values for the large properties set is smaller

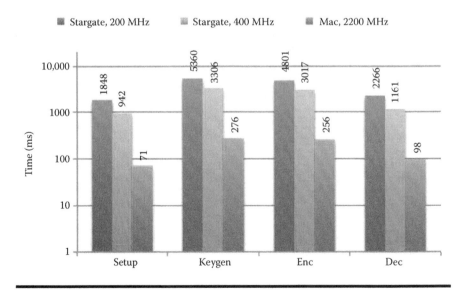

Figure 10.10 CP–ABE execution time for 512-bit curves, large set.

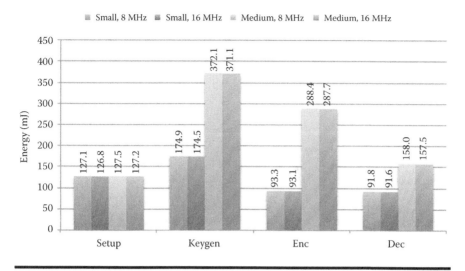

Figure 10.11   CP–ABE energy consumption. Z1, ENERGEST results.

than the values for Z1s on the medium set. On the other side, the current draw in sleep mode for the Stargates microcontroller is 100 times higher than the one from the Z1s. What is also nicely reflected here is the same performance of decryption in the medium and large sets as in the runtime measurements. Here you also have roughly the same energy consumption.

After evaluating the runtime characteristics, Figures 10.13 and 10.14 show the ROM sizes of the implementation. The data shown is for all platforms except the

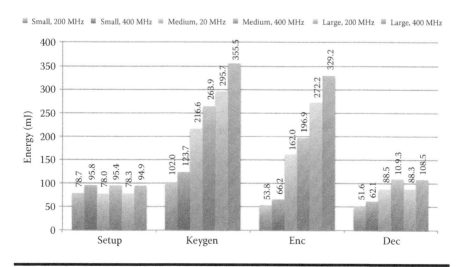

Figure 10.12   CP–ABE energy consumption. Stargate, runtime-based measurement.

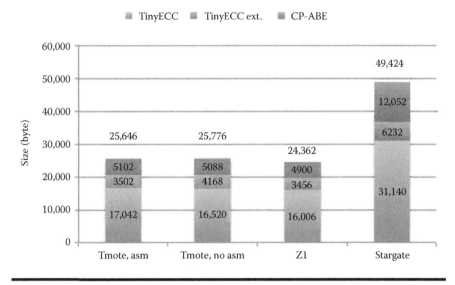

Figure 10.13 CP–ABE ROM size, 192-bit curve parameters.

PC where a tool to get the sizes was missing. Also compilation for Z1s with the 512-bit curve did not work out because of the static RAM demands. Three sizes are shown per entry. These are TinyECC for all code that comes from it. TinyECC extension includes all extensions done to TinyECC; these are minor additions for CP–ABE and also the Tate pairing. Finally, CP–ABE is all code implementing

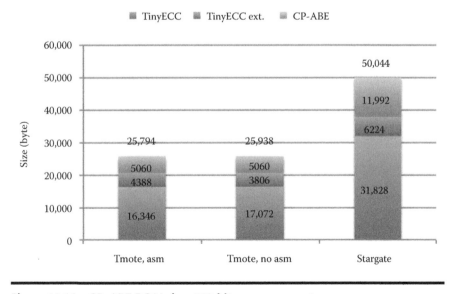

Figure 10.14 CP–ABE ROM size, 512-bit curve parameters.

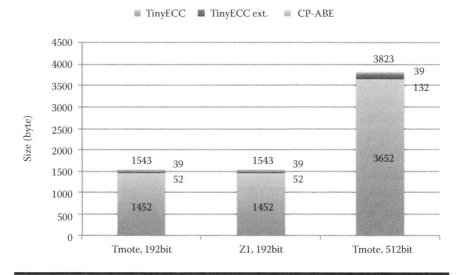

Figure 10.15 CP–ABE RAM size.

this algorithm. This categorization is also used for static RAM evaluation (Figure 10.15). First thing to mention about ROM size is, compiling the code for 192 bit or 512 bit introduces no big changes, so the size is stable. In general, ROM sizes for Z1 and Tmote should be the same; however, the code for Z1s is compiled using a more recent major version of the compiler that employs better code optimizations, which results in slightly smaller code. With the Stargate we change from a 16-bit to a 32-bit architecture that also introduces a doubling in ROM size. The 25 kB for the Tmotes occupy more than half of its ROM, on the Z1s the ratio is better, and here only a quarter is used. For the Stargate with 32 MB flash the 50 kB are no problem.

For the evaluation of RAM we decided to split it to static RAM, which is included in the sensor node image, and dynamic RAM, which contains only the keys used. Actually, all static memory is allocated by TinyECC and used for the optimizations. CP–ABE or the Tate pairing use only a small amount of static RAM. In Table 10.3, we show the different key sizes for the different elliptic curves and platforms that have to be held in RAM. Besides the MK and public parameters, which have a static size, the SK and ciphertext depend on the number of properties or elements of the policy tree. For example, the SK in the large set, for the 512-bit curve on the Stargate has a size of 2384 bytes.

10.6.5 SecScopes Framework

The tests we perform reflect the stability of scopes, the reliability of creating them, and also show the influence of the introduced security measures in the results. Therefore, we create a scope over all nodes in our testbed and keep it active for 70 s.

Table 10.3 CP–ABE Key Structure Size

Size in Byte	192-Bit Tmote, Z1	192-Bit Stargate, PC	512-Bit Stargate, PC
MK	78	84	204
Public parameters	208	224	544
SK	56	64	144
Property elem.	108	120	280
Ciphertext	108	120	316
Policy node	130	164	324

The refresh interval is set to 6 s and the timeout for the scope is set to 20 s. This means, if a scope does not receive a refresh for this time it is assumed dead and is removed. After the 70 s the scope is deleted via a deletion message. If this message is not received the scope is deleted after the timeout interval. Of course, these tests omit the time delay introduced by the CP–ABE operations. The delay is the time for the first encryption of the data for the scope-creation message and the decryption of it at the nodes in the network for evaluation.

The scopes creation test is shown in Figure 10.16a and b. The first figure refers to the scopes framework without the security measures. It can easily be seen that the minimum and maximum values are very close. So we have a very stable creation mechanism. Also, the deletion message is received by all nodes, as all of them close exactly after the deletion message is issued. Compared to Figure 10.16b, with the security measures, the difference can clearly be seen. The minimum and maximum curves are separated in the creation and deletion phases. This means, with the introduction of the larger scope creation and refresh messages (in this test a refresh is built by three messages compared to one in the first case), the reliability of the creation mechanism is reduced. Also the success ratio of transferred data can be slightly influenced, because of the higher load of messages in the network. As

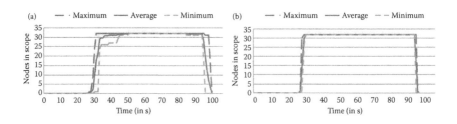

Figure 10.16 Scope creation tests using tree routing. (a) No security measures, and (b) with security measures.

can be seen in the deletion phase, the transfer of the deletion message is delayed by some seconds. Still, it is received by all nodes, and they all close the scope right after receiving it.

When using the key-refresh mechanism, the time needed for execution is the result of three steps. First, a new random scope key is created at the root node and the data for the scope refresh is then newly encrypted. This is the most expensive operation, as shown above. After that it is just sending the new data in a new refresh cycle, which takes for our test with three hops roughly 60 ms per hop including sending and processing the messages. Finally, the time for decrypting the message on member nodes has to be added for the complete key refresh cycle.

The routing tree we observed in our tests for the used tree routing protocol is a tree with a very high fan out. This means a node has many children and only some connections to other hops. Here we have, for example, a root node with 10 child nodes and one that is the next hop. This one has also 8 children and one that is connected to one other. So most nodes are in one or two hop distance and only some require three hops.

10.7 Related Work

The concepts of scopes were originally conceived to provide structuring capabilities in the Rebeca Publish/Subscribe system [13]. The scopes framework offers a powerful tool to select nodes in a sensor network, which is also pursued by other approaches. In generic role assignment [15], a set of roles is distributed throughout the nodes, which must decide to which of these roles they belong. A node choosing a role may trigger other nodes to reevaluate their role selection. Hood [20] provides a neighborhood programming abstraction through which physical neighbors may interact with each other. By leveraging the omnidirectional broadcast medium, Hood efficiently supports a node's one-hop neighborhood. Our scheme for selecting nodes is most similar to that of logical neighborhoods and its Spidey language [12]. Here, nodes export attributes which are comparable to node properties in scopes. While with Spidey the user has to decide how much effort the system should employ in searching for matching nodes, this is transparent in scopes since a scope is network-wide. Also, logical neighborhoods have a transient character, since they have to be specified in each message, while scopes have a more permanent character and are stored locally. Finally, logical neighborhoods resort to a tight integration with the routing algorithm, which we intentionally avoided in scopes in order to offer alternatives depending on the desired traffic type.

In [21] an access control scheme (FDAC) for data in WSNs based on ABE is proposed, but without using a structured network. It uses KP–ABE and therefore, the data is tagged with properties and the users have an access structure in their private keys that limits their ability to decrypt data. The approach we are following in this chapter is not just selecting nodes, but also managing and communicating in a secure way. SM–Sens [6] is a middleware that provides security through different roles and a cluster-based routing scheme. Nodes are assigned one of the roles as regular node, cluster head, gateway, or guardian. Guardian nodes represent a second channel to a location in the network; they periodically listen to the data traffic and send this data to the base station where the regular data and the one from a guardian node is compared and then decided if nodes have to be excluded from the network. In [7] a group-based security scheme is proposed, with intra- and intergroup security, but the groups are distinct and based on node locations, so it creates a set of cluster-like groups. It covers the whole network and does not provide a way to influence the clustering by an application. In the scopes framework a core element is the description of the group. In this approach there is no description what node may join which group. SMEPP Light [19] was derived from an embedded P2P middleware. It provides groups and has group- and network-level security. Both are secured via symmetric encryption. Inside these groups nodes interact via a publish/subscribe mechanism. A node can subscribe for events of another node of the group. Relevant events are then automatically received and processed. Groups can be open or closed. Another approach is RiSeG [3]. The grouping is based on a ring-topology for each group, where a joining node is appended to the end of the group. This seems to ease grouping, but for networks with multihop communication the unordered addition of nodes may lead to a significant overhead for sending messages. The security in RiSeG is centered on the base station and the group controller, the needed cryptographic keys are deployed in a bootstrap phase, but cannot be managed afterwards like in the scopes framework. It is the closest match to our approach so far.

10.8 Conclusions and Future Work

In this chapter, we have presented SecScopes, an extension to the scopes framework to enable access control, secure key exchange, and rekeying via insecure wireless media. We provide methods for a secure structured WSN that can easily be restructured using a declarative language to define scopes.

We also ported CP–ABE to the Contiki platform and showed its performance in an extensive evaluation. As a result we can confirm that CP–ABE can work in low-power sensor nodes, but with a limited set of attributes and smaller elliptic curve sizes. These limits result from the small amount of RAM available on these nodes. If the high-execution times observed can be tolerated depends on the scenario. On sensor network platforms with more RAM and more capable processors, for example, ARM/X-Scale, CP–ABE can reach its full potential at the cost of higher energy consumption in the long term. As we have shown the computation itself is more efficient on these platforms than on the low-power sensor nodes.

The evaluation of the combined SecScope framework shows that introducing the proposed security measures does not interfere with the stability of a created scope, but introduces a higher load on the WSN as the scope refreshes are much larger compared to using no security. This also means a higher possibility of collisions when sending or forwarding messages. Still, these costs are acceptable compared to the gain in security. The inclusion of security and structure in a sensor network requires additional resources on the sensor nodes. The challenge here is the lack of enough resources (RAM and flash space) on the low end of today's low-power sensor network nodes to accommodate the cryptography and the structuring in addition to the application. However, this is possible on slightly more powerful sensor nodes that are expected in the near future. Our results show that these future large-scale deployments are feasible.

Open issues are the support for multiple key authorities or at least multiple property issues that we plan to investigate further, like [14].

References

1. J. Bethencourt, A. Sahai, and B Waters. Ciphertext-policy attribute-based encryption. In *Proceedings of SP*, Berkeley, CA, June 2007.
2. D. Boneh and M. Franklin. Identity-based encryption from the weil pairing. In Joe Kilian, (Ed.), *Advances in Cryptology (CRYPTO 2001)*, Springer LNCS 2139, Berlin, Heidelberg, pp. 213–229, 2001.
3. O. Cheikhrouhou, A. Koubaa, G. Dini, and A.bid, M. Riseg: A ring based secure group communication protocol for resource-constrained wireless sensor networks. *Personal and Ubiquitous Computing*, 15, 783–797, 2011.
4. D. E. Culler and H. Mulder. Smart sensors to network the world. In *Scientific American*, Scientific American, Inc., pp. 84–91, June 2004.
5. A. Dunkels, F. Österlind, N. Tsiftes, and Z. He.Software-based on-line energy estimation for sensor nodes. In *Proc. of Emnets IV*, Cork, IR, June 2007.
6. L. H. Freitas, K. A. Bispo, N. S. Rosa, and P. Cunha. Sm-sens: Security middleware for wireless sensor networks. In *Proc. of GIIS*, Hammamet, Tunisia, June 2009.
7. M. Garcia, J. Lloret, S. Sendra, and R. L. Gilaberte. Secure communications in group-based wireless sensor networks. *IJCNIS*, 2(1), 8–14, 2010.
8. V. Goyal, O. Pandey, A. Sahai, and B. Waters. Attribute based encryption for fine-grained access control of encrypted data. In *Proc. of ACM CCS*, New York, USA, 2006.
9. D. Jacobi, P. E. Guerrero, I. Petrov, and A. Buchmann. Structuring sensor networks with scopes. In *Adj. Proc. of EuroSSC*, Zurich, CH, October 2008.
10. P. T. Kampanakis. Identity-based cryptography: Feasibility & applications in next generation sensor networks. Mastersthesis, NCSU, Raleigh, NC, 2007.
11. A. Liu and P. Ning. TinyECC: A configurable library for elliptic curve cryptography in wireless sensor networks. In *Proc. of IPSN*, St.Louise, USA, April 2008.
12. L. Mottola and G. P. Picco. Logical neighborhoods: A programming abstraction for wireless sensor networks. In *Proc. of DCOSS*, San Francisco, CA, June 2006.
13. L. Fiege, M. Mezini, G. Mühl, and A. Buchmann. *Engineering Event-Based Systems with Scopes*, ECOOP 2002, Springer LNCS 2374, Malaga, Spain, pp. 309–333, June 2002.

14. S. Müller, S. Katzenbeisser, and C. Eckert. On multi-authority ciphertext-policy attribute-based encryption. *Int. Conf. on Information Security and Cryptology (ICISC)*, Hong Kong, 2008.

15. K. Römer, C. Frank, P. J. Marron, and C. Becker. Generic role assignment for wireless sensor networks. In *Proc. of SIGOPS*, Leuven, BE, September 2004.

16. A. Sahai and B. Waters. Fuzzy identity based encryption. In *Advances in Cryptology—Eurocrypt*, volume 3494 of LNCS, Springer, pp. 457–473, 2005.

17. M. Scott and P. S. L. M. Barreto. Compressed pairings. In *In Advances in Cryptology—Crypto 2004*, Springer-Verlag, pp. 140–156, 2004.

18. A. Shamir. Identity based cryptosystems and signature schemes. In *Advances in Cryptology—CRYPTO*, volume 196 of LNCS, Springer, pp. 37–53, 1984.

19. C. Vairo, M. Albano, and S. Chessa. A secure middleware for wireless sensor networks. In *Proc. of MobiQuitous*, Brussels, BE, 2008.

20. K. Whitehouse, C. Sharp, E. Brewer, and D. Culler. Hood: A neighborhood abstraction for sensor networks. In *Proc. of MobiSYS*, Boston, MA, June 2004.

21. S. Yu, K. Ren, and W. Lou. FDAC: Toward fine-grained distributed data access control in wireless sensor networks. *TPDS*, 22:673–686, 2011.

Chapter 11

Anomaly Detection in Wireless Sensor Networks
Challenges and Future Trends

Muhammad Usman, Vallipuram Muthukkumarasamy,
Xin-Wen Wu, and Surraya Khanum

Contents

11.1 Introduction

Over the course of recent years more and more sophisticated threats have emerged for wireless sensor networks (WSNs) due to innovations and advancements in the technology. Therefore, prevention-based techniques cannot serve as a stand-alone security solution for WSNs. Appropriately designed detection-based techniques can complement the prevention-based techniques to offer more stable and robust security solutions to WSNs. Anomaly detection is one of the detection-based techniques that can not only be employed for providing security services such as intrusion detection, it can also be used for other application areas such as fault/error detection and events/objects detection [1]. Anomaly detection is a well-researched area in traditional wired and wireless networks, but due to unique characteristics of WSNs, these security solutions cannot be readily deployed in these low-resource networks. In this section, first we define basic terminologies of anomaly detection, then describe types of anomalies in WSNs, and finally illustrate the performance evaluation metrics.

11.1.1 Terminologies

An anomaly is also known as an outlier, exception, peculiarity, surprise, irregularity, difference, discordant, and containment. The two most commonly used terms are anomaly and outlier, which are interchangeably used in existing literature [2]. Anomaly detection is the process of discovering nonconventional data patterns. In this subsection, we briefly describe a few basic terminologies of anomaly detection [3].

■ *Normal profile:* The normal profile (also known as profile) is the combination of behavior characteristics of subjects with respect to objects in terms of statistical metrics and models of observed actions. Profiles can be automatically generated when initialized from templates. Profiles can also be manually generated or defined on the basis of initial behavior. The anomaly-detection profiles have two types: static and dynamic. The static profile is usually established by an administrator and it remains constant unless updated or changed. The dynamic profile is changed as new events are observed. The network can be trained using any of the three types of machine-learning algorithms (i.e., unsupervised, semi-supervised, and supervised) to define normal profiles at the time of network deployment. The choice of algorithm is based on factors such as size of the network, available node and network resources, and anomaly-detection methodology. Defining a profile which encompasses all possible normal activities is a challenging task. In many applications the normal behavior keeps evolving with the passage of time. Therefore, sometimes it is very difficult to characterize the current notion of normal behavior of the subject.

■ *Subject:* The subject is initiator of activity on a target system. Typically, subjects are nodes and users.

■ *Object:* These are the resources managed by subject. It may include but not limited to resources such as memory, battery, processor, and transceiver.

■ *False attack stimulus:* An event that triggers an alarm when actually there is no anomaly.

■ *True attack stimulus:* An event that triggers an alarm when there is an actual anomaly.

■ *Alarm filtering:* Verifying alarms for removing false positives from the system. This process is usually carried out at the testing phase.

■ *Evasion:* The process in which intruder changes timing and/or pattern of its activities to avoid being detected. Evasion makes it very difficult to detect anomalies.

■ *Noise:* Alarm events that are accurate but not harmful, like scanning.

■ *Tuning:* The process of updating and/or adjusting an anomaly-detection system.

■ *Confidence value:* The measure of the anomaly-detection system's ability to correctly detect and identify anomalies in the network.

11.1.2 Types of Anomalies

Anomalies can be classified into three broad categories in WSNs: data, node, and network anomalies [4]. The data anomalies occur in data that is gathered by one/multiple sensors of a node or multiple sensor networks (SNs). The sensor readings collected from a specific proximity exhibit some sort of similarity and/or consistency. The inconsistency in collected readings is considered as data anomaly. There can be several reasons for data anomalies such as errors, different types of attacks, sudden node restart, nonsynchronizations, and hardware/software faults. Data anomalies are further categorized into: temporal, spatial, and spatiotemporal anomalies. The temporal anomalies are the inconsistencies that are caused with respect to the time continuum. These anomalies are typically discovered on

Table 11.1   Types of Anomalies with Potential Causes

Anomalies	Potential Causes
Data	Errors, different types of attacks, sudden node restart, nonsynchronizations, and hardware/software faults.
Node	Resource degradation, node failures and restarts, physical attack, and malfunctioning.
Network	Loss of connectivity, intermittent connectivity, routing loop, failure of network components, and broadcast storms.

the basis of time-series analysis. The spatial information is of paramount importance in some application domains such as target monitoring. The inconsistencies in spatial information may lead toward spatial anomalies in sensor data. The spatiotemporal anomalies are discovered by analysis of both spatial and temporal aspects of data.

The node anomalies may be caused due to resource degradation, node failures and restarts, physical attack, and malfunctioning. In physical attacks, nodes are captured and they are replicated through cloning. The network anomaly is instigated by unavailability or interference in transmission frequency (jamming attack), failure of network components such as gateway, node failure, or malicious attacks. The network anomalies are also caused by exploiting weaknesses in routing protocols [5,6]. The anomalies caused due to loss of connectivity, intermittent connectivity, routing loop, and broadcast storms are also known as network anomalies. The types of anomalies along with few potential causes are listed in Table 11.1 [4].

11.1.3 Performance Evaluation Metrics

Training and testing are two key phases that are performed before deployment of an anomaly-detection system. Typically, a dataset is used as an input for an anomaly-detection system to learn about normal and anomalous behavior of node/network. The testing is performed to evaluate the potential performance of an anomaly-detection system. The performance evaluation of anomaly-detection schemes is generally based on time complexity, energy analysis, and detection accuracy analysis. The time complexity is an important factor for performance analysis. Generally, the anomaly-detection algorithms with $O(n)$ time complexity are considered as good. Energy analysis is also a vital performance measure. The communication is a most expensive operation in WSN, where energy may fall up to the fourth power of the distance [7,70]. Therefore, energy consumption by processing operations in general and communication in particular should be optimized.

The receiver operating characteristic (ROC) curve is a graphical way to show the detection accuracy of an anomaly-detection system. Typically, ROC is based on true positive rate (TPR) and false positive rate (FPR) [8]. TPR and FPR can be derived from Equations 11.1 and 11.2 respectively.

$$TPR = \frac{TP}{(TP + FN)} \tag{11.1}$$

$$FPR = \frac{FP}{(FP + TN)} \tag{11.2}$$

where TP is true positive, FP is false positive, TN is true negative, and FN is false negative. TP is the anomaly correctly classified as an anomalous value. FP is the normal activity that is incorrectly classified as an anomaly. TN is the normal activity that is correctly classified as normal. FN is the anomalous activity that is incorrectly classified as normal.

The rest of the chapter is organized as follows. Section 11.2 describes methodical taxonomy of anomaly-detection schemes for WSNs. We highlighted limitations of existing schemes and possible future research directions in Section 11.3. Finally, Section 11.4 concludes the chapter.

11.2 Taxonomy of Anomaly-Detection Schemes for WSNs

Over the years, the research community has adopted concepts from several disciplines such as machine learning, statistics, and artificial intelligence (AI) and applied them in the field of anomaly detection in WSNs. We have classified existing literature into four broad categories: machine learning, AI, statistical, and other schemes. Figure 11.1 illustrates the anomaly-detection taxonomy in WSNs. Now we describe the functionality of each category and also present the critical review of some representative schemes.

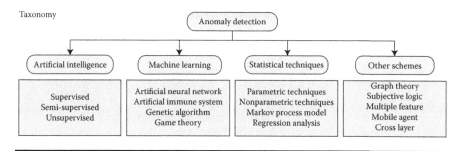

Figure 11.1 Anomaly detection in WSNs: A taxonomy.

11.2.1 Machine Learning-Based Anomaly Detection

Machine learning algorithms allow machines to learn from data. This learning process extracts useful patterns from data and makes predictions on new data based on previously obtained knowledge [9]. The machine learning-based techniques are mainly used for training of anomaly-detection schemes. Three classical approaches are used in training for anomaly-detection systems: supervised, semisupervised, and unsupervised [2]. In this subsection, we critically review a few machine learning-based anomaly-detection schemes for WSNs.

11.2.1.1 Supervised Learning

Supervised learning system labels training set on the basis of pairs (feature, label), denoted as $[(a_i, b_i),...,(a_n, b_n)$ where $i = 1$ to $n]$. The objective of supervised learning is to label "b_i" of each instance of new input with feature "a_i."

A process of supervised learning can be categorized either as regression or classification. It is called regression if $b \in R$ if b has "real" values. On the other hand, it is known as classification if "b" has "whole" values [10]. Though the probability of classifying data on the basis of supervised learning is high and it gives accurate results, it is expensive and a time-consuming approach. Furthermore, it is not always possible to obtain labeled datasets in advance.

Support vector machine (SVM) is one of the supervised learning anomaly-detection methods that is based on classification and regression analysis. SVM is a nonprobabilistic binary linear classifier. SVM algorithm creates a hyperplane that splits data into two classes with the maximum possible distance. The support vectors are the main point of interest that lies near to the separating lines. Figure 11.2 represents a typical SVM graph in a two-dimensional plane that classifies two

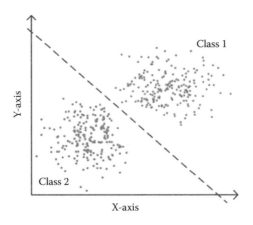

Figure 11.2 A linear SVM in two-dimensional plane.

categories in a dataset. An SVM-based distributed anomaly-detection scheme is proposed by Rajasegarar et al. [11]. In this scheme, the local quarter sphere is calculated by each child, SN, and then sent to their respective parents. The respective parents are responsible for local anomaly detection. Then parent nodes calculate the global radius on the basis of received local radii. After that, the parent node sends a global radius to all of its children to discover anomalies in their proximities. Though this scheme is capable of discovering both local and global anomalies, the computation of only a quarter sphere may miss some important information that can be used in estimating anomalies.

Another supervised approach for anomaly detection is based on Naïve Bayes algorithm [12]. This scheme uses the concept of Bayes classifier to identify anomalous nodes in the network. The Bayes classifier assumes that the presence of a specific feature of a class (either normal or anomalous) is unrelated to the presence of any other feature in a specific class. The anomalous behavior of node is discovered using several parameters such as power, communication, and computational capability. The simulation result demonstrates the Naïve Bayes classifier offers high-detection rate and low FPR. Nevertheless, in this scheme, authors have not suggested the way to overcome/compensate the inherent limitations of supervised learning scheme.

11.2.1.2 Semi-Supervised Learning

The semi-supervised learning approach uses large amount of unlabeled data with small amount of labeled data to construct good classifier [13]. This approach is useful in applications where availability of abnormal data is not easily available, such as medical and aeronautical engineering applications. However, the major problem with the semi-supervised approach is that it is not easy to incorporate new type of normal and/or abnormal data in already running training phase. The semi-supervised techniques are not commonly used in WSNs for anomaly detection because of the difficulty in obtaining the training data that encompasses all the possibilities of normal and anomalous behavior of node/network in advance.

11.2.1.3 Unsupervised Learning

The unsupervised learning scheme uses nonlabeled dataset having features $(a_i, \ldots, a_n)$ for building classifier [14]. This approach requires a threshold or certain criteria in order to discover anomalies in the dataset. The selection of an appropriate threshold is a challenging task especially in dynamic WSN applications where the notion of normal behavior of a system keeps evolving. A typical approach for unsupervised learning is clustering the observations and then detecting anomaly on the basis of distance from either the centroid or boundary of the cluster [15]. This approach is suitable for those applications where labeled data are unavailable.

One of the prominent unsupervised learning schemes that is used for anomaly detection in WSN is based on K-means clustering. In this approach, K-means of each cluster are created. Then these means or centroids are associated with the training set. Each individual point in the training set is associated with the nearest centroid. One anomaly-detection scheme based on K-means clustering is proposed in [16]. In this scheme, each node accumulates the local data to make a norm profile. Then these norm profiles are sent to respective cluster heads (CHs). The CH collects all local norm profiles and creates a global norm profile. Then CH sends this global profile to all SNs in its proximity. The anomaly detection is performed on the basis of a global profile. The K-mean approach is used to optimize clustering. This approach involves significant communication overhead as CH sends global profiles to all member nodes.

Miao et al. [17] proposed an unsupervised anomaly-detection scheme based on principal component analysis (PCA) [17]. PCA is a mathematical process that converts correlated variables into linear and uncorrelated variables through orthogonal transformations. The converted linear, uncorrelated variables are known as principal components (PCs). It should be noted that in the PCA-based method, the number of PCs must be equal or less than the original variable. The researchers used a distance-based approach for reduction in the feature size from multivariate to univariate to optimize the anomaly-detection process. A coefficient of error is introduced to take into account the information loss that incurred during conversion from correlated multivariate to linearly uncorrelated variables. Though this scheme reduces the training overhead, it incurs additional cost for converting interrelated multivariate features into univariate data.

11.2.2 AI-Based Anomaly Detection

The experts of AI define it as a field of studying and designing intelligent agents. They further elaborate the concept of intelligent agents as a system that perceives from its surroundings and takes appropriate action to increase the probability of the agent's success. The overall theme of AI is to simulate human intelligence in machines in order to carry out different tasks [18]. Over the years, several AI-based anomaly-detection schemes were suggested for WSNs. The basic theme of these schemes is to borrow human intelligence and other biological systems in an anomaly-detection process. We now perform the critique of a few representative AI-based anomaly-detection schemes in WSNs.

11.2.2.1 Artificial Neural Networks

Adaptive resonance theory (ART) is a type of neural network. The key idea behind ART is to learn new things without forgetting those that are previously learned. ART has a sequential learning capability [19]. ART neural network is able to update already existing labels or creates a new category for new data samples if it does not belong to any of the existing classes. Walchli et al. [20] presented an anomaly-detection mechanism using fuzzy ART neural network. Fuzzy ART is a descendent of the ART neural network. Unlike traditional ART, fuzzy ART is also capable of handling analogous input samples. The fuzzy ART system is based on two layers: comparison layer *F1* and recognition layer *F2* with *n* and *m* neurons, respectively. Here *n* represents input attributes and *m* denotes categories. A sensitivity threshold/vigilance factor ρ evaluates the similarity among given input and learned categories. A typical fuzzy ART neural network is depicted in Figure 11.3 [20]. This scheme uses fuzzy ART to process, classify, and compress data using time-series data model of SNs. Nevertheless, this scheme is capable of updating already existing label class but it is resource hungry, especially in terms of memory and processing. Limiting memory use by the fuzzy ART may put an adverse effect on efficiency of the scheme.

Li et al. [21] suggested fuzzy ART-based methods to predict missing observations in WSNs. The suggested imputation scheme takes into account the spatial–temporal information of WSN. The idea is based on the hypothesis that the network is highly interrelated in terms of time and space. The R-squared and Pearson correlation coefficient are used for the verification of space correlation. Then a modified fuzzy ART algorithm is used to estimate missing data. The spatial–temporal imputation algorithm is shown in Figure 11.4 [21]. Though this scheme is simple yet effective, it has a limited scope as it only focuses on the spatial–temporal anomalies. Furthermore, in this scheme, the use of fuzzy ART just identifies one type of anomaly, which is resource consuming and may not be suitable for low-resource WSNs.

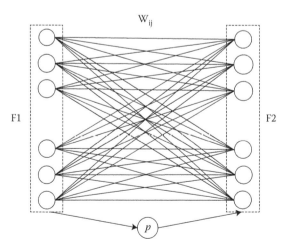

Figure 11.3 A fuzzy ART neural network.

Algorithm 2.1 — (Spatial – Temporal Imputation Algorithm)
1. **for** *each missing sensor node* SN_m **do**
2.   **for** *every sensor node* **SN_n** *within one – hop* **SN_m** **do**
3.   **if** *sensing distance among* SN_m *and* SN_n < α **then**
4. a_m^x *is the most common reading of* a_n^x
5.   **else**
6. $a_m^x \, a_m^{x-1}$
7.   **end if**
8.   **end for**
9.   **end for**

Figure 11.4 A spatial–temporal imputation algorithm.

11.2.2.2 Artificial Immune System

De Castro and colleagues define artificial immune system (AIS) as [22],

> AIS are adaptive systems that are inspired by theoretical immunology and observed immune functions, principles, and models that are applied for problem solving.

The use of AIS for anomaly detection in WSN is a recent approach. Fu and colleague recently suggested a biologically inspired anomaly-detection model for hierarchical WSNs [23]. This anomaly-detection model fuses the advantages of AIS and the fuzzy theory. The model is based on three components: local danger sensing, cosimulation, and global recognition. In case a hostile node launches an attack, the member nodes in the proximity of the antagonist node sense the danger and send signal to the personal area network (PAN) coordinator that acts as a decision maker (DM). The DM announces danger zone as per received danger signals. Then danger zone nodes cooperatively form antigens. The DM establishes and maintains a receptor pool. The receptor pool generates antibodies as per antigen. Then using the principle of negative selection, anomalies are discovered in the network traffic. This model uses physical and media access control (MAC) layer parameters for anomaly detection. Authors used fuzzy logic to generate danger signals. Although this scheme is adaptable and flexible as compare to approach, it is resource hungry and may not be capable of distinguishing among anomalies caused by attacks and faults/errors.

Salmon et al. [24] suggested an anomaly-detection system for WSNs using danger theory immune-inspired technology. The basic idea of the study is to use dendritic cell algorithm (DCA) for anomaly detection in WSNs. This model is based on four phases: (a) collection phase, (b), analysis phase (c), decision phase, and (d) reaction phase. Authors mapped several computational components into their immune-inspired counterparts, shown in Table 11.2 [24]. During the first phase, signals and antigens are captured. In second phase, the analysis is performed on both captured signals and antigens in order to generate output signals. The output

Table 11.2 Computational and Biological Mapping

Computational Components	Biological Components
Attack	Pathogen
Sensor node	Tissue
Node coverage	Danger area
Antagonism identity information	Antigens
Decision manager component in sensor lymph	Lymph node
Intrusion-detection manager and context manager component	DCs
Intrusion-countermeasure component	B and T Cells
Countermeasure triggered by network defense system	Antibody

signals identify the maturation state of dendritic cells (DC). In the third phase, the DC classifies the antigens as normal or adversary and also identifies their degree of anomaly. In the fourth and final phase, B and T cells are triggered to counter the antagonist node. This scheme works on the principle of the human immune system and has the capability of self-defense, but the quality and quantity of antigens and antibodies need to be further verified. Further study is also required in order to analyze the performance of this model against different types of attacks/anomalies.

11.2.2.3 Genetic Algorithm

A genetic algorithm (GA)-based anomaly-detection scheme is proposed by Rahul et al. [25]. The suggested scheme employs GA for an anomaly-detection mechanism on the basis of battery status, packet information, data utilization, and quality of service (QoS) conformity. SNs are categorized as: CH, intercluster router, common SN, and inactive node (sleep mode). The base station (BS) employs GA-based competing fitness functions to optimally select a CH or intercluster router which works as a local monitoring node. The local monitoring node observes the activities of neighbor nodes on the basis of different parameters: (a) packet modification or dropping, (b) received signal strength value, (c) packet transmission rate, (d) delay in response, and (e) fake transmission from compromised nodes. This scheme optimizes the existing anomaly-detection methodology, but the scalability of solution requires further study.

The gene expression programming (GEP) is a biological evolution mimic for computer programming. A GEP-based anomaly-detection scheme for WSNs is proposed by Honglei et al. [26]. In this scheme, a GEP-based network traffic-prediction model is employed for time-series analysis of normal network traffic. The anomaly-detection model is comprised of five tuples: (P_m, F_s, MF, O_p), where,

P_m = parameter set, F_s =function set, M= variable set, F= fitness function, and O_p = operation set. The parameter set is made up of population size, P_{size}. The gene may contain either head H, or tail T, if and only if, Equation 11.3 is satisfied.

$$\text{Gene} = \begin{cases} H \\ T \\ |H| = h \\ |T| = h + 1 \end{cases} \tag{11.3}$$

where $|H|$ is an element number of H and $|T|$ is an element number of T. The function set F_s is based on simple operators that are +, , /, and *. The variable set M, is based on two factors: time t, and length of traffic i. The fitness of nth individual can be estimated from following equation:

$$F_n = R^2 = 1 - SSE/SST \tag{11.4}$$

Here SSE and SST are the sum of square of error and sum of square of total respectively. Note that the fitness function F is based on the standard R-square that returns square of Pearson product correlation efficient. The operation set O_p is based on several functions such as, mutation, insert sequence, root insert sequence, gene transposition, and one/two-point recombination functions. *GEP*-based anomaly-detection algorithm is depicted in Figure 11.5 [26]. This scheme improves efficiency of traditional time-series models by eliminating the need of prior knowledge of network traffic parameters. However, the energy consumption of such an intensive process may have an adverse impact on the low-energy budget of nodes.

11.2.2.4 Game Theory

In game theory, the anomaly-detection mechanism is considered as a game between the anomaly-detection module and antagonist. One such scheme is suggested by

---

Algorithm 2.2 – (GEP based Anomaly Detection Algorithm)
Input: Dataset M_p, where the length of data is i.
Output: Time series mode, relative error, and correlation coefficient.
1. *As per i, transform data into i + 1 time series column data.*
2. *Intitialize population based on Genes that are developed using anomaly detecton j process*
3. *Evaluate individual fitness.*
4. *stop: if maximum threshold of number of generation is achieved, otherwise continue.*
5. *Choose the best.*
6. *Perform every operation in Op.*
7. *Goto step 3.*

Figure 11.5   GEP-based anomaly-detection algorithm.

Agha et al. in [27]. The risk factors such as past behavior of adversary, types of known attacks, and so on, are identified in this scheme and, on the basis of these factors, anomaly detection is carried out. Having a fixed cluster, the adversary can have three options: attack cluster, do not attack cluster, and attack another cluster. On the other hand, the detection module has only two types of responses: defend cluster or defend a different cluster. This makes a matrix of 2 * 3 between two players. Maximizing the profit of each player (i.e., Nash equilibrium) is a challenging task in game theory-based anomaly-detection schemes. Furthermore, it is not always possible to list all the possible states of the system due to dynamic nature of WSN topology.

Reddy [28] suggested another game theory-based anomaly-detection mechanism for identifying anomalous nodes in WSNs. This approach uses a zero sum game for anomaly detection in the forwarding data path. Suppose a network cluster with n nodes in a forwarding path, from which n (2) are malicious nodes. Suppose there is a nonmalicious node q (1) between two malicious nodes. Let σ be the percentage of the nodes that are randomly chosen as check point and ϵ (2) be the designated acknowledgment points in the path of packet. Then the probability of detecting malicious node P_d can be derived from following equation:

$$P_d = P_y = 1 - P_{ack} \tag{11.5}$$

where, P_y is the likelihood of dropped packets from acknowledgment points and P_{ack} is the likelihood of acknowledgments at source from acknowledgment points. The P_{ack} can be evaluated from Equation 11.6.

$$P_{ack} = \sum_{i=1}^{m} P_{em}(i) \tag{11.6}$$

where $P_{em}(i)$ is the likelihood of the packets that are dropped by malicious nodes. The likelihood of packets dropped between chosen check points can be estimated from Equations 11.7 and 11.8.

$$P_{m,n} = P_{ack}(m) - P_{ack}(n) \tag{11.7}$$

$$P_{m,n} = \sum_{j=1}^{m} P_{ack}(j) - \sum_{j=1}^{n} P_{ack}(j) \tag{11.8}$$

By using Equations 11.5 through 11.8, the number of antagonist nodes in a specific path can be determined from following equation:

$$xC_{P_{ack}} = \frac{x!}{(P_{ack}!) * (x - P_{ack})!} = P_d \tag{11.9}$$

This scheme is helpful in determining the suspicious node that is the source of anomalies in the network traffic. This scheme is based on a cooperative zero sum game. However, the appropriate selection and localization of a nonmalicious node between two or more malicious nodes is not an easy task without prior information of the current network state. Furthermore, in a cooperative setting, including even one adversary may spoil the whole process of identifying the malicious node.

11.2.3 Statistical-Based Anomaly Detection

The statistical anomaly-detection techniques in WSNs are generally model-based approaches. In a statistical model, the relationship between variables is formalized by the means of mathematical equations. A typical statistical model must hold two key properties: systematic variation and randomness [29]. The model is supposed to be statistical, if it is stochastically but not deterministically related. A typical classification of statistical-based anomaly-detection schemes for WSN is parametric versus nonparametric schemes [1].

11.2.3.1 Parametric Techniques

The parametric schemes are based on the assumption that the knowledge about underlying data distribution is available. Another key assumption regarding parametric techniques is the normal distribution of data [30]. Typically, mean and standard deviation are used to distinguish between normal and anomalous data. This model is subject to update with aging data. Parametric techniques can be categorized into two types: Gaussian and non-Gaussian-based approaches.

Gaussian models assume that the data distribution is based on Gaussian (bell shape or normal) distribution. An ecological-based anomaly-detection scheme is proposed to detect events, recognize measurement errors, and infer missing readings in application data [31]. The basic idea of the scheme is to discover spatio-temporal correlation in sensor readings. This scheme uses prior information of measurements to discover anomalies in data, as each node compares its current and prior measurements with the neighboring nodes. The major drawback of the scheme is that it only focuses on one-dimensional anomalies that are caused due to irregularities in spatiotemporal data. Furthermore, this technique is application oriented which limits its scope.

Wu et al. [32] presented another Gaussian-based statistical anomaly-detection technique. This technique determines anomalies in two dimensions: (a) discovering an anomalous sensor, and (b) identifying the event boundary. The suggested technique identifies spatially correlated anomalies in sensor readings by comparing them with the median of neighboring node readings. The basic idea is to localize algorithms in order to discover anomalous SNs and event sensors along the event boundary. The anomalous sensor readings and signals indicating an event could not be distinguishable, but anomalous readings are geographically independent.

Furthermore, the sensors that are observing similar phenomenon should be spatially correlated. On the basis of this hypothesis, authors suggested two algorithms for identifying anomalous SNs and detecting event boundaries. In the first algorithm, each SN calculates the difference among its readings and the median of neighboring nodes readings. In case of a large deviation, SN is declared as an anomalous. The second algorithm is based on simple observation that for each event, there must be two zones, each corresponding to a respective sensor. Then the deviation of both zones is calculated individually as per the first algorithm. If the median reading of one zone is significantly different from that of other nodes then SN is declared as anomalous. Both algorithms entail threshold values for comparison. The authors suggested an adaptive mechanism to calculate threshold. The adaptive mechanism is based on ROC curve analysis. Apparently the use of adaptive ROC curve analysis looks very useful to maintain adaptive threshold values, but the iterative process of determining threshold is energy consuming. Therefore, it may not be viable for low-energy sensors. Furthermore, the precision of both anomaly-detection approaches is not high because authors did not consider the temporal correlation among the readings.

In non-Gaussian approaches data is not normally distributed. Jun et al. [33] designed a non-Gaussian anomaly-detection model. This scheme detects spatio-temporal anomalies in SNs. The anomalies are assumed as noncorrelated both in terms of time and space. The outliers are modeled as impulsive noise using symmetric α-*stable* ($S\alpha S$) distribution. Each member node of a cluster performs local anomaly detection and rectifies temporally anomalous data, transmitting it to the respective CH. Then CH receives the rectified data to further discover and correct spatial anomalies. This anomaly-detection model provides two advantages: (a) reduction in communication cost, and (b) enhancing the overall quality of aggregated data. Despite these advantages of the suggested model, $S\alpha S$ distribution may not be viable for most of the SNs due to their dynamic topology.

11.2.3.2 Nonparametric Techniques

The nonparametric statistical techniques do not take prior assumptions about underlying data distribution [34]. These techniques make fewer assumptions, therefore, these are more suited to the dynamic nature of WSNs. The nonparametric techniques typically employ either of two approaches: Histogramming and Kernel functions [1]. Histograms provide probability distribution of a specific variable for a certain range [35]. The height of the bar in a histogram is directly proportional to the frequency of data at a specific range. The histogram uses continuous data, unlike bar graphs where category-based data is used.

A histogram-based anomaly-detection approach in data-centric WSN applications is presented by Sheng et al. [36]. Their suggested approach collects histograms of sensor readings at the sink node instead of collecting raw readings in order to reduce the communication cost. The sink node collects data from the network and extracts the normal pattern. Then anomalies are discovered by collecting more histograms and comparing them with predefined fixed threshold. The main flaw of the scheme is that it only considers univariate data. Furthermore, recollecting histograms incurs network communication overhead that is not suitable for low-resource SNs.

Alternatively in nonparametric techniques, statistical anomaly-detection kernel function-based approaches are used. Typically, a kernel is a weighting function that is used for estimation techniques in nonparametric statistics [37]. Over the years, several kernel function-based anomaly-detection schemes were proposed for WSNs. Palpanas et al. [34] suggested a kernel function-based scheme to identify anomalies in sensor data. The suggested scheme requires no prior knowledge about basic distribution in sensor reading. Each SN locally detects an anomaly by employing kernel density estimator. The reading is said to be anomalous if it is less than the node's specified bounds. The main drawback of the scheme is its only suitable for a one-dimensional data domain, as single threshold is not suitable for multidimensional data. Furthermore, this approach does not consider variations in data.

Subramaniam et al. [38] extended the work of Palpanas et al. [34]. Their suggested model is specifically designed for anomaly detection in multidimensional data. The model is based on two steps to compute global anomalies. In the first method, each SN discovers local anomalies, as suggested in [34]. Then the anomaly report is forwarded to its parent node to determine anomaly. This process continues until the sink determines global anomalies. Whereas, in a second method each node uses the scheme reported in [39] for detecting global anomalies. In this technique, each node discovers global anomalies by using a global estimator that is obtained from the sink. The empirical evaluation of this work demonstrates high-detection rates by optimally utilizing resources. However, this scheme is incapable of detecting spatial anomalies.

11.2.3.3 The Markov Process Model

A few state-of-the-art schemes may either be a parametric or nonparametric anomaly detection. Typically, the Markov process model-based anomaly-detection scheme employs both parametric on nonparametric data models. The Markov process model considers events as state variables. The Markov process model uses transition matrix to define the frequencies of transition between states [40]. The anomalous behavior is determined by comparing the output and input between two consecutive states, respectively. This approach is useful in sequential models. Sometimes a variant of the Markov process model known as a hidden Markov model (HMM) is also used. . In this model, the states and transitions are hidden and only the supposed productions are visible [41]. A typical HMM with observable and hidden states is shown in Figure 11.6.

Over the years, a few Markov process-based anomaly-detection schemes were suggested in WSN literature. Paschalidis and Chen [42] put forward an anomaly-detection model for WSNs to discover both temporal and spatial anomalies. The suggested model used tree-indexed Markov chains to characterize spatial structure of the

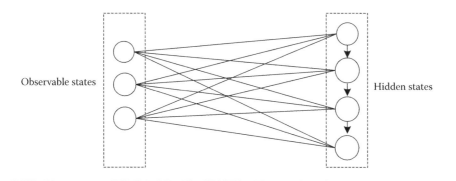

Figure 11.6   A hidden Markov model.

network. In Markov chains, which are indexed by a tree, the value of time replaces integers at vertices of the tree. Then the large-deviation technique is employed to differentiate among previous anomaly-free traces with that of current measures. This model employs decision rules to distinguish among normal and anomalous activity. This approximation leads to discover both spatial and temporal anomalies. However, due to dynamic nature of the WSN's topology, it is not possible to define decision rules that encompass all possible states, which limits the scope of this model.

Gao et al. [43] designed an indirect correlation-based anomaly-detection system for WSNs. This work is based on the hypothesis that the indirect correlation between multiple attributes of sensor data can be employed to model normal behavior of SN. The Markov chain is applied for computing the state transition probability matrix. Then this matrix is used to discover anomalies. This model is implemented in a testbed of 17 TelosB motes. Although this approach is capable of simultaneously detecting multiple types of anomalies and demonstrates high-detection rates, it incurs significant computational overhead that is not suitable for low-resource SNs.

In recent years, the research community has also suggested a few HMM-based anomaly-detection schemes for WSNs. Zheng and Baras [44] proposed sequential anomaly detection for mobile SNs in WSNs. This model is based on multiscale analysis of the traffic generated by SNs. The authors argue that the length of time duration in which network traffic is captured can influence results, therefore, analysis of data in different timescales is imperative. The discrete wavelet transform (DWT) is employed for multiscale analysis of data. DWT is also used for decorrelating stochastic processes. The probabilistic model for wavelet coefficient is constructed using multiple-level HMM to improve detection rates. The expectation maximization (EM) algorithm is used to estimate parameters of the model. The EM algorithm also discovers changes in predicted HMM by analyzing variation score. The variation score is computed as symmetric relative entropy among current and previously estimated HMM. Then anomalies are discovered by identifying changes in variation score. This process is suitable for discovering timescale anomalies in network data, but it is an energy-expensive approach which limits its scope.

11.2.3.4 Time-Series Model

Time series is a nonrandom order of data measurements. Unlike other statistical approaches time-series models work on the assumption that successive measurements are taken at equal intervals of time [45]. In general, time-series analysis has two objectives: (a) discovering the nature of model behavior on the basis of sequence of observations, and (b) predicting future behavior of the model. Singh et al. [46] presented autoregressive integrated moving average (ARIMA)—(p, d, q)-based anomaly-detection model. Here p, d, and q are the order of autoregressive terms, integrated, and number of lagged predicted errors in the forecasting equation respectively. The p, d, and q must be non-negative integers. ARIMA is used for forecasting stationary time series. Statistical functions such as mean, constant, and variant remain constant over the time in ARIMA. This model is based on two algorithms. The first algorithm corrects anomalous data of each SN at BS using ARIMA forecast values at any point in the time series, as illustrated in Figure 11.7 [46]. Then the second algorithm is employed for anomaly detection in SN's data with a 95% confidence interval, as depicted in Figure 11.8 [46]. The anomaly detection is carried out in noncooperative fashion. This scheme has two key flaws. First, typically WSNs have a dynamic topology, therefore, forecasting stationary behavior over the time series is unsuitable for them. Second, temporal and contextual internode relationship can also be considered in order to make anomaly-detection process more robust. Chuah and Fu [47] proposed an anomaly-detection method using time-series analysis in electrocardiogram (ECG) applications. In their suggested model, physiological sensors are deployed to monitor different physiological

---

Algorithm 2.3—(Finding appropriate ARIMA model and forecasting)
1. *Determine existing autocorrelation structure.*
2. *Compute AR(p), order p is calculated on the basis of AIC criterion*
3. *Computer residual of step 2 MA(q) by the order of q*
4. *Perform residual analysis.*
5. *Predict future values and correct anomalies*

Figure 11.7   Finding appropriate ARIMA model and forecasting algorithm.

Algorithm 2.4—(Anomaly Detection)
1. *Run Algorithm 2.3*
2. *Locate 95% confidence interval as [μ ± 1.96 a, here /μ is forecast value and σ is standard error.*
3. *Verify null hypothesis.*
4. *reject or accept values on the basis of step 3.*
5. *The rejection in step 4 demonstrates anomalous value.*

Figure 11.8   Anomaly-detection algorithm.

measures such as pulse rate, heartbeats, and oxygen saturation of elderly people. Sensors send periodic readings through a wireless link to a personal computer for analysis. The scheme is based on adaptive window-based discord discovery (AWDD) method. This method highlights the abnormal pulse rate, heartbeats, and oxygen saturation level. After discovering anomalies, the abnormal data is sent to the remote station where a physician can further diagnose the problem and take appropriate action(s). This scheme is proposed at a high level. Furthermore, the synchronization among received data and submitted data is not considered.

11.2.3.5 Regression Analysis

Regression analysis is a statistical process to find relationship among different variables. Typically, regression analysis is carried out in order to discover causal effects of one variable on other variable(s) [48]. The variables could either be interrelated and/or nonrelated. An autoregression-based malicious node-detection scheme in WSNs is proposed by Curiae et al. [49]. This model employs detectors that are installed at BS to filter SN's measurements. For anomaly detection, SN's current and previous measurements are compared by employing autoregressive predictors. The variation of current values from a certain threshold is considered as anomalous which results in activation of the decision block. The decision block then takes appropriate action against the anomalous node. Although this model is useful in forecasting associations among different variables; however, selection of an appropriate prediction criterion is not an easy task.

Kim et al. proposed a nonparametric regression-based anomaly detection method for heterogeneous SN [50]. A living environment (office room) is used as a case study in which irregular day-to-day events are predicted by employing Bayesian network and regression analysis. The suggested method learns normal behavior of SNs and evaluates the degree of outlier for each data instance as per the estimated variance derived from results of the learning stage. The performance evaluation of this model is carried out in an office room environment in different conditions. The experimental location equipped with light and motion sensors to collect data with little or no prior knowledge. Then outliers are identified on the collection data as per the learning stage outlier threshold. This scheme is simple yet effective for small-scale networks. However, the scalability of such a scheme is questionable, especially when the diverse nature of SNs form a large-scale network.

11.2.4 Other Schemes

In this section, we discuss several other anomaly-detection schemes that may not belong to any of the above-mentioned categories. The graph algorithms such as tree construction, breadth first search, and depth first search are being effectively used for intrusion/anomaly detection. Ngai et al. [51,52] proposed novel algorithms for detecting anomalous node by using graph theory. This scheme is based on two steps: first,

BS collects list of suspicious nodes and second it localizes the exact location of anomalous node by employing network flow graph. This approach is useful for identifying colluding nodes. The voting method is used to declare suspicious node as anomalous or normal on the basis of their previous behavior. This scheme is effective only until majority of nodes are normal. Though this approach is effectual in tracing node anomalies, but it consumes lot of resources for building graph, especially for worst-case scenarios where anomaly lies at the end of either breadth or width of the graph.

Ho et al. [53] suggested a detection scheme based on incorporating a group knowledge factor before deployment. The detection is carried out at SN, when it receives a request from a neighboring node to forward a message. Each group is identified on the basis of a unique group deployment position. The SN in each group is placed at location (a, b), as per Equation 11.10.

$$ f(a,b) = \frac{1}{2\pi\sigma^2} e^{((a-a_G)^2 + (b-b_G)^2)/2\sigma^2} \tag{11.10} $$

Here (a, b) are the location coordinates of SN in a group. (a_G, b_G) denotes the deployment position of group G and σ represent standard deviation. If a node i, member of group Gi, receives a request from neighboring node j, then i verifies the distance between group deployment position and neighboring node. If the distance is less than already-defined distance then j is considered as a valid node, otherwise it is treated as antagonist node. This scheme is effective for identifying anomalies caused by replica nodes. However, the performance of this scheme is highly dependent on accurate deployment and prior knowledge of network behavior which may not be always possible.

Yuan et al. [54] suggested subjective logic-based anomaly detection (SLAD) framework. The subjective logic is suitable to those situations where uncertainty and incomplete knowledge is involved. By applying this concept to anomaly detection, authors suggested SLAD framework. The framework is based on two algorithms. The first algorithm fuses the judgment of neighboring nodes to decide SN as normal or anomalous. However, this approach has three disadvantages: (a) the judgment process also involves anomalous nodes which may affect the anomaly-detection process, (b) this process does not discriminate among anomalous and normal data, and (c) the threshold for the neighborhood is set as 0.5 which may not be suitable in all the situations. In order to overcome these problems, authors extended the idea and proposed a second algorithm. The second algorithm has the following refinements in overall anomaly-detection process: (a) removal of judgment of suspicious node, (b) the spatial correlation among data is exploited to discriminate among anomalous and normal data, and (c) the historical information is considered to appropriately weigh the opinion of neighbor nodes. Despite the improvement in anomaly-detection process, the involvement of historical data may require more memory space. Furthermore, consideration of temporal correlation can increase the robustness of SLAD framework.

Li et al. [55] present a quantitative approach for anomalous node detection in WSN. This approach is based on data transmission quality (DTQ) functions. The WSN is divided into several groups. Each SN maintains DTQ table of neighboring nodes. The DTQ function is given in the following equation.

$$DTQ = k\frac{D}{E} * \frac{STB(\ )}{P(\ )} \quad \text{where } k > 0 \qquad (11.11)$$

Here k is an integer. D/E denotes the number of data packets transmitted in unit energy. $STB()$ represents data transmission stability and $P()$ is the probability of successfully transmitted packets. The value of DTQ remains constant or changes smoothly for normal nodes and keeps changing for antagonist nodes. The final fate of the node as normal or anomalous is decided on the basis of voting among group members. However, the voting approach is susceptible to colluding attack.

Krontiris et al. [56] suggest a rather simple anomaly-detection approach for WSNs. In this approach, selective nodes perform the job of monitoring neighbor nodes. These nodes are known as watchdog nodes. The watchdog nodes are selected based on the certain criteria. Consider a scenario where node "*A*" has a communication link with "*B*". Then the node "*A*" and all other nodes that are positioned within the intersection of the radio range of "*A*" and "*B*", can act as watchdog. The watchdog node monitors neighboring nodes based on the following: (a) if a particular node drop n number of packets in t unit time then an alarm is generated by the watchdog node, and (b) in a specific location, if more than half of the watchdog nodes generate alarms then a suspected node is announced as an anomalous node. Although this approach is simple and efficient, it is highly susceptible to FPR. Moreover, the requirement of half of the node's vote to declare a suspicious node as anomalous may not be suitable for some WSN applications where a high-detection rate is desired. Kumarage et al. [57] recently suggested a distributed anomaly-detection model for industrial WSNs based on fuzzy data modeling. In this approach, distributive data partitioning is performed by employing fuzzy c-means clustering in an incremental model. The anomalies are detected through fuzzy membership evaluations and threshold on intercluster distance. The performance evaluation of this scheme was carried out on multiple datasets based on different distributions. The comparative study is performed on the basis of sensitivity and specificity analysis. Despite the viability of this approach for industrial WSN applications, it requires a specific number of expected clusters at node level. The false specification by a sensor node may result in poor clustering.

11.2.4.1 Multiple Feature-Based Anomaly Detection

Two or more metrics are used for anomaly detection in multiple feature-based anomaly-detection models. The selection of appropriate number of features for anomaly detection is quite imperative in WSN. This selection is mainly dependent on two factors: (a) available network resources and (b) security requirements. Over the years, the research community has proposed several anomaly-detection methods for WSNs that are based on multiple features. One such anomaly-detection technique is proposed by Onat and Miri [58] in which each node maintains a simple normal profile of neighboring nodes. The normal profile is based on two features (parameters): packet arrival rate and average of received power. Since anomalies are discovered on the basis of only two parameters, this model is incapable of detecting more sophisticated anomalies and attacks.

Dai et al. [59] suggested a multivariate classification algorithm for detection of anomalous nodes in large-scale WSNs. This method employs a multivariate classification method. The multivariate classification method extracts SN's preferences pertaining to malicious nodes. Then it establishes a sample space of all SNs that are part of the network. Then the classifier categorizes normal and anomalous nodes on the basis of given criterion. The results demonstrate false-detection rates under 0.5%. Despite such a low-false detection rate, the efficiency of this scheme heavily relies on detection criterion. The selection of detection criteria is not an easy task in some WSN applications where nodes have dynamic behavior.

M. Becker et al. [59] compared and evaluated different learning algorithms using multiple features [60]. The algorithms include k-nearest neighbors, SVMs, Bayes classifier, neural network, AIS, and decision trees. The simulation study shows that simple approaches such as Bayes classifier and decision trees offer better performance as compared to other classification approaches. In an experimentation setup, multiple features are used for anomaly detection (mainly for MAC and routing layer). These features include (a) MAC layer ratio: a ratio of the MAC layer handshake and ready to send (RTS) packets that are transmitted between SN_i and $SN_i + 1$ and are used as normal profile metrics. (b) Forward ratio: the proportion of data packets that are forwarded between SN_i and $SN_i + 1$ and then $SN_i + 2$ are computed. (c) Average delay packet: a time delay to forward data from $SN_i + 1$ to $SN_i + 2$. (d) Forward ratio rout error (RERR): ratio of RERR routing packets is calculated as per (b). (e) Average delay RERR: ratio of RERR routing packet is computed as per (c). However, the performance of anomaly detection is heavily dependent upon the feature selection. The straightforward approach of using multiple features may not be suitable for low-resource WSNs. Therefore, multiple feature-based anomaly-detection processes needs to be optimized.

11.2.4.2 Cross-Layer Anomaly Detection

The cross-layer approach is employed to efficiently utilize overall resources of sensor nodes [61]. One of the recent trends is to detect cross-layer anomalies in WSNs. In this approach, anomalies are detected on the basis of multiple layer-wise network protocol parameters. Bhuse proposed one such anomaly-detection scheme [62]. In this scheme, anomaly-detection parameters for physical, MAC, network, and application layers are used. At the physical layer, the use of a received signal strength indicator (RSSI) value is suggested for anomaly detection. The RSSI value of neighbor nodes is calculated and then significant deviation from that value is considered as an anomaly. At the MAC layer, time division multiple access (TDMA) and sensor-media access control (S-MAC) protocols are considered for anomaly detection. In these MAC layer protocols, time slots are allocated to nodes for communication. The communication outside designated slots indicates the intrusion. For the network layer, protocol is called as information authentication for sensor networks (IASN) is suggested. The idea behind the protocol is to use the forwarding tables that are generated by network layer protocols for anomaly detection. This process requires integration of high-level information in routing tables to construct anomaly-detection table (ADT). Then anomalies are discovered on the basis of ADT. At the application layer, round trip time is considered as an anomaly- or intrusion-detection parameter.

Similarly, Boubiche et al. proposed another scheme for layer-wise anomaly detection in WSNs [63]. The common feature of these schemes is that they encourage detecting layer-wise attacks and/or anomalies at respective layers. It is an expensive approach, considering the cost associated with detecting anomalies at each layer. This approach also has a drawback that each layer's protocol requires amendment for integration of normal profile parameters, which is quite a cumbersome task, if not impossible.

11.2.4.3 Mobile Agent-Based Anomaly Detection

Over the last decade, mobile agents are employed in various capacities in WSNs, such as data diffusion, localization, parallelism, application update, distributed computing, and data dissemination [64]. Likewise, they are also employed in security applications such as intrusion or anomaly detection. In existing literature, one such proposal suggested the use of mobile agents for distributed-intrusion detection [65]. In this methodology, agents are dispatched as guards that roam among different nodes to perform random sampling. If an anomaly is observed during a random examination, then a comprehensive detection is initiated. This proposal reduces the cost associated with dispatch and arrival of mobile agents, but in the absence of guards, nodes are vulnerable to threats.

Kachirski et al. [66] suggested a distributed anomaly-detection system based on mobile agents. Their suggested model employs agents at various levels: monitoring, decision-making, and action agents. To optimize the cost of the system, few agents are stationed on all nodes while others are located at a particular node or group of nodes. The monitoring agents are positioned at each node and they observe the node activities. Whenever monitoring agents observe an anomaly, they report it to the respective decision agent. The decision agents are installed at every CH. After analysis of the anomaly report, CH informs the decision agent regarding the anomaly. Then the decision agent decides the fate of the local agent. This scheme uses various agents which increase the computational and memory cost of the system.

M. Ketel proposed an architecture for using a mobile agent paradigm for distributed anomaly detection in cluster-based WSNs [67]. This scheme employs static and mobile agents in order to detect anomalies: static agents (SA), mobile agents (MA), and nodal agents (NA). Moreover, the architecture is comprised of a mobile agent server (MAS) and victim node list (VNL). The SAs are located at each CH. The SA sends a report to the MAS after observing an anomalous activity. The MAS triggers MAs to the monitoring nodes whenever an attack is detected. The MAs are further divided into thick and thin agents. Thick and thin agents are associated with resource rich and resource low-sensor nodes, respectively. The NAs are positioned at monitoring nodes to detect local anomalies. Finally, VNL contains a list of victim nodes at each CH. The VNL is responsible for generating the itinerary of MA. The suggested idea is at an abstract level. The installation of multiple components/agents on nodes increases the overall cost of the system. The internal details of MAS repository and VNL are not discussed.

Eludiora et al. [68] presented distributed intrusion/anomaly-detection system for WSNs. In this model, SN directly communicates with BS instead of CH. The MAs are employed to communicate among different BSs. The designated role for MAs is to roam among multiple nodes and BSs in order to perform anomaly-detection tasks. The authors suggested two algorithms for detecting anomaly and analyzing data. The first algorithm detects the DoS attacks and consequently updates the status of SN. On the other hand, the second algorithm calculates the probability of BS failure to discover an anomaly. The authors have made an assumption that SN is only a one-hop distance from BS, which is not a realistic for most of the WSN applications. Though single-hop communication reduces energy consumption, it can result in a communication bottleneck on BS. Furthermore, this approach is unsuitable for large-scale networks. In addition, this scheme only focuses on DoS attacks.

An architecture of mobile agent-based hierarchical Intrusion Detection System (IDS) for WSN is presented by Khanum et al. [69]. The authors employed three agents: analyzer agent, management agent, and coordinating agent. The analyzer agent is mobile while management agent and coordinating agent are static in nature. The sensor nodes are deployed in a cluster-based topology. The suggested architecture is installed at each CH. The anomalies are detected at two levels: node and network. The CH detects network-level anomalies while node-level anomalies are discovered by analyzer agent. The architecture is proposed at the abstract level. The authors do not provide any internal details regarding anomaly-detection mechanism. Furthermore, the cost associated with a roaming mobile agent is also not carefully considered. Khanum et al. suggested another anomaly-detection method for WSNs using a Muhammad Usman Surraya Khanum (MUSK) agent [70]. In this scheme, the MUSK agent is installed on each sensor node. Each MUSK agent works independently and there is no collaboration among them. Each MUSK agent is based on three components: monitoring unit, management unit, and coordination unit. This is an architecture-level scheme and its underlying details of anomaly detection are not describer. Therefore, viability of this scheme for WSNs is not established.

Another mobile agent-based anomaly-detection scheme for wireless smart home sensor networks (WSHSN) is proposed by Usman et al. [71]. This scheme takes advantage of the heterogeneous nature of devices in smart homes for effectively detecting anomalies. The anomaly-detection module is installed at resource-rich nodes that are CHs and is based on three components: coordination unit, repository, and anomaly agent, as illustrated in Figure 11.9. The coordination unit is in charge of management and coordination inside the anomaly-detection module, and also with other entities of the network. Repository stores six tuples as shown in Equation 11.12.

$$N, R, Au_{rec}, Prf, An_{rec}, Act_{rul} \tag{11.12}$$

Here N denotes SN identities, R represents resources of corresponding nodes, Au_{rec} identifies audit record, Prf shows normal profiles, An_{rec} express anomaly record, and Act_{rul} represents action rules. The anomalies are discovered after analysis of every received data packet. If required, the anomaly (mobile) agent may travel to the suspicious node for verification of its behavior. The analysis and comparative study showed several advantages of this scheme, such as efficient utilization of memory, reducing network load, and reduction in overall computational cost of the network.

It is important to note that the approach of employing a mobile agent for verification of suspicious nodes may have associated infrastructural costs, particularly at the sensor node. However, the benefit of utilizing mobile agents for a variety of roles such as localization, data fusion, parallelism, and so on, may reduce the impact of that cost. Scalability is another important design factor for articulating any security solution for WSNs as they are generally low-resource networks. This solution is easily scalable, as CHs are assumed as resource-rich nodes, thus, they can accommodate more number of SNs. Nevertheless, timing is an important factor for anomaly detection, therefore, the cluster size must not be too large as it may create a communication bottleneck at CH. Regardless of the comprehensive nature of the model, it cannot serve as a stand-alone security solution. However, it

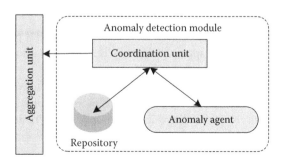

Figure 11.9   Components of a cluster head.

can be employed as a security layer to enhance the probability of detecting more sophisticated anomalies.

11.3 Limitations and Future Directions

In the previous section, we reviewed several anomaly-detection schemes in WSNs. We now sum up major limitations in existing schemes and then describe some potential future research directions.

11.3.1 Limitations

On the basis of the schemes we reviewed in the last section, we realized that existing schemes have several shortcomings. Below we highlight a few major limitations in existing schemes.

- One of the basic focuses of application developers for WSNs is to optimally utilize energy budget of nodes. The same principal applies for anomaly-detection schemes. Though this approach may prolong the overall node and network lifetime, it may not be able to discover more sophisticated anomalies. The incapability of detecting sophisticated anomalies may have an even worse effect, not only on network functionality but also on its lifetime because of unnoticed anomalies. Therefore, there should be a fine balance between lightweight anomaly-detection mechanisms and its capability of detecting more sophisticated anomaly.
- During our review of existing literature, we found that most of the schemes are simply proposed as strategies and their underlying detailed anomaly-detection mechanism and analysis are missing. The absence of detailed anomaly-detection mechanisms raises an important question, whether proposed models are effective for resource-constrained WSNs?
- A considerable portion of existing anomaly-detection schemes are application oriented, thus, those schemes have limited scope. These schemes may not be readily viable to other applications.
- Some existing schemes are based on either single or dual parameters to discover anomalies. This approach ignores the fact that multiple parameters together can also exhibit anomalies.
- Mobility is an important feature for many WSNs. However, the majority of existing anomaly-detection schemes are proposed for static sensor nodes. These solutions are not readily viable for mobile sensor nodes.
- Many existing schemes are based on static bounds on normal profile parameters. Setting static bounds for normal profile parameters may not be a good approach because the normal notion of the node/network may change with the passage of time.

■ The performance evaluation of existing anomaly-detection schemes is generally based on detection accuracy and energy efficiency. Some schemes are evaluated by examining the false alarm rate. However, there is no benchmark in existing literature that can serve as an evaluation standard.

11.3.2 Future Direction

Over the years, due to advancements in technologies and application requirements, the WSN applications have become more and more complex. Similarly over the years, the nature of faults/errors and attacks has also advanced. Therefore, there are several potential research areas that can be further explored. In this section, we briefly describe these research areas of anomaly detection in WSNs.

■ *Network anomaly detection:* Only a minimal amount of work is done on network intrusion/anomaly detection. More focus is on neighbor node monitoring (i.e., node anomalies). The approach of detecting network anomalies is particularly useful in cluster-based network topology. It is a challenging task to detect network anomalies, particularly when multiple nodes are relaying data to cluster head and/or sink node.

■ *Mobile agent-based anomaly detection:* The mobile agents are being used in different capacities in anomaly-detection systems. However, the common shortfall of existing schemes is that they have overlooked the infrastructure and transmission cost that is associated with mobile agents when they roam among multiple nodes. Moreover, security threats that emerged for WSNs due to deployment of mobile agents are also ignored. Therefore, these areas can also be further investigated.

■ *Multivariate anomaly detection:* In some applications such as area monitoring, sensor readings may be spatially and/or temporally correlated. A typical approach to detect anomalies in such applications is to employ first-order statistics where spatial and temporal statistical characterization of sensor data is performed to establish normal profile of the node. Then deviation from that profile is considered as anomaly/outlier. In such cases, advance multivariate covariance and correlation techniques can be investigated to discover more robust anomalies.

■ *Mobility of nodes and anomaly detection:* Another potential research area can be integrating mobility factor for nodes in anomaly-detection systems. Though there are several such schemes in existing mobile ad hoc network (MANET) literature but due to scarce resource capability of sensor nodes, these solutions are not readily viable for WSNs.

■ *Adaptive threshold for anomaly-detection system:* An appropriate threshold is quite imperative in order to accurately identifying anomalies. Setting an appropriate threshold is a challenging task in anomaly detection. The notion of WSN may change with the passage of time and a static threshold may become invalid. Therefore, adaptable threshold mechanisms need to be built for robust anomaly detection in WSNs.

■ *Bio-inspired anomaly detection:* One of the recent tendencies of the research community is to adopt biological methods for anomaly detection in WSNs. In such approaches, researchers try to mimic living organism's defense systems as anomaly-detection methods. Nevertheless, suitability of such approaches for low-resource WSNs requires further investigation.

■ *Nonparametric and nonprobabilistic anomaly detection:* A considerable portion of existing literature is based on either parametric or probabilistic approaches. However, recently, the research community has started to integrate concepts such as fuzzy data modeling into WSN anomaly-detection research. Such approaches can be further investigated as a counterpart of traditional parametric and probabilistic anomaly-detection approaches.

11.4 Conclusions

Anomaly detection is one of the most researched areas in WSNs over the recent years. This research ranges from machine learning to AI and statistical to several other schemes. Throughout this chapter, we first described a few basic concepts and types of anomalies in WSNs. We then highlighted taxonomy of anomaly-detection techniques in WSNs along with its application domains. We observed, in existing literature, most of the machine-learning algorithms are being employed at training and testing phases of the anomaly-detection process. The AI-based schemes replicate human intelligence and other biological methods in anomaly detection. The considerable portion of existing schemes is based on statistical methods. We also observed that the recent trend is to hybrid other applications with anomaly detection, either as a whole or partially, for optimally utilizing overall resources of WSNs. Mobile agent-based anomaly detection is one of the prominent examples of such approaches. In this approach, mobile agents are not only used for anomaly detection but also for several tasks such as data diffusion, localization, parallelism, application update, distributed computing, and data dissemination. Nevertheless, the research of such approaches is at preliminary stages and needs further investigation.

References

1. Y. Zhang, N. Meratnia, and P. Havinga. Outlier detection techniques for wireless sensor networks: A survey. *IEEE Communications Surveys and Tutorials.* 12(2), 159–170, 2010.
2. V. Chandola, A. Banerjee, and V. Kumar. Anomaly detection: A survey. *ACM Computing Surveys*, 41(3), Article 15, 1–58, 2009.
3. D. E. Denning. An intrusion-detection model. *IEEE Transactions on Software Engineering*, 13(2), 222–232, 1987.

4. R. Jurdak, X. R. Wang, O. Obst, and P. Valencia. *Wireless Sensor Network Anomalies: Diagnosis and Detection Strategies. Intelligence-Based Systems Engineering*, Springer-Verlag, Berlin, Germany, pp. 309–325, 2011.

5. C. Karlof and D. Wagner. Secure routing in wireless sensor networks: Attacks and countermeasures. *1st IEEE International Workshop on Sensor Network Protocols and Applications*, 113–127, 2003.

6. Z. Bankovic, J. C. Vallejo, P. Malagon, A. Araujo, and J. M. Moya. Eliminating routing protocol anomalies in wireless sensor networks using AI techniques, 3rd ACM workshop on Intelligence and Security, pp. 8–13, 2010.

7. Y. K. Tan and S. K. Panda. Review of energy harvesting technologies for sustainable wireless sensor network. Book chapter, sustainable wireless sensor networks, INTECH, pp. 15–43, 2010.

8. R. A. Maxion and R. R. Roberts. Proper use of ROC curves in intrusion/anomaly detection. Technical report, School of Computer Science, University of Newcastle upon Tyne, pp. 1–32, November 2004.

9. T. M. Mitchell. *Machine Learning*. McGraw-Hill Science/Engineering/Math; 1st edition, pp. 432, 1997.

10. Xiaojin Zhu. Semi-supervised learning with graphs. PhD Dissertation. School of Computer Science, Carnegie Mellon University, USA, pp. 1–4, 2005.

11. S. Rajasegarar, C. Keckie, M. Palaniswami, and J.C. Bazdek. Quarter sphere based distributed anomaly detection in wireless sensor networks. *IEEE International Conference on Communications*, Glasgow, United Kingdom, 3864–3869, 2007.

12. Z. Xiao, C. Liu, and C. Chen. An anomaly detection scheme based on machine learning for WSN. *IEEE International Conference on Information Science and Engineering (ICISE)*, China, 3959–3962, 2009.

13. T. H. Lim. Detecting anomalies in wireless sensor networks. Qualifying Dissertation, Department of Computer Science, University of York, pp. 23–37, 2010.

14. C. E. Loo, M. Y. Ng, C. Leckie, and M. Palaniswami. Intrusion detection for routing attacks in sensor networks. *International Journal of Distributed Sensor Networks* 2, 313–332, 2006.

15. P.-N. Tan, M. Steinbach, and V. Kumar. *Introduction to Data Mining*. Addison-Wesley, 769, USA, 2006.

16. S. Rajasegarar, C. Keckie, M. Palaniswami, and J. C. Bazdek. Distributed anomaly detection in wireless sensor networks. *10th IEEE International Conference on Communication Systems*, 1–5, 2006.

17. X. Miao, H. Song, and T. Biming. Highly efficient distance-based anomaly detection through univariate with PCA in wireless sensor networks. *In 10th IEEE conference on Trust, Security and Privacy in Computing and Communications (TrustCom)*, China, 564–571, 2011.

18. D. Poole, A. Mackworth, and R. Goebel. *Computational Intelligence: A Logical Approach*. New York: Oxford University Press, pp. 576, 1998.

19. G. Carpenter and S. Grossberg. *Adaptive Resonance Theory*. Bradford Books. MIT Press, United Kingdom, 2002.

20. M. Walchli and T. Braun. Efficient Signal Processing and Anomaly Detection in Wireless Sensor Networks. *6th European Workshop on the Application of Nature-inspired Techniques for Telecommunication Networks (EvoCOMNET 2009)*, Tubingen, Germany, pp. 81–86, 2009.

21. Y. Y. Li and L. E. Parker. Classification with missing data in a wireless sensor network. *IEEE Conference*: Southeastcon, pp. 533–558, 2008.
22. D. Castro, N. Leandro, and T, Jonathan. *Artificial Immune Systems: A New Computational Intelligence Approach.* Springer, USA, pp. 57–58, 2002.
23. R. Fu, K. Zheng, T. Lu, D. Zhang, and Y. Yang. Biologically inspired anomaly detection for hierarchical wireless sensor networks. *Journal of Networks*, Academy Publisher 7, 1214–1219, 2012.
24. H. M. Salmon, C.M.D. Farias, P. Loureiro, L. Pirmez, S. Rossetto, P. H. D. A. Rodrigues, R. Pirmez, F. C. Delicator, and L. F. R. D. C. Carmo. *Intrusion Detection System for Wireless Sensor Network Using Danger Theory Immune-Inspired Techniques.* Springer Science, USA, pp. 1–28. 2012.
25. R. Khanna, H. Liu, and H.-H. Chen. Reduced complexity intrusion detection in sensor networks using genetic algorithm. *IEEE Proceedings of the 2009 IEEE International Conference on Communications*, USA, pp. 598–602, 2009.
26. H. Gao, G. Chen, and W. Guo. A GEP-based anomaly detection scheme in wireless sensor networks. *International Conference on Computational Science and Engineering*, Canada, pp. 817–822, 2009.
27. A. Agah, S. K. Das, and K. Asu. A non-cooperative game approach for intrusion detection in sensor networks. *IEEE 60th Vehicular Technology Conference*, USA, pp. 2902–2906, 2004.
28. Y. B. Reddy. A game theory approach to detect malicious nodes in wireless sensor networks. *IEEE 3rd International Conference on Sensor Technologies and Applications (SENSORCOMM.)*, Greece, 462–468, 2009.
29. A. C. Davison. *Statistical Models*, Cambridge University Press, United Kingdom, 738, 2003.
30. D. J. Sheskin. *Handbook of Parametric and Nonparametric Statistical Procedures.* Third Edition, Chapman & Hall/CRC Press, USA, 1016, 2000.
31. L. M. A. Bettencourt, A. A. Hagberg, and L. B. Larkey. *Separating the Wheat from the Chaff: Practical Anomaly Detection Schemes in Ecological Applications of Distributed Sensor Networks. In Distributed Computing in Sensor Systems*, Springer-Verlag, LNCS, pp. 223–239, 2007.
32. W. Wu, X. Cheng, M. Ding, K. Xing, F. Liu, and P. Deng. Localized outlying and boundary data detection in sensor networks. *IEEE Trans. Knowledge Data Engineering*, 19, 1145–1157, 2007.
33. M. C. Jun, H. Jeong, and C. C. J. Kuo. Distributed spatio-temporal outlier detection in sensor networks, *Proceedings of SPIE*, USA, 2006.
34. T. Palpanas, D. Papadopoulos, V. Kalogeraki, and D. Gunopulos. *Distributed Deviation Detection in Sensor Networks.* ACM Special Interest Group on Management of Data, USA, pp. 77–82, 2003.
35. A. Gibson. *Exposure and Understanding the Histogram.* Peachpit Press, 20, 2011.
36. B. Sheng, Q. Li, W. Mao, and W. Jin. Outlier detection in sensor networks. Proc. ACM 8th International Symposium on Mobile ad Hoc Networking and Computing, pp. 219–228, 2007.
37. B. Miladinovic. Kernel density estimation of reliability with applications to extreme value distribution, Proquest, Umi Dissertation Publishing, 160, 2011.
38. S. Subramaniam, T. Palpanas, D. Papadopoulos, V.Kalogeraki, and D. Gunopulos. Online outlier detection in sensor data using nonparametric models. In 32nd International Conference on Very Large Data bases (VLDB), pp. 187–198, 2006.

39. S. Papadimitriou, H. Kitagawa, P.B. Gibbons, and C. Faloutsos. LOCI: Fast outlier detection using the local correlation integral. *Proceedings of the IEEE International Conference on Data Engineering*, USA, pp. 315–326, 2003.

40. W. Daniel. *Graduate Texts in Mathematics: An Introduction to Markov Processes*. Springer, USA, 178, 2005.

41. P. Dymarski. Hidden Markov Models: Theory and applications. In Tech, 314, 2011.

42. I. C. Paschalidis and Y. Chen. Anomaly detection in sensor networks based on large deviations of markov chain models. *47th IEEE Conference Decision and Control*, Mexico, pp. 2338–2343, 2008.

43. Y. Gao, C. Chen, J. Bu, W. Dong, and D. He. ICAD: Indirect correlation based anomaly detection in dynamic WSNs. *IEEE Wireless Communications and Networking Conference (WCNC)*, Mexico, pp. 647–652, 2011.

44. S. Zheng and J. S. Baras. Sequential anomaly detection in wireless sensor networks and effects of long range dependant data. *Special IWSM Issue of Sequential Analysis (SQA)*, 31, 458–480, 2012.

45. J. D. Cryer and K.-S. Chan. *Time Series Analysis with Applications in R*. 2nd edition, Springer, USA, 2008.

46. A. K. Singh, B. Giridhar, and P. S. Mandal. Fixing data anomalies with prediction based algorithm in wireless sensor networks. *7th IEEE Conference on Wireless Communication and Sensor Networks (WCSN)*, 6, 2011.

47. M. C. Chuah and F. Fu. ECG Anomaly Detection via Time Series Analysis, Frontiers of High Performance Computing, Lecture Notes in Computer Science (LNCS), pp. 123–135, 2007.

48. S. Chatterjee and A. S. Hadi. *Regression Analysis by Example*. Fourth edition, John Wiley & Sons, USA, 408, 2006.

49. D.-I. Curiae, O. Banias, F. Dragan, C. Volosencu, and O. Dranga. Malicious node detection in wireless sensor networks using an auto regression technique. *IEEE 3rd International Conference on Networking and Services*, Greece, 83, 2007.

50. S. Y. Kim, M. Imada, and M. Ohta. Detecting anomalous events in ubiquitous sensor environments using Bayesian networks and nonparametric regression. *1st International Conference IEEE Advanced Information Networking and Applications*, Canada, pp. 236–243, 2007.

51. E. C. H. Ngai, J. Liu, and M.R. Lyu. On the intruder detection for sinkhole attack in wireless sensor networks. *IEEE International Conference on Communications*, Turkey, pp. 3383–3389, 2006.

52. E. C. H. Ngai, J. Liu, and M.R. Lyu. An efficient intruder detection algorithm against sinkhole attacks in wireless sensor networks. *Journal of Computer Communications*, 30(11–12), 2353–2364, 2007.

53. J.-W. Ho, D. Liu, M. Wright, and S. K Das. Distributed detection of replica node attacks with group deployment knowledge in wireless sensor networks. *Ad Hoc Networks*, 7, 1476–88. 2009.

54. J. Yuan, H. Zhou, and H. Chen. Subjective logic-based anomaly detection framework in wireless sensor networks. *International Journal of Distributed Sensor Networks*, 2012, 13, 2011.

55. T. Li. M. Song, and M. Alam. Compromised sensor nodes detection: A quantitative approach. *IEEE 28th International Conference on Distributed Computing Systems Workshops*, China, pp. 352–357, 2008.

56. I. Krontiris, T. Dimitriou, and F. C. Freiling. Towards intrusion detection in wireless sensor networks. *13th European Wireless Conference*, France, 7, 2007.

57. H. Kumarage, I. Khalil, Z. Tari, A. Zomaya. Distributed anomaly detection for industrial wireless sensor networks based on fuzzy data modeling. *Journal of Parallel and Distributed Computing*, 73(6), 790–806, 2013.

58. I. Onat and A. Miri. An intrusion detection system for wireless sensor networks. *Proceedings of the IEEE International Conference on Wireless and Mobile Computing, Networking and Communications*, (WiMob'2005), 3, 253–259, 2005.

59. H. Dai, H. Liu, Z. Jia, and T. Chen. A multivariate classification algorithm for malicious node detection in large-scale WSNs. *IEEE 11th International Conference on Trust, Security, and Privacy in Computing and Communications*, United Kingdom, pp. 239–245, 2012.

60. M. Becker, M. Drozda, S. Schaust, S. Bohlmann, and H. Szczerbicka. On classification approaches for misbehavior detection in wireless sensor networks. *Journal of Computers*, 4(5), 357–365, 2009.

61. N. A. Alrajeh, S. Khan, Jaime Lloret, and J. Loo. Secure routing protocol using cross-layer design and energy harvesting in wireless sensor networks. *International Journal of Distributed Sensor Networks*, 2013, 1–13, 2013.

62. V. S. Bhuse. Lightweight intrusion detection: A second line of defense for unguarded wireless sensor networks. PhD Dissertation, Department of Computer Science, Western Michigan University, 2007.

63. D. E. Boubiche and Z. Bilami. Cross layer intrusion detection system for wireless sensor network. *International Journal of Network Security & Its Applications (IJNSA)*, 4(2), 35–52, 2012.

64. M. Chen, T. Kwon, Y. Yuan, and V. C. M. Leung. Mobile agent based wireless sensor networks. *Journal of Computers*, 1(1), 14–21, 2006.

65. C. Krugel and T. Toth. Applying mobile agent technology to intrusion detection. *Proceedings of ICSE Workshop on Software Engineering and Mobility*, Canada, pp. 1–5, 2001.

66. O. Kachirski and R. Guha. Effective intrusion detection using multiple sensors in wireless ad hoc networks. *Proceedings of 36th Annual Hawaii International Conference on System Sciences*, USA, pp. 57–65, 2003.

67. M. Ketel. Applying the mobile agent paradigm to distributed intrusion detection in wireless sensor networks. *IEEE 40th Southeastern Symposium on System Theory*, USA, pp. 74–78, 2008.

68. S. I. Eludiora, O. O. Abiona, A. O. Oluwatope, S. A. Bello, M. L. Sanni,, D. O. Ayanda, C. E. Onime, E. R., Adagunodo, and L. O. Kehinde. A distributed intrusion detection scheme for wireless sensor networks. *IEEE International Conference on Electro/Information Technology*, China, pp. 1–5, 2011.

69. S. Khanum, M. Usman, and A. Alwabel. Mobile agent based hierarchical intrusion detection system in wireless sensor networks. *International Journal of Computer Science Issues (IJCSI)*, 9(3), 101–108, 2012.

70. S. Khanum, M. Usman, K. Hussain, R. Zafar, and M. Sher. *Energy-Efficient Intrusion Detection System for Wireless Sensor Network Based on MUSK Architecture. Lecture Notes on Computer Science (LNCS)*. Springer-Verlag, Berlin Heidelberg, pp. 212–217, 2010.

71. M. Usman, V. Muthukkumarasamy, X.-W. Wu, and S. Khanum. Wireless smart home sensor networks: Mobile agent based anomaly detection. *IEEE 9th Conference on Ubiquitous Intelligence and Computing (UIC)*, Japan, September 2012.

Chapter 12

Taxonomy of Security Protocols for Wireless Sensor Communications

Jasone Astorga, Eduardo Jacob, Nerea Toledo, and Marivi Higuero

Contents

12.1 Introduction

The aim of this chapter is to provide an overview of current approaches to protect communications with sensor devices. The concept of wireless sensor network (WSN) communications security can cover a broad range of issues, including key management mechanisms, communications confidentiality, source authentication, protection against denial-of-service attacks, secure routing, resistance to node capture, intrusion detection, protection against traffic analysis, secure data aggregation, and so on. In this context, this chapter will focus on security protocols specifically aimed at protecting the confidentiality and integrity of the information transmitted and received by sensor devices, as well as to guarantee that this information is only retrieved by authenticated and authorized third parties.

Traditional security mechanisms, meant for powerful workstations, are not directly applicable to environments that include heavily resource-deprived devices, such as sensors, since these devices present severe limitations regarding processing power, storage capacity and battery. For this reason, the development of security protocols specifically tailored to sensor devices has been essential. In fact, sensor network security is currently a very active research topic, with new protocols and approaches being constantly proposed. Taking into account the vast amount of existing security protocols for sensor communications, in this chapter we present the most relevant ones or those which introduce special characteristics.

The security protocols studied in this chapter have been classified into four main groups, as shown in Figure 12.1.

The first group focuses on protocols aimed at protecting the integrity and confidentiality of data transmissions within sensor networks at the link layer as well as at network or application layers.

The second and third groups include protocols with the same objective: the secure and efficient establishment of cryptographic keys between nodes of the same sensor network. The reason to divide these protocols in two different groups is that they make use of opposite technologies to achieve the common objective. Therefore, the second group of protocols defined in this chapter gathers key exchange protocols that are solely based on symmetric key cryptography, while the third group includes key exchange protocols that make use of some kind of public-key cryptographic algorithm.

These first three groups of protocols have been designed for traditional sensor networks, which are regarded as isolated isles based on proprietary systems and communication protocols. In such scenarios, the access to the information provided by the sensors is restricted to the use of intermediary proxies or centralized information repositories. Therefore, all of these protocols focus on protecting the communications between nodes within the same network, but none of them considers the possibility that an external entity connected to the Internet would directly query a node within the sensor network.

However, the recent development of technologies like 6LoWPAN [1] allows breaking away from this traditional view, giving place to a new concept of sensor networks, which are no longer isolated isles. Instead, sensor networks are integrated in the IP world. Consequently, sensors essentially become tiny information or application servers, directly addressable by any other IP entity in the Internet. As a result, the security services required by these new types of sensor networks are radically different to the ones required by traditional sensor networks. Specifically, strong security features, such as authentication and authorization of remote entities, as well as key exchange mechanisms are required. In the fourth group of protocols defined in this chapter we aim at gathering security protocols designed with

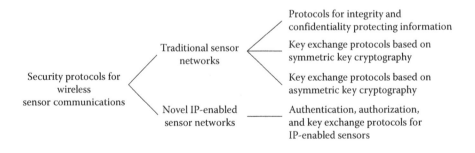

Figure 12.1 **Taxonomy of the security protocols for wireless sensor communications.**

this goal. However, to the best of our knowledge, so far a single-security protocol has been proposed with this aim: Ladon.

Finally, it must be taken into account that due to the severe resource-constraints presented by the targeted devices, most of the security protocols for sensor networks are not general purpose protocols, but are specifically designed to deal with a particular security necessity. For this reason, the description of the security protocols included in this chapter are not fully homogenized.

The rest of the chapter is organized as follows. Section 12.2 provides a review on the most popular approaches to protect the integrity and confidentiality of data exchanges within different entities of the same sensor network. Section 12.3 gathers the most important key exchange protocols for traditional sensor networks which solely rely on symmetric key cryptography, while Section 12.4 includes the most relevant key exchange protocols for the same environments that make use of some kind of asymmetric cryptographic algorithm for their operation. With respect to security mechanisms for novel IP-enabled sensors, the Ladon security protocol is presented in Section 12.5. Finally, Section 12.6 summarizes future research directions and Section 12.7 gathers the most remarkable conclusions of the taxonomy of security protocols for wireless sensor communications presented in this chapter.

12.2 Protocols for Integrity and Confidentiality Protection in WSNs

In this section, we present protocols aimed at protecting the integrity and confidentiality of the packets exchanged between members of a WSN and at providing source authentication. These security services can be enforced at different layers of the protocol stack, depending on the nature of the communication to be protected. That is, confidentiality and integrity protection can be applied hop-by-hop to link-layer frames or end-to-end to network or application layer packets. Therefore, different protocols have been designed to operate at each level of the protocol stack.

On the other hand, most of these protocols do not define the mechanisms through which legitimate participants can obtain or calculate the necessary keys to afterwards protect information through encryption or message authentication code (MAC) computation. They assume that the necessary keys are already available at the communicating endpoints.

12.2.1 IEEE 802.15.4 Security Mechanisms

The IEEE 802.15.4 standard [2], which defines physical and link layer communications of resource-deprived devices, also defines some basic security mechanisms, which include access control, data confidentiality, frame integrity, and replay protection. These services are defined at the link layer and they are provided by means of three different security modes:

- Unsecured mode: It does not provide any security service.
- Access control list (ACL) mode: It ensures that a device only accepts frames from those entities included in the ACL, but it does not provide any mechanism to protect the transmission of these frames over the wireless link.
- Secured mode: It provides frame confidentiality, integrity, access control, and replay protection.

12.2.1.1 Basic Operation

The *access control* service implies that each device can select which other devices it accepts to communicate with. The IEEE 802.15.4 standard provides this service by the maintenance of ACLs in each protected device, specifying the devices from which it expects to receive frames.

Regarding the *data confidentiality* service, the IEEE 802.15.4 standard defines the use of the advanced encryption standard (AES) symmetric key algorithm over the payload of data, command, and beacon frames.

Similarly, *data integrity* is achieved by means of message integrity codes (MICs), which also serve to guarantee source authentication: the frame has been generated by the entity that owns the key necessary to generate the MIC and it has not been modified by any third entity. The integrity service is also applied to data, command, and beacon frames.

Finally, *replay protection* is achieved by using a counter, so that the recipient device compares the value embedded in each frame with the last value stored for the given communication and discards frames with an older value of the counter.

12.2.1.2 Frame Format

IEEE 802.15.4 frames convey in their header two security-related fields. Figure 12.2 shows the format of an IEEE 802.15.4 frame, highlighting these specific fields, whose meaning is detailed below.

- *Frame control:* This field specifies if the frame is protected by any security mechanism or not. The field *auxiliary security header* is only processed when the *security-enabled* bit is set to 1.
- Auxiliary security header: It consists of three subfields:
  - Security control (1 byte): It specifies the used security suite.
  - Frame counter (4 bytes): Unique sequence number is used to implement replay protection.
  - Key identifier (0–9 bytes): Identifier of the key used to protect the frame.

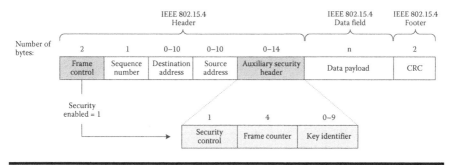

Figure 12.2   IEEE 802.15.4 frame format.

12.2.1.3 Performance Analysis

The implementation of security functionalities at the link layer implies mainly an increase in the IEEE 802.15.4 header of 5–14 bytes, which results in the corresponding increase of energy consumption due to the use of the radio transceiver to transmit these extra bytes.

Additionally, it also entails a computation cost derived from the encryption and decryption of data when data confidentiality is desired, as well as a storage cost associated with the storage of ACLs in nodes that implement access control functionalities.

12.2.2 TinySec Protocol

The objective of TinySec [3] is to define the most efficient security primitives to perform message encryption and MAC computation, with the final goal of providing authentication, integrity, and optionally, confidentiality to the packets of the popular TinyOS [4] operating system for sensor devices. This protocol, which is currently integrated in TinyOS, supports two different security options:

■ TinySec-Auth (authentication only): It is based on authenticating messages by the corresponding MAC, but the data field of the messages is transmitted unencrypted.
■ TinySec-AE (authenticated encryption): It is based on encrypting the data field of the messages and authenticating those messages by appending a MAC.

12.2.2.1 Basic Operation

TinySec implements encryption with semantic security, using Skipjack in cipher-block chaining (CBC) mode and an 8-byte initialization vector (IV).

Regarding the encryption algorithm, Skipjack is used because it is one of the fastest block ciphers currently available. The reason to select a block cipher, usually slower than most stream ciphers, is that block ciphers are stronger in the case of repeated IVs. Specifically, if two independent packets happen to be encrypted with the same IV using a stream cipher, then it is often possible to recover both plaintexts.

With respect to the IV, its length is a critical design decision, as too long IVs imply including unnecessary bits in the packets and thus, an increase of the energy consumed during transmission. Too short IVs, in turn, might result in repeated IVs, with the consequent security vulnerabilities. TinySec makes use of IVs composed as a concatenation of the following fields: destination and source addresses, the active message (AM) handler type (a concept similar to the port numbers in TCP/IP packets), the length of the data field, and a 16-bit counter. Therefore, although the used IV has an 8-byte length, it only implies an additional overhead of 4 bytes, as the other 4 correspond to fields already included in the message headers.

With regard to MAC computation, TinySec makes use of the CBC–MAC block cipher construction to compute and verify the authenticity of the exchanged packets. CBC–MAC is efficient and fast, and as it is based on the same algorithm used for encryption, it allows minimizing the number of cryptographic primitives to be implemented by the sensors and thus, it allows saving memory.

The security of CBC–MAC constructions is directly linked to the MAC size. Conventional security protocols, not oriented to resource-deprived devices, make use of 8 or 16-byte MACs. TinySec, in turn, with the objective of maximizing efficiency and minimizing resource consumption, makes use of short 4-byte MACs. The security level provided by 4-byte MACs is not enough for conventional scenarios. In the case of sensor networks, instead, the designers of TinySec argue that they provide an adequate security level, since due to the bandwidth and energy limitations of sensor networks, an attacker would exhaust the node's batteries before being able to forge a valid MAC.

12.2.2.2 Packet Format

The format of TinySec packets is based on the format of the packets of the TinyOS operating system. Figure 12.3 shows the differences between TinyOS and TinySec packets.

As we can observe in Figure 12.3, the common fields are the destination address, the AM handler type, and the length. These fields are transmitted unencrypted in order to enable the implementation of *early rejection* techniques when a node is not the destination of a given message, thus allowing energy saving. On the other hand, TinyOS packets include a 16-bit cyclic redundancy check (CRC) field for error detection. However, CRC codes do not provide protection against malicious modification or forgery of messages. Therefore, to guarantee the integrity and authenticity of messages, TinySec replaces the CRC code with a MAC, which also serves to detect transmission errors. Finally, TinyOS packets include a group field to prevent different sensor networks from interfering with each other. Given that in TinySec access control is guaranteed thanks to the MAC, this group field is no longer necessary.

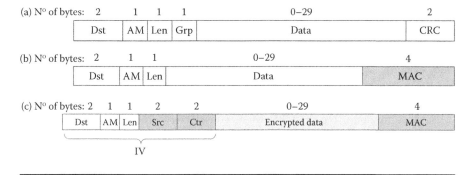

Figure 12.3   TinyOS and TinySec packet formats. (a) TinyOS packet format; (b) TinySec-Auth packet format; and (c) TinySec-AE packet format.

However, TinySec packets do not include any field to provide replay protection. The designers of the TinySec protocol argue that this issue can be more easily dealt with at the application layer.

12.2.2.3 Performance Analysis

The performance of TinySec is assessed in terms of three parameters: (1) additional packet overhead, which is independent from the specific implementation (2) energy cost, and (3) extra latency, which are directly linked to the characteristics of the specific implementation used for the evaluation.

Regarding the increase of packet lengths, TinySec-Auth implies a 1.5% increase compared to raw TinyOS packets and TinySec-AE, an 8% increase. On the other hand, experimental measures using Berkley Mica2 sensor nodes show that the implementation of TinySec-Auth entails a 3% increase on power consumption compared to transmitting the same amount of data using raw TinyOS packets, and TinySec-AE, a 10% increase. It must be noted that TinySec's energy consumption comes from two sources: cryptographic computations and additional transmission needs due to increased packet lengths. However, experimental measures show that the most important contributor is the energy cost due to extra transmissions. Finally, the experimental measures with respect to latency match the expected analytically computed values: TinySec-Auth implies an extra latency of 1-byte time per each IEEE 802.15.4 hop and TinySec-AE an extra 5-byte time per hop.

12.2.3 Link Layer Security Protocol

Link Layer Security Protocol (LLSP) [5] is a protocol based on TinySec, which aims at improving its efficiency. Therefore, the objective of LLSP is to efficiently implement message authentication, access control, confidentiality, and replay protection in TinyOS packets.

12.2.3.1 Basic Operation

LLSP makes use of AES encryption algorithm in CBC mode in order to guarantee the confidentiality of the transmitted messages. This encryption schema provides semantic security, guaranteeing that two consecutive encryptions of the same plaintext message produce different outputs.

Authentication and access control are achieved by including a MAC in each message, computed using the CBC-MAC construction. This way, it is possible to reuse the code related to the encryption algorithm for MAC calculation and consequently, save memory.

On the other hand, the most common way to provide replay protection is to maintain independent counters in the communicating endpoints, so that the receiver rejects packets with a counter value lower than the expected one. However, for the memory-deprived devices of sensor networks, it is unsuitable to maintain a counter for each of the other members of the network. In order to implement an efficient replay protection mechanism, LLSP takes some important assumptions. It assumes that each node has knowledge of the network topology and that the number of neighbors is small and does not change. This way, each sensor does not need to maintain a counter for every member of the sensor network, but only for its neighbors. According to LLSP, each node maintains a 4-byte synchronous counter with each of its neighbors. This counter is updated every time a new message is received using a feedback shift register (FSR). This way energy saving is possible, as it is not necessary to transmit the counter in each message.

However, the fact of not including the counter in each message can result in synchronism errors of the counter value when some packet is lost during transmission. In this sense, one of the problems of LLSP is that it does not define any mechanism to deal with this situation.

12.2.3.2 Packet Format

The LLSP packet format is based on TinySec packets. As it is shown in Figure 12.4, basically both packet formats differ just in the absence of the counter in the LLSP packet. However, like in TinySec, in LLSP the counter is included in the structure of the IV and in the MAC computation.

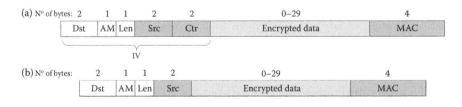

Figure 12.4 LLSP and TinySec–AE packet formats. (a) TinySec-AE packet format and (b) LLSP packet format.

As in the case of TinySec packets, LLSP message headers travel unencrypted, enabling the implementation of early rejection techniques. In this way, it is possible to save energy, just making the sensors switch off the radio transceiver when they detect that they are not the destination of the received message. The same happens with the MAC. As it is not encrypted, a receiver can check the authenticity of a message with low energy consumption.

12.2.3.3 Performance Analysis

The developers of LLSP use the PowerTOSSIM simulation environment to assess the performance of their protocol comparing it to the performance of the TinySec-AE protocol. The energy model used by PowerTOSSIM is based on the Mica2 sensor node. As LLSP packets are 2 bytes shorter than TinySec–AE packets, LLSP implies energy and time saving compared to TinySec-AE due to reduced packet overhead. More specifically, simulation results show that LLSP achieves a latency reduction of about 3% and a reduction of the energy cost of nearly 15% compared to TinySec-AE.

12.2.4 The MiniSec Protocol

MiniSec [6] provides data confidentiality, source authentication, and replay protection. This protocol presents two operating modes: one for unicast communications and another one for broadcast communications.

In order to achieve high efficiency and security levels, MiniSec is based on an offset code book (OCB) block cipher and Bloom filters [7]. The OCB mode of operation provides authenticated encryption, that is, it allows guaranteeing the authenticity and confidentiality of the encrypted information with a single pass of the block cipher. Additionally, MiniSec makes use of long IVs, providing a high security level. The cipher used by MiniSec is Skipjack, with a 64-bit block size. On the other hand, Bloom filters are a space-efficient data structure used for fast probabilistic membership tests. They basically provide two functionalities: membership addition and membership query. These structures can provide false positives, that is, the query may return true even when the element is not in the set. However, it is not possible to have false negatives.

Finally, in order to implement efficient replay protection in broadcast communications, MiniSec requires loose time synchronization between communicating endpoints.

12.2.4.1 Basic Operation of MiniSec-U (Unicast Communications)

MiniSec-U is based on a counter shared between each sender/receiver pair. This counter is incremented with each transmitted message and it is used as IV for OCB encryption. In order to optimize energy consumption, MiniSec-U does not transmit the whole counter in each message. Instead, only the last 3 bits of the counter are transmitted. Thanks to these 3 bits included in each packet, resynchronization is almost implicit in case of packet losses and thus, energetically more efficient than executing a specific resynchronization mechanism.

As a result of using the counter as IV for encryption, semantic security is ensured, as the IV value changes for each transmitted packet. Additionally, once a node has received a packet with a given value of the counter, it rejects all the messages with the same or a lower value, avoiding the possibility of replay attacks being carried out.

12.2.4.2 Basic Operation of MiniSec-B (Broadcast Communications)

MiniSec-U is not directly applicable to broadcast communications, since each node must store a counter value for each remote entity it communicates with, which would result in excessive memory consumption. For this reason, MiniSec-B was designed, mainly based on sliding-windows and Bloom filters.

The sliding-windows approach consists on dividing the time in discrete intervals, known as *epoch*s. Communicating entities agree upon the value of the current epoch (E_i) and use this value as IV for OCB encryption; this way, achieving protection against replay attacks from older epochs. In order to deal with synchronization errors at epoch transitions, two possible epoch values are always used for decryption: E_i and $E_i - 1$ when the message is received at the beginning of an epoch; and E_i and $E_i + 1$ when it is received at the end of an epoch.

However, this approach presents vulnerability since an attacker can replay a valid message for the whole duration of the epoch in which it was originally sent. This problem can be easily dealt with just by storing all the messages received during the vulnerability window. However, such an approach implies a high-memory cost. For this reason, MiniSec-B implements a solution based on storing a local counter in the sender and two Bloom filters in all the recipient entities.

The counter used in MiniSec-B is shorter than the one used in MiniSec-U, because the sender resets it at the beginning of each epoch. Consequently, the counter can be included in each sent message. In addition, this counter is not directly used as IV for OCB encryption, instead, a concatenation of the counter value, the source node identity, and the current epoch value (E_i) is used. On the other hand, the intended recipients maintain two Bloom filters to store the received packets: one for the current epoch and the other for the previous or next epoch. It is important to note that the implementation of rejection policies based on Bloom filters allows detecting all the replay attacks, since they do not provide false negatives. However, as these filters present a positive probability of false positives, a valid message could be identified as a replay and consequently, discarded.

12.2.4.3 Packet Format

MiniSec packets are based on the format of TinyOS packets. Figure 12.5 shows the structure of TinyOS, MiniSec-U, and MiniSec-B packets:

As shown in Figure 12.5, MiniSec packets share most of their fields with TinyOS packets. However, MiniSec replaces the TinyOS CRC field with a 4-byte MIC, which provides protection against any type of modification, intended or not. Similarly, the group field of TinyOS messages is not necessary in MiniSec, where access control is achieved by using different keys to encrypt the messages corresponding to different communications. Finally, MiniSec requires the inclusion of a 2-byte field to convey the source address.

Regarding the counter used for encryption, it is transmitted in a different way depending on whether it is a MiniSec-U packet or a MiniSec-B packet. In the case of MiniSec-U, just the last 3 bits of the counter must be transmitted. As the data field of TinyOS packets is never longer than 29 bytes, the length field can be overloaded and embed the counter information in the three most significant bits of the length field. This way, the counter information is sent without increasing the message length.

In the case of MiniSec-B short 8-bit counters are used. In order to avoid increasing the message length, some message fields are overloaded to transmit the counter bits. First, the first 3 bits of the counter are transmitted by overloading the length field, as in the case of MiniSec-U. The remaining 5 bits can be included in the destination field, since this field has a 2-byte length and it is unlikely to have more than 2048 broadcasting destinations in a network.

12.2.4.4 Performance Analysis

The developers of MiniSec carry out an analytical evaluation in which they compare the energy cost of both versions of MiniSec (MiniSec-U and MiniSec-B) with the

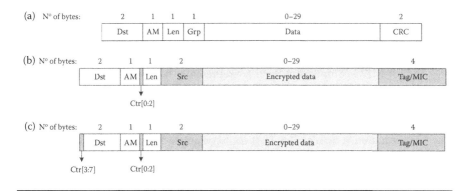

Figure 12.5 MiniSec and TinyOS packet formats. (a) TinyOS packet format; (b) MiniSec-U packet format; and (c) MiniSec-B packet format.

TinySec protocol. As MiniSec (both MiniSec-U and MiniSec-B) reduces the security overhead of 5 bytes introduced by TinySec in each packet to 3 bytes, it achieves an energy saving due to shorter transmission times. Specifically, MiniSec-U reduces the energy cost for each transmitted packet in nearly a 70% and MiniSec-B in about a 60%. However, for very high-packet loss probabilities (beyond 90%), implicit resynchronization is not probable in MiniSec-U and an explicit resynchronization mechanism, consisting of message exchanges, must be performed, resulting in a high increase of the consumed energy. The authors of the protocol argue that such scenarios are very rare in practice.

12.2.5 The SPINS Protocol

SPINS [8] allows implementing authentication, semantic security, replay protection, and freshness in unicast communications, and authentication and integrity in broadcast communications. This protocol is specifically designed to operate in a sensor network where all the communications travel across a more powerful and mains-powered base station (BS). Therefore, sensors establish a routing forest with the BS at the root of every routing tree. This way, every node can forward packets toward the BS, recognize packets addressed to it, and manage message broadcasts.

The SPINS protocol consists of two subprotocols, SNEP and μTESLA, each of them providing different functionalities.

12.2.5.1 Basic Operation of SNEP

The SNEP protocol provides source authentication, replay protection, semantic security, and weak message freshness. It is based on maintaining two counters (one for each direction of the communication) in each of the communicating entities. Therefore, a client C that wants to send a message M to a destination server S protecting it with SNEP, will send the following message:

$$\{M\}K_{CS}\|MAC(K'_{CS}, C_C\|\{M\}K_{CS})$$

where, K_{CS} denotes the symmetric key used for encryption, K'_{CS} the symmetric key used for MAC computation, and C_C represents the counter value in the communication direction $C \rightarrow S$. This counter value provides weak freshness, as the C_C value allows controlling the order in which the messages are sent by the client, but it does not allow associating a response message to its corresponding request. To face this limitation, a nonce value (N_1) could be appended to the previous message. This same nonce value should also be included in the response message, linking this way the response with its corresponding request. The approach followed by SNEP is that the client appends the N_1 value to the request message and the server responds including this same value in the MAC computation. This way, strong freshness would be achieved.

As the protocol operation is based on a counter value not transmitted in each message, it is necessary to implement some mechanism to allow communicating endpoints to agree upon the value of this counter, both at the beginning of a communication or during it, in the case of inconsistencies caused by message losses.

12.2.5.2 Basic Operation of μTESLA

μTESLA is the SPINS subprotocol in charge of guaranteeing authenticity in broadcast communications. Broadcast message authentication requires some asymmetric mechanism to prevent a malicious receiver from impersonating the sender. To avoid using asymmetric key cryptography, μTESLA introduces asymmetry by delaying the transmission of the corresponding symmetric keys. Therefore, all the communicating entities must have synchronized clocks and must know an upper bound on the maximum synchronization error.

μTESLA is based on the properties of public one-way functions. Remind that for a one-way function (F) is relatively easy to compute forward (i.e., to obtain K_j given $K_j + 1$); however, it is computationally unfeasible to compute backward (i.e., to obtain $K_j + 1$ given K_j). Therefore, first, the sender generates a one-way key chain by selecting a random last value (K_n) and successively applies the one-way function F to compute the rest of the values of the key chain: $K_i = F(K_i + 1)$. Additionally, time is divided in fixed intervals and each key is linked to a time interval, so that all the messages sent during that interval are authenticated using the same key. Keys are distributed by the source some time intervals after they have been used for message protection.

Hence, when a node receives an authenticated broadcast message, the key necessary to compute the MAC is only known by the legitimate sender. Therefore, the receiver stores the message until the corresponding symmetric key is distributed by the source. At this moment, the receiver checks the veracity of the distributed key and uses it to assess the integrity and authenticity of the stored message.

In order to verify the keys used by a certain sender, the receiver just needs to obtain in an authenticated way one of the keys of the key chain. Then it can authenticate the rest of the keys on the chain by applying the one-way function. That is, if the recipient obtains K_i in an authenticated way, it cannot compute $K_i + 1$, but once it receives $K_i + 1$, it can verify it by checking that $K_i = F(K_i + 1)$. Therefore, to implement this approach it is essential that first the receiver obtains one of the keys of the key chain (K_i) in an authenticated way. Additionally, the sender and the receiver must be synchronized and the receiver must know the timing used by the sender for key distribution. In order to establish the initial key (K_i), as well as the synchronism between sender and receiver, the message exchange depicted in Figure 12.6 is used, where N_1 is a nonce value used to link the request with its corresponding response, T_S represents the current time in the BS and K_i is the initial key needed to check the rest of the keys of the chain. Similarly, T_i represents the initial instant of the i-th interval, T_{int} the duration of that interval and δ the disclosure delay.

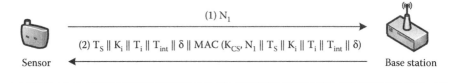

Figure 12.6   Message exchange for the µTESLA protocol initialization.

Due to the computation and storage capacities required by the source entity, µTESLA is commonly used to broadcast messages from the BS to all the members of the sensor network.

12.2.5.3 Performance Analysis

To evaluate the performance of SPINS, the protocol developers propose a specific implementation based on using the RC5 block cipher in counter mode (CTR mode) for encryption and the CBC–MAC construction for message authentication. According to this implementation, the cost of using SNEP consists of a 6 byte increase in the length of each message to provide source authentication and freshness, as well as the computation overhead derived from the execution of the corresponding cryptographic operations. For a 30-byte packet, the transmission of the additional 6 bytes implies an increase of 30% in the energy cost, while the execution of the cryptographic operations related to encryption and MAC computation are just an increase of 2%. In the case of µTESLA, the authentication cost per message is the same than in SNEP, but it also requires the periodic key disclosure. If keys are disclosed along with routing updates, key disclosure is nearly free. In any other case, key disclosure implies an additional message per time interval.

12.2.6 Location-Aware End-to-End Data Security

The objective of LEDS (Location-aware End-to-end Data Security) [9] is to provide end-to-end message confidentiality, authenticity, and availability within a sensor network. LEDS is designed to protect a specific type of communication: the transfer of an event report from the generating node to the BS. To that end, the protocol makes some general assumptions. It assumes a static sensor network consisting of a high amount of nodes uniformly deployed over a large area. The terrain where the sensor network is deployed is virtually divided into multiple cells, forming a virtual geographic grid. In addition, each node is assumed to have a unique identifier and to be able to determine its location within the sensor network.

With respect to the communications to be protected, LEDS envisions a sensor network where each event of interest is detected by multiple nodes and all of them agree upon a joint report. This report is then transmitted toward the BS, which is the final destination of all the communications. In this scenario, LEDS also allows for the implementation of message-filtering mechanisms. This way, intermediate entities can determine the authenticity of the messages they should forward and directly discard forged messages. The approach presented by LEDS is based on a one-to-many forwarding schema, where a message cannot be removed from the network just with one of the next-hop nodes discarding it. To this aim, LEDS makes use of a linear secret sharing (LSS) schema with (t, T) threshold. That is, an intended receiver is able to recover the original message with just t legitimate shares of messages consisting of a total of T shares.

12.2.6.1 Basic Operation

For its operation, LEDS relies on a number of cryptographic keys computed by each node. These keys are computed using geographic location data and a set of preconfigured information, so message exchanges are not required. In fact, each node computes three types of keys:

■ Two *unique secret keys* shared between each node and the BS and used to provide end-to-end source authentication.
■ A *cell key*, shared among all the nodes in a cell. This key is used to confidentiality protect the information exchanged among the nodes within a given cell.
■ A set of *authentication keys*, used to provide authentication in cell-to-cell forwarding of messages and to implement message-filtering mechanisms.

As previously mentioned, the objective of LEDS is to protect the event reports generated by the nodes of the network. In order to provide data confidentiality, the sent data are encrypted with the cell key of the cell where the event took place. In order to guarantee data availability, the encrypted report is divided into T unique shares using an LSS schema, so that the BS can recover the original message from a subset of t shares. In addition, multiple MACs are attached to each report, generated with the authentication keys of the nodes that generated the report. Thus, intermediate-forwarding nodes can detect and discard fake messages.

In order to implement such an approach, each report is generated by T nodes, which agree upon a joint report. Each participating node encrypts the report with the cell key and computes a unique part of the encrypted message by using an LSS schema. Next, each node sends a message with its identity and its share of encrypted message and gathers the $T - 1$ shares computed by the rest of the nodes of the cell. Once a node has all the independent shares of the encrypted message, it computes two MACs over all the shares, using two of its authentication keys. These authentication keys are selected so that they are shared with cells in the path toward the BS. Finally, each node broadcasts its identifier along with the computed MACs. This way, each node can obtain the MACs computed by the rest of the nodes of its cell. Therefore, each node can send a final report consisting of: (1) the identifier of the cell where the event took place, (2) the identifiers of the T nodes taking part in the report generation, (3) its share of the encrypted message, and (4) $T + 1$ MACs. The intermediate nodes that receive the report verify its authenticity and process it, after which, one of them forwards it toward the BS.

Once the report reaches the BS, it checks its authenticity, recovers the original message from t valid shares, and decrypts the received event report.

12.2.6.2 Performance Analysis

The developers of LEDS evaluate its performance mainly in terms of storage and communication overheads. Regarding storage cost, each node maintains two unique secret keys, shared with the BS; a cell key, shared with all other nodes in its cell; and a number of authentication keys, where the specific amount depends on the networks size and the location of the node within the network.

With respect to communication overhead, the main contributor is the $T + 1$ MACs that must be added to each message. This overhead can vary depending on the length of each MAC, finding this way a trade-off between performance and security.

12.2.7 Discussion

In this section, we have dealt with protocols oriented to protect data transmissions within sensor networks. Therefore, the main objective of these protocols is to define efficient mechanisms to guarantee the confidentiality and integrity of the transmitted messages. However, these protocols do not define any mechanism so that the communicating endpoints can agree upon the necessary cryptographic keys and they assume that these keys are preconfigured or obtained by means of appropriate protocols.

This section gathers protocols that operate at different layers of the protocol stack. The IEEE 802.15.4 standard defines some basic security mechanisms to protect frame transmissions between adjacent neighbors of a WSN at the link layer.

On the other hand, TinySec, LLSP, and MiniSec have been designed to protect a specific type of message, that is, the messages generated by the popular TinyOS operating system. Therefore, these protocols are aimed at defining efficient security primitives for encryption and MAC computation, as well as at minimizing the communication overhead derived from the implementation of these security services.

One of the drawbacks of TinySec is that it does not provide any mechanism to protect communications against replay attacks. LLSP, which is heavily based on TinySec, attempts to solve this problem by maintaining an independent counter in the communicating endpoints. However, this counter is not included in each transmitted message, which can give place to synchronism errors of the counter value. MiniSec, in turn, defines specific subprotocols to deal with unicast and broadcast communications separately. In the case of unicast communications, MiniSec tries to achieve a higher security level than TinySec by defining longer IVs. However, not to penalize energy consumption, instead of transmitting the whole IV in each message, MiniSec proposes transmitting just the three least significant bits. For broadcast communications, MiniSec relies for its operation on dividing the time in discrete intervals and using Bloom filters to store the messages received during each interval.

Finally, SPINS and LEDS have been designed to provide end-to-end security to generic network or application layer messages within the WSN. SPINS defines two subprotocols to deal with unicast and broadcast communications independently. In the case of unicast communications, it provides source authentication, semantic security, replay protection, and freshness through encryption and MAC computation. Regarding broadcast communications, its operation is based on the use of one-way key chains.

LEDS is aimed at protecting a very specific type of communication: reports generated by network nodes and sent to the BS. One of its disadvantages is the hurdle it implies for sensor nodes, as it requires that each node stores a high amount of keys and computes a high number of MACs.

Table 12.1 provides a comparison of the protocols presented in this section regarding the most significant security features that characterize these types of protocols, namely, the capacity to provide confidentiality, integrity and replay protections, as well as whether they require clock synchronization or not. In fact, the maintenance of synchronized clocks is a rather undesirable requirement in sensor environments.

12.3 Key Exchange Protocols for WSNs Based on Symmetric Key Cryptography

In this section, we will describe those security protocols oriented to the efficient establishment of symmetric cryptographic keys within a sensor network by solely relying on symmetric key cryptographic algorithms. For their operation, they generally make use of certain information configured in the nodes prior to their deployment that identifies them as legitimate members of the network and of cryptographic material shared with the BS. The established keys can be pair-wise keys, group keys, or network keys, depending on whether they are shared by two or more nodes of the network. The keys agreed upon can then be used to protect the information transmitted by sensors by means of encryption or MAC computation.

12.3.1 Localized Encryption and Authentication Protocol

LEAP (Localized Encryption and Authentication Protocol) [10] is a symmetric key exchange protocol aimed at operating on massively deployed and unattended static sensor networks. One of the objectives of this protocol is to allow in-network processing, for which it is necessary that intermediate nodes are able to decrypt or verify protected messages exchanged between two other network nodes.

Table 12.1 Comparison of Security Protocols for Integrity and Confidentiality Protecting Data Transmissions

Protocol	Preloaded Information	Confidentiality	Integrity	Replay Protections	No Syncronization Needed
IEEE 802.15.4	Two symmetric keys and ACL entries for all potential destinations	Optional	If pair-wise shared MAC key	✓	✓
TinySec	Two symmetric keys for each secure group	Optional	If pair-wise shared MAC key	✗	✓
LLSP	Two symmetric keys for each secure group, identities of all neighboring nodes, and initial counter value for each neighbor	✓	If pair-wise shared MAC key	✓	✓
MiniSec	MiniSec-U: Two symmetric keys and two counters MiniSec-B: Current epoch, counter, and two Bloom filters	✓	✓	✓	✗
SPINS	SNEP: Two symmetric keys and two counters µTESLA: One symmetric key, one key of the key chain, and key disclosure timing information	✓	✓	✓	✗
LEDS	Two symmetric keys, dimensions of the virtual grid, and location information	✓	Only between a sensor and the BS	✗	✓

LEAP supports the establishment of four types of keys:

- *Individual keys*: Keys shared between each node and the BS.
- *Pair-wise shared keys*: Pair-wise keys shared between a node and each of its neighbors.
- *Cluster keys*: Keys shared among a set of neighboring nodes. They are used to protect local broadcast messages and to allow the implementation of in-network processing techniques.
- *Group key*: Key shared among all the nodes of the sensor network. It is used to protect broadcast messages, usually sent by the BS.

12.3.1.1 Key Establishment Mechanisms

Individual keys are computed by applying a pseudorandom function to the master key (MK) shared between each node and the BS. Therefore, no message exchange is necessary.

Regarding pair-wise shared keys, the approach presented by LEAP does not allow a node to share a key with any other node of the sensor network, it is restricted to the establishment of shared keys between neighboring nodes. Therefore, when a node boots up it broadcasts its identity and as a response it receives a message with the identity of all its neighboring nodes. Then, pair-wise shared keys are computed taking as input parameters the identities of the two nodes that will share the key and a globally preset MK.

The mechanism envisioned for the establishment of cluster keys is based on the use of the previously computed pair-wise shared keys. When a node wants to establish a cluster key, it computes the desired key, individually encrypts it with the key it shares with each of its neighbors and finally, it sends the encrypted key to each of the intended neighbors. The destination neighbors, in turn, store the received cluster key and respond to the source node with their own cluster key.

The mechanism proposed by LEAP to establish a group key simply consists on preloading the group key in each node prior to its deployment. However, it is necessary to implement some mechanism by which this key can be updated, in order to avoid cryptanalysis attacks or if a network node is compromised. To this end, the μTESLA protocol is used. As explained in Section 12.2.5, this protocol is based on one-way key chains and it requires that all the nodes maintain synchronized clocks. Additionally, in order to authenticate the keys in a key chain, each node must obtain an initial key in an authenticated way. In LEAP this initial key is directly the group key preloaded in all the network nodes.

12.3.1.2 Local Broadcast Authentication

The objective of the local broadcast authentication is to provide authentication to the messages sent by a node of the sensor network. μTESLA is not suitable for the authentication of local broadcast messages, mainly by the introduced latency and the required storage capacity. The approach proposed by LEAP is based on the use of one-way key chains. Therefore, a node that wants to send authenticated broadcast messages, first generates a one-way key chain and sends the initial key of the chain to each of its neighbors, individually protecting it by using the pairwise shared keys. Then, to authenticate each of the broadcasted messages, the node attaches to each message the next key of the key chain. The neighboring nodes can check the authenticity of the received messages just by applying the one-way function to the key value received with the message and comparing the obtained result with their locally stored key value.

This solution is only valid to authenticate broadcast messages sent by a node to its immediate neighbors, but not to all the nodes of the sensor network. Additionally, it is subject to replay attacks, as an adversary can capture a legitimate message and send it afterwards impersonating the legitimate source node. In order to face such attacks LEAP proposes to use as authentication key, a key derived from the combination of the cluster key and the key of the one-way key chain. This way it is avoided the possibility of an external attacker impersonating one of the nodes of the cluster, but not the possibility that a node of the cluster impersonates one of its neighbors.

12.3.1.3 Performance Analysis

The developers of LEAP evaluate its performance in terms of computational, communication, and storage cost.

Specifically, the establishment of individual keys implies just the execution of a pseudorandom function. The establishment of pair-wise shared keys, in turn, implies the execution of a MAC and a pseudorandom function, the exchange of two short messages and the storage of the identities of each neighboring node and an MK. On the other hand, the establishment of a cluster key entails to each node an encryption and a decryption operation for each neighbor, the transmission and reception of a message for each neighbor and the storage of as many cluster keys as neighbors has the given node. Finally, updating the group key implies for each receiving node the reception of the broadcasted message and a decryption operation. Additionally, the broadcasting node must also store all the keys that constitute the key chain.

12.3.2 Lightweight Security Protocol

LiSP (Lightweight Security Protocol) [11] considers that in sensor network environments the security must be oriented to a group communication model. That is, all the legitimate nodes of the sensor network must share a secret key that allows them to exchange encrypted traffic. Based on the properties of one-way cryptographic functions, LiSP implements efficient mechanisms for the establishment of this group key, as well as for its update when a node leaves the network or periodically as a protection against cryptanalysis attacks.

12.3.2.1 Protocol Architecture

In LiSP, the sensor network is divided into multiple groups, each of them managed by a control node, known also as *group head* (GH). Therefore, the sensor network has a hierarchical architecture where control nodes communicate with each other as well as with entities outside the sensor network. Data collection nodes, in turn, can directly communicate with any other node of their group, but not with nodes of other groups or entities outside the sensor network. These types of communications are carried out through the GH.

On the other hand, LiSP introduces the concept of *key server* (KS), which is implemented by each GH. Additionally, there is a global KS, known as *key server for the network* (KSN), which is in charge of coordinating the different group KSs in order to allow intergroup communications.

Moreover, LiSP makes use of two types of keys: (1) *temporal keys* (TK), used to protect the packets exchanged among members of the same group, and (2) *master keys* (MK), which are pair-wise keys shared between each node and its group GH. These MKs are assumed to be preconfigured.

Related to the management of these types of keys, the protocol assumes the existence of two modules in the network: (1) an *intrusion detection system* (IDS), whose objective is to detect compromised nodes, and (2) a *TK manager*, which is in charge of periodically redistributing the group TK. Therefore, each KS implements an IDS and a TK manager module. Similarly, to avoid the existence of compromised KSs in the network, an IDS located in a more powerful node controls the security of the network KSs.

Finally, it is worth noting that for its operation, LiSP assumes that the clocks of all the nodes that form a group are permanently synchronized.

12.3.2.2 Key Establishment Mechanisms

LiSP relies on the properties of one-way cryptographic functions and on dividing time into discrete intervals, which determine the periodicity with which TK keys are distributed.

Initially, the KS computes a one-way key chain. Thanks to the characteristics of one-way key chains, any entity can compute TK_i if it knows TK_{i+1}, just by applying the one-way cryptographic function ($TK_i = H(TK_{i+1})$). However, nobody that knows TK_i can compute TK_{i+1}. Then, it sends an *InitKey* message to each of the members of the group conveying an initial value of the key chain TK_{t+2}. With this value each node is able to compute the rest of the keys of the chain ($TK_{t+1}, ..., TK_1$). In order to protect *InitKey* messages, they are individually encrypted and authenticated with each destination's MK.

After this initialization mechanism, each node starts using TK_1 for encryption and activates a timer configured with the duration of a time interval. When this timer expires, the node activates the next key of the key chain for encryption and sets the timer again.

Then, periodically the KS distributes a new key of the chain (TK_{t+3}) by means of *UpdateKey* messages. This key is authenticated by executing the one-way function and checking that $TK_{t+2} = H(TK_{t+3})$. If authentication is successful, the received value is added to the chain maintained by each node. Additionally, thanks to the properties of one-way functions, in case that any *UpdateKey* message is lost during transmission, the key chain value embedded in further *UpdateKey* messages is used to compute intermediate lost values, just by applying the one-way function to the last value received. However, as nodes have limited buffers for storing the keys, this recovering mechanism is only valid when the number of consecutively lost *UpdateKey* messages is smaller than the buffer size. To recover from greater amount of consecutively lost *UpdateKey* messages, the affected node sends a *RequestKey* message to the KS, which responds with a new *InitKey* message.

In order to face possible synchronization errors, LiSP defines a set of decryption keys consisting of the current encryption key and the encryption keys corresponding to the previous and the next time intervals. Additionally, to each message sent within a group a MAC is also attached in order to guarantee its authenticity. Finally, to allow secure communications between different groups, the KSs maintain their own TKs, which are distributed and updated by the KSN.

12.3.2.3 Performance Analysis

The developers of LiSP evaluate the performance of the proposed TK management scheme in terms of computational and communication overhead. The obtained results show that on average, the KS must perform a hash computation per TK disclosure and each client less than three hash computations, even when half of the TK broadcasts get corrupted. On the other hand, the communication overhead is determined by the key buffer length. Using a buffer that can store five or six keys, the normalized communication cost is nearly one, even in the case of frequent corruption of TK broadcasts.

12.3.3 Peer Intermediaries for Key Establishment

The basis of PIKE (Peer Intermediaries for Key Establishment) [12] is using peer sensor nodes as trusted intermediaries for key distribution. This way, PIKE allows the establishment of pair-wise keys between any two sensor nodes regardless of the network topology or node density.

For its operation, PIKE assumes that all the network nodes are globally addressable, that is, any node in the sensor network can communicate with any other node in the same network through intermediary hops.

12.3.3.1 Basic Key Establishment Mechanisms

In PIKE, each sensor node is configured with an identifier of the form (x, y), where $x, y \in \{0, 1, 2, ..., \sqrt{n} - 1\}$, n being the maximum number of nodes in the sensor network. Each of these nodes is configured with $2(\sqrt{n} - 1)$ secret keys, shared with those other sensor nodes with which the node shares a common x or y identifier index. That is, the node with the identifier (x,y) is initially configured with secret keys shared with the nodes with identifiers of the form (i, y) and (x, j), where i and $j \in \{0, 1, 2, ..., \sqrt{n} -1\}$. It must be noted that the identifier indexes x and y are not related to the geographic location of the nodes.

Therefore, two nodes A and B with identifiers (x_A,y_A) and (x_B,y_B) respectively, can always find two intermediary nodes that have shared keys with both A and B. The identifiers of these intermediary nodes will be (x_A,y_B) and (x_B,y_A). Once the identifiers of the two possible intermediaries have been found, the node that initiates the key establishment process (for instance, A), selects one of them (C), and sends to it the key to be shared with B, encrypted with the key shared between A and C. The intermediary node decrypts the key received from A, encrypts it again with the key it shares with the destination node (B), and sends it to B. Finally, in order to acknowledge the reception of the key, the destination node (B) sends to the source node (A) a nonce value encrypted with the new key established between them and a MAC computed with this same key. This message exchange can be represented as shown in Figure 12.7.

Therefore, in PIKE, it is assumed that the established keys are secure, as long as the endpoint nodes (A and B) and the intermediary node (C) are not compromised.

Regarding the network deployment, nodes are to be deployed following the order of their identifiers, that is, for example in the following order: $(0,0)$, $(0,1)$,... $(0, \sqrt{n})$,$(1,0)$,..., $(1, \sqrt{n})$, and so on. This way it is guaranteed that the space of the identifiers of the deployed nodes has always a rectangular form and thus, it is highly probable that two intermediaries will be available, or at least one.

12.3.3.2 Extension of the Basic Key Establishment Mechanisms to Three or More Dimensions

The basic PIKE protocol makes use of two-dimensional identifiers. However, the same concepts can be extended to identifiers with three or more dimensions.

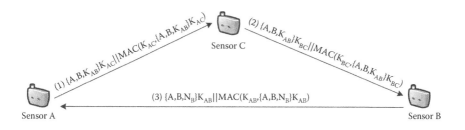

Figure 12.7   Basic message exchange of the PIKE protocol operation.

In the case of identifiers with three dimensions, each node a will have an identity of the form (a_1, a_2, a_3) and will maintain secret keys shared with those nodes that have two common identifier indexes, that is, with the nodes with identifiers of the form (i, a_2, a_3), (a_1, i, a_3), and (a_1, a_2, i), being $i \in \{0, 1, 2, \ldots, \sqrt[3]{n}\}$. With such a configuration, the messages exchanged for the key establishment must travel through at most two intermediaries.

Thanks to increasing the number of dimensions of the used identifiers, it is possible to save memory, as the number of secret keys to be stored by each node is reduced. However, it implies an additional cost regarding the communication necessities and the security of the established keys. In fact, the established keys will usually depend on the security of two intermediaries, instead of just one.

12.3.3.3 Using Newly Established Keys to Enhance the Basic Key Establishment Mechanisms

During the execution of PIKE, new keys are established between pairs of nodes. These keys can also be used, as the originally stored secret keys, to help establishing new keys between other pairs of nodes. This way, it is possible to reduce the number of intermediaries involved in a key establishment process, and consequently, its communication cost.

Nevertheless, the fact of using newly established keys to help establish further keys has a negative impact from the security viewpoint. While a first-generation key depends on the security of one intermediary, a second-generation key usually depends on two first-generation keys and thus, on three intermediaries. In the same way, a third-generation key will typically depend on seven intermediaries. Consequently, it is essential to limit the use of newly established keys for the establishment of additional keys, since the compromise of one of the first-generation keys implies that all the keys that depend on the compromised key are also compromised. Therefore, if no limitations are enforced, the compromise of a single key could derive in the compromise of the whole sensor network.

12.3.3.4 Performance Analysis

The PIKE protocol entails high-memory consumption in the network nodes in order to store the initially deployed keys as well as the new keys established during the execution of the protocol. In a basic scheme using two-dimensional node identities, each node must initially store $2(\sqrt{n} - 1)$ keys. Similarly, using three-dimensional node identities, each node must initially store $6(\sqrt[3]{n} - 1)$ keys. These figures can be reduced to half using a specific key derivation method. However, this method implies additional computation overhead to derive the keys which are not initially stored in the node. Regarding communication overhead of each network node, the protocol developers show through simulation that it is proportional to $\sqrt{n}$ and grows with the network size.

12.3.4 The Protocol Proposed by Liu and Ning

Liu and Ning [13] present a general framework for the establishment of pair-wise cryptographic keys, based on a polynomial algorithm and on a probabilistic distribution of keys. Following this approach, two specific schemas are proposed: one based on the random assignment of polynomial subsets and another based on grid architecture.

The fundamental idea is to use t-degree polynomials generated as $f(x,y) = \sum_{i,j=0}^{t} a_{ij} x^i y^j$, over a finite field F_q, q being a large enough prime as to accommodate a cryptographic key so that $f(x,y) = f(y,x)$. Therefore, for each sensor i, the configuration server computes a subset $f(i,y)$ of $f(x,y)$. This way, in case that nodes i and j want to establish a shared key, i node will calculate it by evaluating $f(i,y)$ in the point j, and j node will obtain the same key by evaluating $f(j,y)$ in the point i, since $f(i,j) = f(j,i)$.

Such an approach is proven to be secure as long as the number of compromised nodes in the network is not greater than t. However, the previous mechanism is not directly applicable to a sensor network, since the storage cost of a polynomial subset grows exponentially with the size of the group.

12.3.4.1 Basic Key Exchange Mechanism

On the basis of the previously introduced polynomial system, Liu and Ning propose a general key distribution framework tailored to the characteristics of sensor networks. To that end, first each sensor is initialized with a polynomial subset. Then, when two nodes want to establish a shared key, they check if they share a common polynomial. In a positive case, they use the previously explained mechanism to compute the shared key.

Therefore, some mechanism must be implemented so that two nodes are able to determine whether they share a polynomial or not. One solution is that the configuration server predistributes certain information to each node so that, with the identifier of the targeted endpoint, it can be determined whether a shared polynomial exists or not. Another solution is real-time discovering, which is based on the communicating nodes exchanging the identifiers of the polynomials they own and searching for coincidences. These communications must also be protected by appropriate mechanisms, thus involving a large communication overhead.

Nevertheless, it can happen that the given i and j nodes do not share a common polynomial. In this case, a mechanism known as *path key establishment* must be executed, which requires collaboration from other network nodes. This mechanism consists of finding a path between i and j through intermediate network nodes with which pair-wise keys can be directly established, thanks to shared polynomials. Then, either of the nodes, i or j, can send a request to establish a key with the other node through the previously established path. The problem here is how to find the path between nodes i and j. As before, two mechanisms are possible: predistribution of information by the configuration server or real-time discovery.

12.3.4.2 Key Exchange Mechanism Based on Random Polynomial Subset Assignment

This schema is based on an instantiation of the general procedure described before. First, the configuration server generates a polynomial set and assigns a random subset of it to each node. Then, each node broadcasts a list with the identifiers of the polynomials it owns. In this way, it is possible to determine if a key between two nodes can be directly established.

If the direct key establishment is not possible, a path discovery mechanism must be carried out. To that end, one of the nodes broadcasts a request message conveying two lists of polynomial identifiers: one for the polynomials owned by the source node and another for the polynomials owned by the destination node. If any of the entities receiving this message is able to establish a direct key with both the source and the destination nodes, it responds with a message that contains two instances of a randomly generated key, one encrypted with the direct key shared with the source node and the other encrypted with the direct key shared with the destination node.

12.3.4.3 Key Exchange Mechanism Based on a Grid Architecture

This approach is based on dividing the sensor network in a grid so that each row i has an associated polynomial $f_i^r(x, y)$ and each column j an associated polynomial $f_j^c(x, y)$. Additionally, the configuration server assigns to each node an identifier that allows identifying its location in a specific row and column within the grid.

Direct key establishment can be carried out when both nodes have a common row or a common column. In any other case, a path key must be established. To that end, both nodes use their row and column identifiers to search an intermediate node that can directly establish a shared key with both of them. This schema gets more and more complex when the number of intermediate hops increases. For this reason, the protocol designers limit the distance between the two endpoints to two intermediate hops.

12.3.4.4 Performance Analysis

The protocol developers evaluate its performance in terms of the number of processor operations that a sensor must perform in order to evaluate a t-degree polynomial, which typically implies large integer operations. In order to minimize this computation overhead, the protocol designers propose a scheme based on suitably choosing the node IDs, which are in fact, the points in which the polynomials must be evaluated.

However, apart from this high-computational cost, the protocol implies also an important storage cost, associated with the polynomial subset initially assigned to each node. Additionally, it also entails a communication cost, derived from the broadcast messages that must be sent during the execution of the path discovery mechanisms when the endpoint nodes do not initially share a common polynomial.

12.3.5 Discussion

This section briefly describes the most relevant key establishment protocols specifically designed for sensor environments that solely make use of symmetric key cryptographic algorithms. All these protocols base their operation on information preloaded in the sensor nodes and usually shared with the BS.

LEAP implements a complete framework for key establishment in sensor networks. However, the procedures proposed by this protocol are complex and they require that the sensor nodes maintain a large amount of information regarding the state of neighboring nodes and the network topology, such as shared keys, lists of neighboring nodes, and so on. In fact, each node must store $3d + 2 + L$ keys, d being the number of neighbors and L the length of the one-way key chain.

On the other hand, LiSP is specifically designed for the efficient distribution of group keys in sensor environments, and it is not suitable for the establishment of pair-wise keys. Additionally, LiSP requires some stringent characteristics from the nodes that execute the protocol: they must be able to maintain synchronized clocks with every other member of their group and they need to store permanently $t + 2$ keys, where t defines the length of the key buffer used to face potential packet losses.

PIKE, in turn, is based on using a third node of the sensor network as a trusted third party for the key establishment between two other nodes of the network. For its operation, this protocol requires that each node be loaded with $2(\sqrt{n} - 1)$ secret keys. Additionally, one of the main disadvantages of PIKE is that the secret key established between the communicating endpoints is also known by the intermediary entity. Therefore, the compromise of a node does not involve just the compromise of the node's communications, but also the compromise of the communications of the rest of node pairs for which it has acted as intermediary.

Finally, Liu and Ning's approach provides some general notions regarding the use of polynomials for pair-wise key establishment in sensor networks, but some specific procedures, such as the introduction of new nodes in the network or the revocation of compromised nodes are missing. On the other hand, the overhead introduced by path discovery mechanisms when a pair-wise key cannot be directly established between two nodes grows quickly with the number of intermediate hops. Additionally, in this case, the key to be shared is generated by a third entity, which will also have knowledge of it.

A comparison of the functionalities of the protocols gathered in Section 12.3 is provided in Table 12.2, where the most relevant features of these types of protocols

Table 12.2   Comparison of Symmetric Key-Based Key Exchange Protocols for Sensor Networks

Protocol	Preloaded Information	Pair-Wise Keys	Group Keys	Replay Protection	No Syncronization Needed
LEAP	Identities of neighboring nodes and one symmetric key	Not secret	✓	✗	✗
LiSP	One symmetric key	✗	✓	Only of *UpdateKey* messages	✗
PIKE	$2(\sqrt{n} - 1)$ symmetric keys	✓	✗	✗	✓
Liu and Ning's approach	Polynomial set	✓	✗	✗	✓

are considered. Specifically, it is detailed whether each protocol can be used for the establishment of pair-wise and group keys, if replay protection is provided and whether the protocol can operate without the need of synchronized clocks.

12.4 Key Exchange Protocols for WSNs Based on Asymmetric Key Cryptography

Although in general terms, public key cryptography is not considered a suitable alternative to operate on sensor devices due to the high computational and storage requirements that it entails, recently several efficient public key-based approaches are emerging. The basis of such approaches is to make use of a reduced set of functionalities of this type of technology, and mainly of elliptic curve cryptography (ECC), to efficiently implement authentication and key establishment mechanisms in sensor networks. In this regard, it is worth noting that the use of public key cryptography is always restricted to the authentication phase. From then on, a symmetric key is generated and shared between the communicating pair and further data encryption and authentication is always performed by means of symmetric key cryptographic procedures.

12.4.1 The TinyPK Protocol

The objective of the TinyPK protocol [14] is to provide the necessary functionalities so that a sensor network and any other third entity can mutually authenticate and securely transfer the sensor network's session key to the external party. This key can then be used to protect the traffic exchanged between the sensor network and the external entity. Therefore, TinyPK assumes that all the members of the sensor network share a session key that identifies them as legitimate members of the network and that is used to protect the traffic exchanged within the sensor network.

For its operation, TinyPK requires the introduction in the sensor network of a certification authority (CA), which is an entity with a pair of public/private keys that are trusted by all the legitimate network members. Any entity willing to interact with a sensor will need its own public/private key pair and will need to have its public key signed with the CA's private key, thus establishing its identity in a reliable way. Finally, each sensor will have the CA's public key installed before network deployment. The most common way to distribute public keys is by means of certificates. However, taking into account the limitations of the targeted sensor devices, TinyPK does not use certificates.

TinyPK makes use of a Rivest, Shamir and Adleman (RSA) system and of key distribution techniques based on the Diffie–Hellman protocol. Regarding the RSA system, TinyPK uses e = 3 as the public exponent, which is a rather low value for the implementation of security features in conventional environments and can lead to security vulnerabilities. The designers of the TinyPK protocol argue that by making suitable use of the random padding during data encryption it is possible to face those vulnerabilities. Additionally, in order to make RSA feasible for the targeted resource-deprived devices, TinyPK is designed so that those devices must only perform encryption and signature check operations, which in the case of the RSA system have the advantage of being very fast, compared to other public key technology computations.

12.4.1.1 Mutual Authentication Mechanisms

The mechanism introduced by TinyPK to authenticate any third entity to a sensor network is based on a challenge/response approach. First, the remote third entity sends a challenge to the sensor, which consists of its public key signed by the CA and a certain text signed with its own private key. The signed text corresponds to a nonce value and a message checksum. The nonce value provides replay protection, while the checksum serves to guarantee message integrity.

When a sensor receives such a challenge, it uses the preconfigured CA's public key to verify the public key sent by the third entity. Then, it uses this public key to check the nonce value and checksum included in the message. If the verification ends up successfully, the sensor relies on the identity of the third party and transmits the sensor network's session key. To this end, the sensor encrypts the session key and the received nonce value with the public key of the remote entity.

When the authentication takes place in the opposite direction, it is not possible to implement a mechanism based on signatures, since sensors do not have enough capacity to perform full RSA operations. Alternatively, TinyPK proposes that each sensor uses as a credential of its identity, a pair of Diffie–Hellman static keys along with a text string encrypted with the CA's private key. This text string contains some identity features of the given sensor, such as its sequence number, manufacture date, and so on.

To verify the identity of a sensor, the remote third party must carry out a semistatic key exchange with the sensor. Specifically, the remote third party sends a challenge to the sensor that consists, as in the previous case, of its public key signed by the CA and a nonce/checksum pair signed with its own private key. This communication is protected by using the sensor network's session key, previously obtained during the authentication of the remote third party against the sensor. The destination node verifies the request and responds with a message including its credential and the received nonce value, encrypted with the third party's public key. The remote third entity checks the response message sent by the sensor and if it is correct, it stores the sensor's ID number along with the session key. Therefore, the next authenticated messages sent by the sensor will include the ID number along with a MAC computed using the appropriate session key.

12.4.1.2 Key Establishment Mechanism

TinyPK considers a key exchange mechanism based on the Diffie–Hellman protocol. Specifically, one of the nodes generates a random number $R1$ and computes the function $g^{R1} \bmod p$, whose output sends to the remote node. The remote node, in turn, generates also a random number $R2$ and computes the function $g^{R2} \bmod p$, whose output sends to the first node. Therefore, the shared key is computed by both endpoints as: $(g^{R1} \bmod p)^{R2} \bmod p = (g^{R2} \bmod p)^{R1} \bmod p = g^{R1*R2} \bmod p$.

12.4.1.3 Performance Analysis

In order to implement TinyPK, a sensor must permanently store the public key of the CA, a pair of Diffie–Hellman static keys and its identity information signed with the CA's private key.

Additionally, for the authentication of the remote entity, the sensor must perform a transmission and a reception operation, two decryption operations and an encryption operation. The cost of authenticating the sensor identity to the remote entity is similar, but it entails an additional decryption operation.

Finally, the key establishment mechanism implies a message reception and transmission, the generation of a prime number and the computation of the functions $g^{R1} \bmod p$ and $(g^{R1} \bmod p)^{R2} \bmod p$.

12.4.2 Lightweight Security Protocol

LSec (Lightweight Security protocol) [15] implements a simple key exchange mechanism that allows providing node authentication and authorization functionalities as well as data confidentiality and protection against intrusions and anomalies.

12.4.2.1 Protocol Architecture

The operation of the LSec protocol is based on a number of modules implemented by the BS.

- Key management module (KMM): The module in charge of storing the public keys of all the nodes and the secret symmetric keys each node shares with the BS.
- Token generator module (TGM): It is responsible for generating tokens, which are the credentials that requesting nodes must present to requested nodes as proof of their identity.
- Authorization module (AzM): This module determines whether a given requesting node is authorized to communicate with a specific targeted node or not.
- Intrusion detection system (IDS): It is in charge of detecting intrusions or network anomalies.

Additionally, LSec assumes that the sensor network is divided into groups, known as *clusters*, and that each cluster is managed by a special node known as *cluster head*. In LSec the cluster heads are responsible for detecting network intrusions or anomalies and for sending the corresponding alerts to the IDS module.

12.4.2.2 Key Establishment Mechanisms

The operation of the LSec protocol consists of three phases: (1) authentication and authorization phase, (2) key distribution phase, and (3) data transmission phase.

The authentication and authorization phase solely relies on the utilization of symmetric key cryptography. When a client C wants to communicate with a server S, first it sends a *Request* message to the BS, specifying the identity of the destination node it wants to communicate with and including a nonce value, used to match the current request with its corresponding response. This *Request* message is encrypted with the symmetric key the client node shares with the BS. The BS checks if the requesting client is allowed to communicate with the desired server and only in a positive case, it generates the corresponding token and sends it to the requesting client in a *Response* message. The generated token is encrypted with the secret key shared between the BS and the destination server (S), so the client node cannot access or modify the information contained in it. The *Response* message, in turn, is encrypted with the key shared between the BS and the requesting node and it contains the targeted server's public key and the same nonce conveyed in the *Request* message.

The key distribution phase is based on the use of public key cryptography and its objective is that the client and server nodes end up sharing a symmetric secret key. For that purpose, the client node sends an *Init* message, including a new nonce, its own public key and the token provided by the BS. Additionally, all this information is encrypted with the intended server's public key. In order to assess the identity of the requesting client, the server checks the token and if it is correct, it generates a random key and sends it to the client in an *Ack* message, which is encrypted with the client's public key.

The symmetric key conveyed within the *Ack* message, is used afterwards, during the data transmission phase, to protect the communications between the given client/server node pair.

12.4.2.3 Performance Analysis

LSec requires that each node stores five keys (its public/private keys, the secret key shared with the BS, the public key of the remote node, and a shared secret key). Additionally, for every connection, the protocol implies the transmission of four control packets (Request, Response, Init, and Ack) during the authentication, authorization and key exchange phases, and the corresponding encryption/decryption operations over the protected fields of those messages.

The protocol developers show by means of the SENSE sensor network simulator and emulator system that most of the energy consumed during the execution of the protocol comes from transmission/reception operations, while cryptographic operations are less expensive from the energy cost point of view.

12.4.3 Protocol Presented by Lin and Tseng

The protocol presented by Lin and Tseng [16] consists of a key exchange mechanism based on the identifiers of the participating entities. This protocol allows a sensor to establish shared cryptographic keys with all its one-hop neighbors. One of the main advantages of this protocol is that it scales well, since the amount of node memory required remains constant regardless of the number of immediate neighbors it establishes a key with. The proposed mechanism is based on ECC, which requires shorter keys and lower computation capacity than traditional public key cryptography to achieve the same security level.

12.4.3.1 Basic Concepts

LSec assumes the existence in the network of a registration server (RS), which is responsible for generating each node's public/private key pairs and assigning them to the corresponding node.

The basic protocol consists of two main phases: (1) the key issuing phase and (2) the key agreement phase.

Initially, the RS creates an elliptic curve group EC and selects a generator G of EC. Then, the RS selects the master private key (K_{RS}^-) and computes the corresponding public key as: $K_{RS}^+ = K_{RS}^- \cdot G$.

Afterwards, when a node wants to establish shared keys with its neighbors, it starts the key issuing phase. The node i sends its identity (ID_i) to the RS, which generates a random number t_i and computes i node's public and private keys as $K_i^+ = t_i \cdot G$ and $K_i^- = t_i + K_{RS}^- \cdot h(ID_i \,\|\, x(K_i^+))$ *mod q* respectively. Here $h()$ denotes a one-way key function, $x(Q)$ the x coordinate of a point Q in EC and q the order of the generator G.

To end this phase, the RS sends to the node i the values G, q, K_{RS}^+, K_i^+, and K_i^-.

The key agreement phase takes place between the node i and its neighbor j, whose public and private keys are denoted as K_i^+/K_i^- and K_j^+/K_j^- and have been already obtained by the execution of the key issuing phase. To establish a shared key, node i generates a random number r_i and computes $V_i = r_i \cdot G$. Then, it makes use of its secret key to compute $w_i = r_i + K_i^- \, x(V_i)$ *mod q* and sends to node j the values V_i, K_i^+, and ID_i. Upon receiving these values, node j executes the symmetric operations. That is, it generates a random number r_j and computes $V_j = r_j \cdot G$. Then, it uses its secret key to compute $w_j = r_j + K_j^- \, x(V_j)$ *mod q*, and sends the values V_j, K_j^+, and ID_j to the node i. Thanks to the exchanged information and to the properties of the ECC, both nodes are able to independently compute a shared key.

12.4.3.2 Key Establishment Mechanisms

This protocol is based on the basic concepts introduced in the previous section, but it is specifically tailored to the characteristics of the sensor networks. The resulting protocol consists of four phases.

In the *manufacture phase*, the sensor manufacturer performs the tasks corresponding to the key issuing phase, providing each sensor with a public/private key pair. Additionally, in order to optimize resource consumption in the sensors, the manufacturer computes several sets of (r_i, V_i, w_i) values on behalf of each sensor and installs them in the corresponding node.

In the *network deploying phase,* the node chooses an instance of (r_i, V_i, w_i) from its stored set and broadcasts the values (V_i, K_i^+, ID_i) to its neighboring nodes. Each neighboring node responds with its own (V_j, K_j^+, ID_j) values. Consequently, nodes i and j already have the necessary information to compute a shared key.

Afterwards, a *rekeying phase* might be necessary if a node wants to establish new shared keys with its neighbors, for example, because current keys have expired or have been compromised. To that end, node i simply selects new (r_i, V_i, w_i) values from its stored set and broadcasts (V_i, ID_i).

Finally, the *adding new nodes phase* allows including new nodes in an already deployed network. This phase simply consists on the new node broadcasting its

(V_N, K_N^+, ID_N) values and the adjacent nodes proceeding as in the network-deploying phase.

12.4.3.3 Performance Analysis

This protocol presents a low-communication cost, as it just implies one broadcast message. However, it entails an important computation overhead, since costly operations must be performed, such as point scalar multiplications, point additions, one-way hash function executions, and modular multiplications. Nevertheless, the protocol designers argue that the protocol is well suited for the targeted sensor devices because it makes use of ECC. Additionally, it also entails a high-storage cost due to the storage of (r_i, V_i, w_i) values.

12.4.4 Discussion

Section 12.4 includes key establishment protocols for WSNs that make use of asymmetric key cryptography.

Among these approaches, TinyPK implements mutual authentication between a sensor network and an external third entity. To that end, the protocol assumes the existence of a CA trusted by all the members of the sensor network. However, TinyPK does not specify any procedure so that the CA can distribute its public key, and it assumes that it is preconfigured in each sensor prior to the network deployment.

LSec is a simple and lightweight security protocol. However, it presents some important vulnerabilities. First, a crucial assumption of the protocol is that only the BS has knowledge of the public keys of the sensor network nodes. Nevertheless, these keys are distributed during the protocol operation, so a malicious node could store them and use an old token to impersonate some other network entity. Similarly, revocation of permissions is not straightforward. Once a client node has obtained a valid token and a public key to communicate with a given server, it can afterwards reuse those same values to access the server, even if its permissions have now been revoked in the BS.

Finally, the protocol proposed by Lin and Tseng provides simple key exchange mechanisms, but it is restricted to the establishment of shared keys between each node and its adjacent neighbors. Therefore, it cannot be used to establish a shared key between any two nodes of the sensor network. On the other hand, in order to make public key cryptography available to sensor devices, this protocol is based on performing the expensive cryptographic operations offline in a more powerful third entity and storing the computed values in each sensor prior to its deployment. This implies an important storage cost for the sensor devices if it must be guaranteed that they do not run out of cryptographic material.

Table 12.3 summarizes the main functionalities of the protocols presented in this section. These functionalities are the same as the ones used in the comparison

Table 12.3   Comparison of Asymmetric Key-Based Key Exchange Protocols for Sensor Networks

Protocol	Preloaded Information	Pair-Wise Keys	Group Keys	Replay Protection	No Syncronization Needed
TinyPK	CA's public key, static Diffie–Hellman key pair, one symmetric key	✓	✓	Weak	✓
LSec	One symmetric key and one public/private key pair	✓	✗	✗	✓
Lin and Tseng's approach	Public/private key pair and (r_i, V_i, w_i) point information sets	✓	✗	✗	✓

of symmetric key-based key exchange protocols in Table 12.2. That is, the capacity to establish pair-wise and group keys, whether replay protection is provided and whether clock synchronization is required.

12.5 End-to-End Authentication, Authorization, and Key Exchange Protocols for IP-Enabled Sensors

This section aims at completing the previous study with protocols oriented to protect the access to a new type of sensor that has recently emerged, that is, IP-enabled sensors. The security services required by this type of sensor are radically different to the security requirements of traditional sensor networks and they are similar to the security needs of any other Internet-connected information or application server. Although the protection of Internet-connected servers is a long-researched issue, traditional security mechanisms, meant for powerful workstations, cannot be directly implemented in sensor devices, due to the stringent resource and energy limitations these devices pose and thus, specific security protocols for sensor environments must be designed. To the best of our knowledge, currently the only existing protocol designed with this specific goal is the Ladon [17] protocol.

12.5.1 The Ladon Protocol

Ladon is a security protocol aimed at allowing low-capacity devices to implement end-to-end authentication, authorization, and key establishment functionalities at the application level.

For the authentication purpose, Ladon is based on the well-known Kerberos protocol. Although not specifically developed for resource-deprived environments, Kerberos is very well-suited for the requirements that these scenarios pose, mainly due to its centralized user account management. However, raw Kerberos does not address all the security requirements presented by IP-enabled sensors, basically due to two reasons. First, Kerberos uses timestamps to determine the freshness of messages. Having the clocks of all possible communicating entities synchronized can be a feasible solution when the operational environment is a controlled scenario. However, it is completely unrealistic that all entities in the Internet will permanently maintain synchronized clocks. Second, Kerberos does not support authorization functionalities. When sensors can be queried by any entity in the Internet, it is crucial to implement reliable and flexible authorization mechanisms. Therefore, Ladon presents two big modifications with respect to Kerberos: (1) it does not require clock synchronization, and (2) it supports authorization functionalities.

12.5.1.1 Protocol Architecture and Basic Operation

The design of Ladon implies the modification of the Kerberos key distribution center (KDC) to include two new information stores: (1) an *active connections information base*, used to assess the freshness of tickets and protocol messages, and (2) an *authorization information base*, used to store the authorization-related policies. In addition, three new messages (LDN_AP_IND, LDN_AP_IND_REQ, and LDN_AP_IND_REP) have been defined and the original meaning of some Kerberos message fields has been altered. Basically, special nonces have been introduced to avoid the necessity for synchronized clocks. Although the replacement of timestamps with nonces is a classic technique to avoid time synchronization, the difficulty lies in doing it without increasing the number of messages exchanged by the protocol. In Ladon, this is achieved by the use of one-way key chains.

Figure 12.8 shows the basic interactions of the Ladon protocol, where LDN_AP_IND_REQ and LDN_AP_IND_REP have not been represented, as these two messages are not sent during the normal operation of the protocol, they are only rarely sent to face high-packet error situations.

The operation of Ladon is organized into three different phases: the authentication phase, the authorization phase, and the service access phase. In the

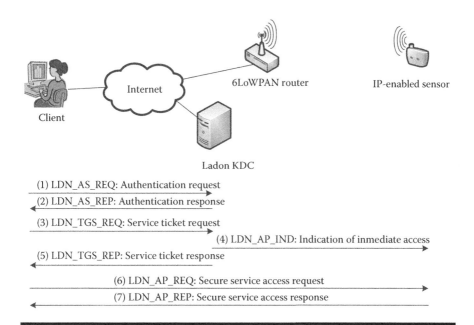

Figure 12.8   Basic message exchange of the Ladon protocol operation.

authentication phase (messages 1–2), the Ladon authentication server (AS) verifies the identity claimed by the requesting client. As a result of a successful authentication, the client obtains a ticket granting ticket (TGT), which allows him to prove his identity to the Ladon ticket granting server (TGS) in order to obtain as many service tickets as he may need during the validity period of the TGT. During the authorization phase (messages 3–5), the Ladon TGS checks if a legitimately authenticated client is entitled to access some specific data stored within a given sensor. The targeted sensor and data are defined as a service principal. If the verification is successful, both the targeted service principal and the requesting client are provided with the necessary information so that the communication can be securely established. Finally, during the service access phase (messages 6–7), the client presents the credentials provided by the Ladon TGS to the desired service principal, who checks them against the information provided by the Ladon TGS. If they can be positively validated, the sensor responds with the requested information.

12.5.1.2 Authorization Support

The proposed authorization model is based on a combined design of role-based access control (RBAC) and attribute-based access control (ABAC), integrating attributes with pure RBAC. Each service principal is preconfigured with the role value that should be granted access to the given service and only accept service tickets with that value embedded.

After authenticating an incoming service ticket request, the TGS queries the *authorization information base* in order to determine if the requesting client is authorized to access the desired service and only in a positive case, it generates a service ticket with the corresponding role embedded. Therefore, Ladon service tickets assert both, the veracity of the identity claimed by the client and his right to access the requested service.

12.5.1.3 Avoiding Permanently Synchronized Clocks

Ladon implements a nonce-based solution to probe the freshness of messages and avoid replay attacks.

In order to limit the validity period of TGTs, a special nonce is embedded in each TGT at creation time and it is also stored in the *active connections information base*. After a given lifetime expires, the entry is deleted and thus, the ticket containing the deleted nonce value is no longer valid.

Service tickets are understood as instantaneous credentials and thus, their validity period is much shorter than the one of the TGTs. This validity period is communicated by the TGS to the target service principal within a LDN_AP_IND message.

On the other hand, to avoid the possibility of an attacker replaying legitimate LDN_TGS_REQ messages, Ladon implements a counter-based mechanism. On the other hand, to provide replay protection to LDN_AP_REQ messages, the nonce value embedded in the service ticket used to authenticate the message is used. Finally, in order to avoid the replay of LDN_AP_IND messages, a mechanism based on a one-way key function is implemented. Thus, each time the TGS sends a LDN_AP_IND message to a given service principal (S), it embeds a value of the key chain, following the order: $K^0_{S,TGS}, K^1_{S,TGS}, \ldots, K^L_{S,TGS}$. Therefore, once the service principal owns a value $K^{i-1}_{S,TGS}$, it is enough to check that $F\left(K^i_{S,TGS}\right) = K^{i-1}_{S,TGS}$ to assert the freshness of the message conveying the $K^i_{S,TGS}$ value. However, the first time a given service principal receives a LDN_AP_IND message, it does not own a $K^{i-1}_{S,TGS}$ value to validate this message. In this case, the service principal directly queries the TGS about the next value of the key chain by means of a LDN_AP_IND_REQ/_REP message exchange.

12.5.1.4 Performance Analysis

We evaluate the performance of the Ladon protocol in terms of delay for a secure session establishment and energy cost for each service principal. Additionally, in order to show that the mechanisms designed to deal with message losses are suitable for the targeted environments, we analyze these parameters depending on the packet loss probability in the IEEE 802.15.4 network.

For the evaluation purpose, we assume a cluster-tree sensor network, with 54 service principals and a 3-hop depth from the BS to each of the leaf nodes where service principals reside. Additionally, we assume N_C clients generating requests to each of these service principals according to a Poisson distribution with a mean rate of 1 request per minute. On the other hand, for the parameterization of the Ladon protocol and the IEEE 802.15.4 link layer, we select $L = 100$ as the length of the one-way key chain used to implicitly authenticate LDN_AP_IND messages and default values for the IEEE 802.15.4 backoff process. For the energy model, we consider the instantaneous power consumptions specified in the datasheet of the MEMSIC TelosB mote (TPR2420CA) powered with a 3 V power supply. We also consider that the client principal executes cryptographic operations at a rate of 50 Mbps, the TGS at a rate of 100 Mbps and the service principal at a rate of 30 Kbps. Figure 12.9a and b shows the evolution of the delay for a secure connection establishment and energy cost for each service principal with respect to the packet loss probability in the IEEE 802.15.4 network and different network loads (N_C = 1, 50, and 100).

Figure 12.9a shows that the delay introduced by the Ladon protocol is within an acceptable limit, even in situations of very heavy load conditions. More specifically, for typical packet loss rates in sensor networks (40–60%), the delay for a secure connection establishment varies from 360 to 550 ms, in the most demanding of the evaluated situations. These delays can be considered acceptable for the envisioned applications, as they are significantly shorter than the double of the maximum allowed end-to-end delay for traditional IP applications (1 s) and also for interactive data transactions (400 ms) as specified by the ITU-T.Y.1541 recommendation (network performance objectives for IP-based services). It must be noted that end-to-end delay considers the time elapsed since a packet leaves the client entity until it reaches the destination, while in Figure 12.9a, we measure the time needed for the whole establishment of a secure connection, with all the message exchanges it implies. On the other hand, for very high-packet loss rates (above 80%), the delay for a secure connection establishment does not tend to infinity, since the number of retransmissions allowed for each type of message is bounded.

Similarly, Figure 12.9b shows that the energy cost of Ladon is acceptable for the targeted devices. For packet loss rates between 40% and 60%, the energy cost of a secure session establishment varies between 1.6 and 1.9 mJ. On the other hand, when the packet loss rate grows too much (beyond 60%), very few of the transmitted messages reach their destination and therefore, the energy consumed by the sensor is drastically reduced. In such cases, the probability of the secure session being actually established is also very low.

12.6 Future Research Directions

The research program of the authors includes performance evaluations of the presented protocols. In order to obtain comparable results, the protocols to be assessed

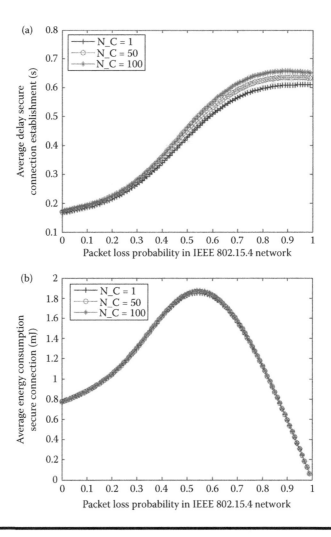

Figure 12.9 Performance of the Ladon protocol for a secure connection establishment depending on the packet loss rate in the IEEE 802.15.4 network and the number of simultaneously requesting client principals: (a) average delay and (b) average energy consumption.

are planned to be implemented using the same methodology and evaluated according to the same benchmark.

On the other hand, the authors also plan to extend their research on security protocols for wireless sensor communications with protocols aimed at protecting the privacy of such communications. In this regard, authors do not only consider content privacy, that is, protecting the privacy of the transmitted data, but also context privacy, including issues such as privacy of location or timing-related data and anonymous queries.

12.7 Conclusions

This chapter provides a review of the broad research area of security in wireless sensor communications. Traditionally, sensors have been organized in networks that implement IEEE 802.15.4 at the link and physical layers and proprietary protocols on top of it. These traditional sensor networks require security mechanisms to protect the integrity and confidentiality of the traffic exchanged between legitimate members of the network from potential attackers. To that end, specific protocols have been designed which aim at including efficient mechanisms to perform encryption and MAC computation in sensor devices as well as to minimize the communication overhead derived from the protection of the exchanged messages. However, these protocols do not define how the necessary cryptographic keys are obtained by the communicating endpoints. To address this issue, key establishment protocols for sensor networks have been developed. Most of these protocols are based on the use of symmetric key cryptographic algorithms, as traditional public key cryptography is regarded as prohibitive to be implemented in sensor devices, due to the high computational requirements it imposes. However, in the last few years, some alternatives have emerged that make use of specific public key cryptographic mechanisms to allow key establishment in sensor networks. In this chapter, the most relevant security protocols designed for traditional sensor networks have been introduced highlighting their main benefits and deficiencies.

However, this view of sensor networks radically changes with the emergence of IP-enabled sensors. Thanks to technologies like 6LoWPAN, sensors can be integrated in the Internet as any other IP-enabled device and thus, they become subject to the same potential attacks as any other Internet-connected server. This results in a drastic change of the security services required by sensors. In order to address this problem, the Ladon security protocol allows resource-deprived devices to implement strong security properties such as end-to-end authentication, authorization, and pair-wise key establishment. To the best of our knowledge, this is the first and only protocol specifically designed to protect the data collected by sensors from illegitimate accesses, when these accesses can be originated outside the sensor network and in the most general case, by any Internet-connected entity.

Taking into account the latest advances in electronic technologies and the benefits of implementing standardized communication protocols, it is reasonable to assume that IP-enabled sensors will become popular in a near future. Therefore, the development of security protocols for sensor devices should be channeled to this new paradigm of sensor networks where the devices are IP-enabled.

Acknowledgments

The work described in this chaper has been produced within the training and research unit UFI11/16 funded by the UPV/EHU. This work was supported by the Basque Government through the strategic project Future Internet II and Saiotek project SensITS (Seguridad y movilidad en redes de sensores para entornos ITS).

References

1. Montenegro, G., N. Kushalnagar, J. Hui and D. Culler. 2007. Transmission of IPv6 packets over IEEE 802.15.4 networks. *RFC 4944*.
2. IEEE Computer Society. 2006. IEEE 802.15.4 Standard: Wireless Medium Access (MAC) and Physical Layer (PHY) Specifications for Low-Rate Wireless Personal AreaNetworks (WPANs).
3. Karlof, C., N. Sastry and D.Wagner. 2004. TinySec: A link layer security architecture for wireless sensor networks. In *Proceedings of the 2nd ACM Conference on Embedded Networked Sensor Systems: SenSys*, Baltimore, Maryland, USA, 2004, November 3–5,162–175.
4. Hill, J., R. Szewczyk, A. Woo, S. Hollar, D. Culler and K. Pister. 2000. System architecture directions for networked sensors. *ACM SIGPLAN Notices* 35(11): 93–104.
5. Lighfoot, L. E., J. Ren and T. Li. 2007. "An energy efficient link-layer security protocol for wireless sensor networks. In *Proceedings of the IEEE International Conference on Electro/Information Technology, Chicago, IL, USA*, 2007, May 17-20, 233–238.
6. Luk, M., G. Mezzour, A. Perrig and V. Gligor. 2007. MiniSec: A secure sensor network communication architecture. In *Proceedings of the 6th International Conference on Information Processing in Sensor Networks: IPSN, Cambridge, Massachusetts, USA*, 2007, April 25–27, 479–488.
7. Bloom, B. H. 1970. Space/time trade-offs in hash coding with allowable errors. *Communications of the ACM* 13(7): 422–426.
8. Perrig, A., R. Szewczyk, J. D. Tygar, V. Wen and D. E. Culler. 2002. SPINS: Security protocols for sensor networks. *ACM Wireless Networks* 8(5): 521–534.
9. Ren, K., W. Lou and Y. Zhang. 2008. LEDS: Providing location aware end-to-end data security in wireless sensor networks. *IEEE Transactions on Mobile Computing* 7(5): 585–598.
10. Zhu, S., S. Setia and S. Jajodia. 2003. LEAP: Efficient security mechanisms for large-scale distributed sensor networks. In *Proceedings of the 10th ACM Conference on Computer and Communications Security: CCS, Washington, DC, USA*, 2003, October 27–30, 62–72.
11. Park, T. and K. G. Shin. 2004. LiSP: A lightweight security protocol for wireless sensor networks. *ACM Transactions on Embedded Computing Systems* 3(3): 634–660.
12. Chan, H. and A. Perrig. 2005. PIKE: peer intermediaries for key establishment in sensor networks. In *Proceedings of the 24th Annual Joint Conference of the IEEE Computer and Communications Societies: INFOCOM, Miami, FL, USA*, 2005, March 13–17, 524–535.
13. Liu, D. and P. Ning. 2003. Establishing pairwise keys in distributed sensor networks. In *Proceedings of the 10th ACM Conference on Computer and Communications Security: CCS, Washington, DC, USA*, 2003, October 27–30, 52–61.

14. Watro, R., D. Kong, S. Cuti, C. Gardiner, C. Lynn and P. Kruus. 2004. TinyPK: Securing sensor networks with public key technology. In *Proceedings of the 2nd ACM Workshop Security of Ad Hoc & Sensor Networks, Washington, DC, USA,* 2004, October 25–29, 59–64.

15. Shaikh, R. A., S. Lee, M. A. U. Khan and Y. J. Song. 2006. LSec: Lightweight security protocol for distributed wireless sensor network. In *Proceedings of the 11th IFIP International Conference on Personal Wireless Communications: PWC, Albacete, Spain,* 2006, September 20–22, 367–377.

16. Lin, H. and Y. Tseng. 2007. A scalable ID-based pairwise key establishment protocol for wireless sensor networks. *Journal of Computers* 18(2): 13–24.

17. Astorga, J., E. Jacob, M. Huarte and M. Higuero. 2012. Ladon: End-to-end authorisation support for resource-deprived environments. *IET Information Security* 6(2): 93–101.

SECURITY IN OTHER AD HOC NETWORKS

Chapter 13

Securing the Control of Euler–Lagrange Systems in Networked Environments with Model-Free Sliding Mode Control

Shuai Li, Yunpeng Wang, and Long Cheng

Contents

13.1 Introduction

The development in mechanics, electronics, sensing technology, microelectrome-chanical systems (MEMS), and so on, have greatly advanced the progress of auto-mation in the past decades, which makes automatic systems consisting of sensors for perception and actuators for action more widely used in applications [1–3]. Besides the proper choices of sensors and actuators and an elaborate fabrication of mechanical structures, the control law design also plays a crucial role in the implementation of automatic systems, especially for those with networked inter-connections of sensing and actuating elements. The signal processing problem in wireless sensor networks or mobile phone networks, which are typical networked systems, has been widely investigated [4–6]. The actuation problem in manipula-tor networks, which is a typical actuator network, also has attracted many studies [7,8]. Networked systems as shown in Figure 13.1 sometimes are more fragile to

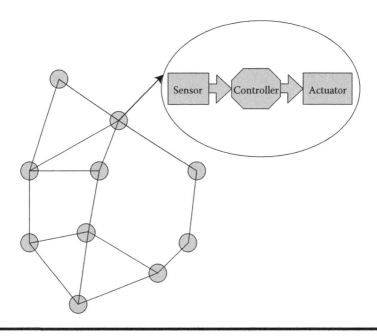

Figure 13.1   The schematic of a sensor–actuator network, where the circles rep-resent sensor–actuator nodes and the black lines represent the communication connection.

the failure of a single node in the network. The failure of one node may lead to the failure of the whole system due to the damage propagation. In such a situation, a robust control law with guaranteed stability and adaptive learning capability is becoming desirable. On the other hand, for most mechanical sensor–actuator systems, it is possible to model them in Euler–Lagrange equations [3,9]. In this chapter, we focus on the investigation of control strategies for sensor–actuator systems modeled by Euler–Lagrange equations.

Euler–Lagrange equations are widely studied as it models many real sensor–actuator systems. According to the type of constraints, the Euler–Lagrange system can be categorized into a system without nonholonomic constraints and the system subject to nonholonomic constraint [10]. Examples of nonholonomic-free Euler–Lagrange systems include fully actuated manipulators [7,11] and omnidirectional mobile robots [12]) while nonholonomic-constrained Euler–Lagrange systems include the cart–pole system [13], the underactuated multiple body system [14]), and so on. For the Euler–Lagrange system without nonholonomic constraints, the dimension of inputs is often equal to the dimension of outputs and the system is often able to be transformed into a double integrator system by employing feedback linearization [15]. Other methods, such as control the Lyapunov function method [16], passivity-based method [17], optimal control method [18], and so on, are also successfully applied to the control of the Euler–Lagrange system without nonholonomic constraints. In contrast, as the dimension of inputs is lower than that of outputs, it is often impossible to directly transform the Euler–Lagrange system subject to nonholonomic constraints to a linear system and, thus, feedback linearization fails to stabilize the system. To tackle the difficulty, sliding mode control (SMC)-based method [19], backstepping-based control [20], optimal control-based method [21], discontinuous control method [22], and so on, are widely investigated and some useful design procedures are proposed. However, due to the inherent nonlinearity and nonholonomic constraints, most existing methods [19–22] are strongly model-dependent and the performance is very sensitive to model errors. Inspired by the success of human operators for the control of Euler–Lagrange systems, various intelligent control strategies, such as fuzzy logic [23], neural networks [24], evolutionary algorithms [25], to name a few, are proposed to solve the control problem of Euler–Lagrange systems subject to nonholonomic constraints. As demonstrated by extensive simulations, these types of strategies are indeed effective to the control of Euler–Lagrange systems subject to nonholonomic constraints. However, rigorous proof on the stability is difficult for this type of method and there exists some initializations of the state from which the system cannot be stabilized.

Approximate dynamic programming (ADP), which is a modified version of the general reinforcement learning method to resolve the curse of dimensionality problem—an unavoidable problem for direct dynamic programming (DP) methods, has attracted researchers from different communities, such as control engineering [26] and neural networks [27] as ADP opened an avenue to real-time reinforcement learning control with theoretically guaranteed convergence. Applying ADP to SMC of Euler–Lagrange systems, the relaxation of the exact model information is expectable. In this chapter, we propose an ADP-based intelligent SMC method for Euler–Lagrange systems including the ones without nonholonomic constraints and the ones subject to nonholonomic constraints. In contrast to existing work on intelligent control of Euler–Lagrange systems, the stability of the closed-loop system with the proposed method is proven in theory. On the other hand, different from model-based design strategies, such as backstepping-based design [20], sliding mode-based design [19], and so on, the proposed method does not require information of the model parameters and therefore is a model-independent method. We formulate the problem from an optimal control perspective. In this framework, the goal is to find the input sequence to minimize the cost function defined on infinite horizons under the constraint of the system dynamics. The solution can be found by solving a Bellman equation according to the principle of optimality [28]. Then an adaptive DP strategy [26,27,29] is utilized to numerically solve the input sequence in real time. Note that the optimal control problem, as widely studied in model predictive control, can be numerically solved using finite horizon-control techniques by approximately considering the cost function on a finite horizon [30,31]. However, due to the curse of dimensionality, such a strategy is computational intensive and is not suitable for the online control of rapid evolving systems [32]. More importantly, model information is necessary for the design of a finite horizon controller [33] and, therefore, this technique cannot be applied directly to model-free control. As a review and extension of our previous work on ADP-based SMC of Euler–Lagrange systems [34], more rigorous theoretical proof on the convergence is presented and more numerical simulations are performed to investigate the validity of the proposed method.

The remainder of this chapter is organized as follows: in Section 13.2, preliminaries on Euler–Lagrange systems and SMC are briefly given. In Section 13.3, the problem is formulated as a constrained optimization problem. In Section 13.4, the model-free control strategy is presented and analyzed in theory. Section 13.5 considers issues in the practical implementation and introduces the critic model and the action model to facilitate the control. In Section 13.6, simulations are given to show the effectiveness of the proposed method. The chapter is concluded in Section 13.7.

13.2 SMC of the Euler–Lagrange System

In this chapter, we are concerned with the following Euler–Lagrange system,

$$D(q)\ddot{q} + C(q,\dot{q})\dot{q} + \phi(q) = u \tag{13.1}$$

where $q \in \mathbb{R}^n$, $D(q) \in \mathbb{R}^{n \times n}$ is the inertial matrix, $C(q,\dot{q}) \in \mathbb{R}^{n \times n}$, $\phi(q) \in \mathbb{R}^n$, and $u \in \mathbb{R}^n$. Note that the inertial matrix $D(q)$ is symmetric and positive definite. There are three terms on the left side of the above equation. The first term involves the inertial force in the generalized coordinates, the second one models the Coriolis force and friction, the values of which depend on $\dot{q}$, and the third one is the conservative force, which corresponds to the potential energy. The control force u applied on the system drives the variation of the coordinate q. It is also noteworthy that we assume the dimension of u is equal to that of q here. This definition also admits the case for u with a lower dimension than that of q by imposing constraints to u, for example, the constraint $u = [u_1, u_2, ..., u_n]$ with $u_1 = 0$ restricts u in an $n-1$ dimensional space. Defining state variables $x_1 = q$ and $x_2 = \dot{q}$, the Euler–Lagrange Equation 13.1 can be put into the following state-space form

$$\dot{x}_1 = x_2$$
$$\dot{x}_2 = -D^{-1}(x_1)(u + C(x_1, x_2)x_2 + \phi(x_1)) \tag{13.2}$$

Note that the matrix $D(x_1)$ is invertible as it is positive definite. The control objective is to asymptotically stabilize the Euler–Lagrange system (13.2), that is, design a mapping $(x_1, x_2) \rightarrow u$ such that $x_1 \rightarrow 0$ and $x_2 \rightarrow 0$ when time elapses.

As an effective design strategy, SMC finds applications in many different types of control systems including the Euler–Lagrange system. The method stabilizes the dynamics of a nonlinear system by steering the state to an elaborately designed sliding surface, on which the state inherently evolves toward the zero state. Particularly for the system (13.2), we define $s = s(x_1, x_2)$ as follows:

$$s = c_0 x_1 + x_2 \tag{13.3}$$

where $c_0 > 0$ is a constant. Note that $s = c_0 x_1 + x_2 = 0$ together with the dynamics of x_1 in Equation 13.2 gives the dynamics of x_1 as $\dot{x}_1 = -c_0 x_1$ for $c_0 > 0$. Clearly, x_1 asymptotically converges to zero. Also, we know $x_2 = 0$ when $x_1 = 0$ according to $s = c_0 x_1 + x_2 = 0$. Therefore, we conclude the states x_1, x_2 on the sliding surface $s = 0$ for s defined in Equation 13.3 converge to zero with time. With such a property of the sliding surface, a control law driving the states to $s = 0$ definitely guarantees the ultimate convergence to the zero state. Accordingly, the stabilization of the system can be realized by controlling s to zero. To reach this goal, a positive definite control Lyapunov function $V(s)$, for example, $V(s) = s^2$, is often used to design the control law. For stability consideration, the time derivative of $V(s)$ is required to be negative definite. In order to guarantee the negative definiteness of the time derivative of $V(s)$, exact information about the system dynamics (13.2 is often necessary, which results in the model-based design strategies.

Remark

In this chapter, we are only concerned with the systems represented by Euler–Lagrange equations. The proposed method in this chapter is the ability to control this type of model since Euler–Lagrange equation-modeled systems can be controlled by using the variable structure control methodology. Also, the dependence of the control law on accurate model parameters can be relaxed by using ADP for parameter estimations. Therefore, other nonlinear systems, which do not fall into the class of Euler–Lagrange models, may also be solved in a similar way if the SMC law applies. ■

13.3 DP-Based Formulation of SMC

Without losing generality, we stabilize the system (13.1) by steering it to the sliding surface $s = 0$ with s defined in Equation 13.3. Different from existing model-based design procedures, we design a self-learning controller, which does not require accurate knowledge about $D(q)$, $C(q,\dot{q})$, and $\phi(q)$ in Equation 13.1. In this section, we formulate such a control problem from the optimal control perspective.

In this chapter, we set the origin as the desired operating point, that is, we consider the problem of controlling the state of the system (13.1) to the origin. For the case with other desired operating points, the problem can be equivalently transformed to the one with the origin as the operating point by shifting the coordinates. At each sampling period, the norm of $s = c_0 x_1 + x_2$, which measures the distance from the desired sliding surface $s = 0$, can be used to evaluate the one-step performance. Therefore, we define the following utility function associated with the one-step cost at the ith sampling period,

$$U_i = U(s) \tag{13.4}$$

$$U(s) = \begin{cases} 0 & |s_1| < \delta_1, |s_2| < \delta_2,..., |s_n| < \delta_n \\ 1 & \text{otherwise} \end{cases} \tag{13.5}$$

where s is defined in Equation 13.3 and $s = [s_1,s_2,...,s_n]^T$, $|s_i|$ denotes the absolute value of the ith component of the vector s, the parameter $\delta_i > 0$ for $i = 1, 2, ..., n$. At each step, there is a value U_i and the total cost starting from the kth step along the infinite time horizon can be expressed as follows:

$$J_k = J(x(k),\bar{u}(k)) = \sum_{i=k}^{\infty} \gamma^{i-k} U_i \tag{13.6}$$

where $x(k)$ is the state vector of system (13.1) sampled at the kth step with $x(k) = [x_1^T(k), x_2^T(k)]^T$, γ is the discount factor with $0 < \gamma < 1$, $\bar{u}(k) = (u_k, u_{k+1},..., u_\infty)$ is the control sequence starting from the kth step. Note that for the deterministic system (13.1), the preceding states after the kth step are determined by $x(k)$ and the control sequence $\bar{u}_k$. Accordingly, J_k is a function of $x(k)$

and $\bar{u}(k)$ with $J_k = J(x(k), \bar{u}(k))$. Also note that both the cost function J_k and the utility function U_k are defined based on the discrete samplings of the continuous system (13.1). Now, we can define the problem of controlling the sensor–actuator system (13.1) in this framework as follows:

$$\min_{u(0),u(1),\dots,u(\infty)\in\Omega} \quad J_0 = \sum_{i=0}^{\infty} \gamma^i U_i \tag{13.7a}$$

subject to:
$$\begin{cases} \dot{x}_1(t) = x_2(t) \\ \dot{x}_2(t) = -D^{-1}(t)(x_1(t))\Big(u(t) + C(x_1(t), x_2(t))x_2(t) + \phi(x_1(t))\Big) \end{cases}$$

$$\tag{13.7b}$$

$$u(t) = u(i) \quad \text{for } i\tau \le t < (i+1)\tau \tag{13.7c}$$

where U_i is defined by Equations 13.4 and 13.5, $\tau > 0$ is the sampling period, the set Ω defines the feasible control actions, J_0 is the cost function for $k = 0$ in Equation 13.6. It is worth noting that J_0 is a function of $\bar{u}(0) = (u_0, u_1, \dots, u_\infty)$ and $x(0)$ according to Equation 13.6. The optimization in Equation 13.7 is relative to $\bar{u}(0)$ with a given initial state $x(0)$. Also note that in the optimization problem (13.7), the decision variable $u(0)$, $u(1)$, ..., $u(\infty)$ are defined in every sampling period. The control action keeps the value in the duration of two consecutive sampling steps. This formulation is consistent with the real implementations of digital controllers.

Remark

There are infinitely many decision variables, which are $u(0)$, $u(1)$,..., $u(\infty)$, in the optimization problem (13.7). Therefore, this is an infinite dimensional problem. It cannot be solved directly using numerical methods. Conventionally, such these types of problems are often solved by using a finite dimensional approximation [35]. In addition, note that the dynamic model of the system appears in the optimization problem (13.7) and it will also show up in the finite dimensional relaxation of the problem, which means the resulting solution requires model information and, thus, is also model-dependent. In contrast, in this chapter we investigate the model-independent SMC of sensor–actuator systems on the infinite time horizon. ■

13.4 Model-Free SMC of the Euler–Lagrange System

In this section, we present the strategy to solve the constrained optimization problem efficiently without knowing the model information of the chaotic system.

We first investigate the optimum condition of Equation 13.7 and present an iterative procedure to approach the analytical solution. Then, we analyze the convergence of the iterative procedure and the stability with the derived control strategy.

13.4.1 Optimality Condition

Denoting J^* the optimal value to the optimization problem (13.7), that is,

$$J^* = \min_{u(0),u(1),...,u(\infty)\in\Omega} J_0$$
$$\text{subject to: } (13.7\text{b}),(13.7\text{c})$$

(13.8)

According to the principle of optimality [28], the solution of Equation 13.7 satisfies the following Bellman equation

$$J^*(y) = \min_{u_k\in\Omega}(U_k + \gamma J^*(z))\forall x, \quad \forall k = 0,1,2,...$$

(13.9)

where z is the solution of Equation 13.7b at $t = k + 1$ with $x(k) = y$ and the control action $u(t) = u_k$ for $k\tau \le t < (k + 1)\tau$. Without introducing confusion, we simply write Equation 13.9 as follows:

$$J^* = \min(U_k + \gamma J^*)$$

(13.10)

Define the Bellman operator $\mathcal{B}$ relative to function $h(z)$ as follows:

$$\mathcal{B}h(z) = \min(U_k + \gamma h(z))$$

(13.11)

Then, the optimality condition (13.10) can be simplified into the following with the Bellman operator

$$J^* = \mathcal{B}J^*$$

(13.12)

Note that the function U_k is implicitly included in the Bellman operator. Equation 13.12 constitutes the optimality condition for problem (13.7). It is difficult to solve the explicit form of J^* analytically from Equation 13.9. However, it is possible to get the solution by iterations. We use the following iteration to solve J^*

$$\hat{J}(n + 1) = \mathcal{B}\hat{J}(n)$$
$$\text{subject to: } (13.7\text{b})(13.7\text{c})$$

(13.13)

The control action keeps constant in the duration between the kth and the $k + 1$th step, that is, $u^*(t) = u^*_k$ for $k\tau \le t < (k + 1)\tau$. u^*_k can be obtained from Equation 13.9 based on Equation 13.13

$$u^*_k = \text{argmin}_{u_k \in \Omega}(U_k + \gamma J^*) \tag{13.14}$$

13.4.2 Convergence of the Iteration

In this part, we investigate the convergence of the iterative procedure given by Equation 13.13.

Before presenting the convergence result, the following lemma about the contraction property, the Bellman operator, is stated below

Lemma 13.1 ([28])

The Bellman operator is a γ contraction mapping in the infinity norm, that is,

$$\| \mathcal{B}h_1 - \mathcal{B}h_2 \|_\infty \le \gamma \| -h_2 \|_\infty$$

The following lemma, known as the contraction mapping principle, is useful for the convergence analysis of the iteration (13.13). ■

Lemma 13.2 ([36])

Let (S,d) be a complete metric space and $T{:}S \to S$ be a mapping such that $d(T(x),T(y)) \le cd(x,y)$ for some $0 \le c < 1$ and all x and y in S. Then T has a unique fixed point in S such that $x^* = T(x^*)$. Moreover, for any $x_0 \in \mathbb{S}$ the sequence of iterates x_0, $T(x_0)$, $T(T(x_0))$, … converges to the fixed point x^*.

Now, we are ready to state the convergence result on Equation 13.13. ■

Theorem 13.1

For any $\hat{J}(0)$, the iteration (Equation 13.13) converges to J^* such that $J^* = \mathcal{B}J^*$ for $0 < \gamma < 1$.

Proof
The result directly follows Lemmas 13.1 and 13.2. According to Lemma 13.1, the Bellman operator $\mathcal{B}$ is a γ contraction mapping in the infinity norm, which yields

$$d(\mathcal{B}(x),\mathcal{B}(y)) \le \gamma d(x,y)$$

with $d(x, y) = \| x - y \|_\infty$ and $0 < \gamma < 1$. With Lemma 13.2, we conclude that the sequence $\hat{J}(0)$, $\mathcal{B}(\hat{J}(0))$, $\mathcal{B}(\mathcal{B}(\hat{J}(0)))$, … converges to the fixed point J^*, which is the solution of $J^* = \mathcal{B}J^*$. Clearly, the sequence can be generated by the iteration (13.13) initialized with any $\hat{J}(0)$. This completes the proof. ■

13.4.3 Control Performance

Stability is often the first consideration in most control designs. It requires all trajectories generated by the dynamic system to travel to the origin eventually. In real applications, this requirement is often relaxed to ultimate boundedness due to the presence of additive noises, perturbations, modeling errors, and so on. In this part, we consider ultimate boundedness of the control strategy (13.5) with the utility function (13.14), which means all trajectories eventually converge to a small vicinity of the origin.

About the boundedness of the eventual states, we have the following theorem.

Theorem 13.2

If there exists a control law $u = u(x_1, x_2)$, such that the system (13.1) converges to the sliding surface $s = 0$ asymptotically, where s is defined in Equation 13.3, the system (13.1) under the control input (13.14) with the utility function (13.5) ultimately converges to the set, where $\delta = \sqrt{\delta_1^2 + \delta_2^2 + \cdots + \delta_n^2}$ with δ_1, δ_2^2,..., δ_n defined in Equation 13.5, $c_0 > 0$ defined in Equation 13.3.

Proof

Since the system (13.1) converges to the sliding surface $s = 0$ with the control law $u = u(x_1, x_2)$. We conclude that for any positive constant $\delta_1 > 0$, $\delta_2 > 0$, …, $\delta_n > 0$, there exists a time t_1, such that $|s_1(t)| < \delta_1$, $|s_2(t)| < \delta_2$,…, $|s_n(t)| < \delta_n$ for $_t > t_1$ [37]. As $\|s\| = \sqrt{s_1^2 + s_2^2 + \cdots + s_n^2} < \sqrt{\delta_1^2 + \delta_2^2 + \cdots + \delta_n^2} = \delta$, we have $\|s(t)\| < \delta$ for $t > t_1$. Define k_1 as the least integer greater than t_1/τ. Then, for any $n > k_1$, we have $U_n = 0$ for the utility function with the definition (13.5). Therefore, the cost function for this control law is $J_1 = \sum_{i=0}^{k1} \gamma^i U_i = \text{constant}$. ■

Now we consider the cost function J_2 under the control input (13.14). As J_2 is the minimum one under the constraints, $J_2 \le J_1$. Also note that J_1 is upper bounded and J_2 is the summation of nonnegative values, we conclude that $\lim_{n \to \infty} U_n = 0$ for the control law (13.14) (otherwise if $\lim_{n \to \infty} U_n \ne 0$, J_2 will keep increasing and be ultimately unbounded). As there are only two possible values for U_n, there must exist an integer k_2, such that $U_n = 0$ for all $n > k_2$. Recalling the definition of the utility function (13.5), we conclude that $\|s\| < \delta$ for $n > k_2$ (i.e., $t > k_2\tau$). As for $t > k_2\tau$, $x_2 = -kx_1 + s$ with $\|s\| < \delta$, we have the following dynamics of x_1

$$\dot{x}_1 = -kx_1 + s \tag{13.16}$$

where $\|s\| < \delta$ for $t > k_2\tau$. Note that Equation 13.16 can be regarded as a linear system with x_1 as state and s as input. Clearly, the system (13.16) is bound-input-bounded-output since its system matrix has all eigenvalues strictly on the left-half-plane. Therefore, x_1 is bounded ultimately as the input s is bounded. To analyze the bound on x_1 in quantity, we construct the measurement $V = x_1^T x_1$. The time derivative of V is

$$\begin{aligned}
\dot{V} &= 2x_1^T \dot{x}_1 = -2kx_1^T x_1 + 2x_1^T s \\
&= -2kx_1^T x_1 + 2x_1^T s \le -2kV + 2\,\|x_1\|\,\|s\| \\
&= -2kV + 2\sqrt{V}\,\|s\| \le -2kV + 2\sqrt{V}\delta \\
&\forall t > k_2\tau
\end{aligned} \tag{13.17}$$

Defining $W = \sqrt{V}$, the differential inequality in the last line of Equation 13.17 changes to

$$\dot{W} \le -kW + \delta \quad \forall t > k_2\tau \tag{13.18}$$

Using the comparison Lemma [37], we find the solution of the differential inequality (13.18) as follows:

$$W(t) \le W'(t) \quad \forall t > k_2\tau \tag{13.19}$$

where $W'(t)$ is the solution of the following differential equation

$$\dot{W}' = -kW' + \delta \tag{13.20}$$

with $W'(k_2\tau) = W(k_2\tau) = \sqrt{V(k_2\tau)} = \|x_1(k_2\tau)\|$. Note that system (13.20) is asymptotically stable and W' converges to δ/k ultimately, that is, $\lim_{t\to\infty} W'(t) = (\delta/k)$. Calculating limits on both sides of the inequality (13.19) yields

$$\lim_{t\to\infty} W(t) \le \lim_{t\to\infty} W'(t) = \frac{\delta}{k} \tag{13.21}$$

According to the definition of W, we get

$$\lim_{t\to\infty} \|x_1\| \le \frac{\delta}{k} \tag{13.22}$$

With the definition of s given in Equation 13.3, we have

$$\lim_{t \to \infty} \|x_2\| = \lim_{t \to \infty} \|-kx_1 + s\| \leq \lim_{t \to \infty} k \|x_1\| + \delta \leq 2\delta \qquad (13.23)$$

From Equations 13.21 and 13.23, we conclude that the system (13.1) under the control input (13.14) with the utility function (13.5) ultimately converges to the set $\mathbb{S} = \{x_1 \in \mathbb{R}^n, x_2 \in \mathbb{R}^n, \|x_1\| < \delta/k, \|x_2\| < 2\delta\}$, which completes the proof.

13.5 Using Neural Network to Approximate the Mapping

In the last section, the iteration (13.13) is derived to calculate J^* and the optimization (13.14) is obtained to calculate the control law. The iteration to approach J^* and the optimization to derive u^* have to be run in every time step in order to obtain the most up-to-date values. Inspired by the learning strategies widely studied in artificial intelligence [27,38], a learning-based strategy is used in this section to facilitate the processing. After a long enough time, the system is able to memorize the mapping of J^* and the mapping of u^*. After this learning period, there will be no need to repeat any iterations or optimal searching, which will make the strategy more practical.

Note that the optimal cost J^* is a function of the initial state. Counting the cost from the current time step, J^* can also be regarded as a function of both the current state and the optimal action at the current time step according to Equation 13.10. Therefore, $\hat{J}(n)$, the approximation of J^*, can also be regarded as a function relative to the current state and the current optimal input. As to the optimal control action u^*, it is a function of the current state. Our goal in this section is to obtain the mapping from the current state and the current input to $\hat{J}(n)$ and the mapping from the current state to the optimal control action u^* using parameterized models, denoted as the critic model and the action model, respectively. Therefore, we can write the critic model and the action model as $J_n(u_n^*, x_n, W_c)$ and $u_n^*(x_n, W_a)$ respectively, where W_c is the parameters of the critic model and W_a the parameters of the action model.

In order to train the critic model with the desired input–output correspondence, we define the following error at time step $n+1$ to evaluate the learning performance

$$e_c(n + 1) = \mathcal{B} \hat{J}(n) - \hat{J}(n + 1)$$

$$E_c(n + 1) = \frac{1}{2} e_c^2(n + 1) \qquad (13.24)$$

Note that $\mathcal{B}\,\hat{J}(n)$ is the desired value of $\hat{J}(n + 1)$ according to Equation 13.13. Using the back-propagation rule, we get the following rule for updating the weight W_c of the critic model

$$W_c(n+1) = W_c(n) + \delta W_c(n) = W_c(n) - l_c(n)\frac{\partial E_c(n)}{\partial W_c(n)}$$

$$= W_c(n) - l_c(n)\frac{\partial E_c(n)}{\partial \hat{J}(n)}\frac{\partial \hat{J}(n)}{\partial W_c(n)} \tag{13.25}$$

where $l_c(n)$ is the step size for the critic model at the time step n.

As to the action model, the optimal control u^* in Equation 13.14 is the one that minimizes the cost function. Note that the possible minimum cost is zero, which corresponds to the scenario with the state staying inside the desired bounded area. In this regard, we define the action error as follows:

$$e_a(n) = \hat{J}_n$$
$$E_a(n) = \frac{1}{2}e_a^2(n) \tag{13.26}$$

Then, similar to the update rule of W_c for the critic model, we get the following update rule of W_a for the action model

$$W_a(n+1) = W_a(n) - l_a(n)\frac{\partial E_a(n)}{\partial \hat{J}(n)}\frac{\partial \hat{J}(n)}{\partial u(n)}\frac{\partial u(n)}{\partial W_a(n)} \tag{13.27}$$

where $l_a(n)$ is the step size for the action model at the time step n.

Equations 13.25 and 13.27 update the critic model and the action model progressively. After W_c and W_a have the model information by learning for a long enough time, their values can be fixed at the final step and no further learning is required, which is in contrast to requiring to solve an optimization problem even after a long enough time (Equation 13.14).

13.6 Simulation Experiments

In this section, we consider the simulation implementation of the proposed control strategy. The dynamics given in (Equation 13.1) model a wide class of sensor–actuator systems. Particularly, to demonstrate the effectiveness of the proposed self-learning sliding mode method, we apply it to the stabilizations of typical benchmark systems: the cart–pole system and the underactuated two-link manipulator system.

13.6.1 Cart–Pole System

The cart–pole system, as sketched in Figure 13.2, is a widely used testbed for the effectiveness of control strategies. The system is composed of a pendulum and a cart. The pendulum has its mass above its pivot point, which is mounted on a cart moving horizontally. In this part, we apply the proposed control method to the cart–pole system to test the effectiveness of our method.

13.6.1.1 The Model

The cart–pole model used in this work is the same as that in [39], which can be described as follows:

$$\ddot{\theta} = \frac{g\sin\theta + \cos\theta[-F - ml\dot{\theta}^2\sin\theta + \mu_c\text{sgn}(\dot{y})] - (\mu_p\dot{\theta}/ml)}{l((4/3) - (m\cos^2\theta/m_c + m))} \tag{13.28}$$

$$\ddot{y} = \frac{F + ml[\dot{\theta}^2\sin\theta - \ddot{\theta}\cos\theta] - \mu_c\text{sgn}(\dot{y})}{m_c + m} \tag{13.29}$$

where

$$\text{sgn}(x) = \begin{cases} 1, & \text{if } x > 0 \\ 0, & \text{if } x = 0 \\ -1, & \text{if } x < 0 \end{cases} \tag{13.30}$$

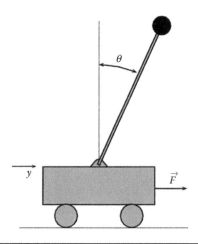

Figure 13.2 The cart–pole system.

with the following values of the parameters,

g: 9.8 m/s²acceleration due to gravity;
m_c: 1.0 kg, mass of cart;
m: 0.1 kg, mass of pole;
l: 0.5 m, half-pole length;
μ_c: 0.0005, coefficient of friction of cart on track;
μ_p: 0.000002, coefficient of friction of pole on cart; and
F: ±10 N, force applied to cart center of mass.

This system has four state variables: y is the position of the cart on track, θ is the angle of the pole with respect to the vertical position, and $\dot{y}$ and $\dot{\theta}$ are the cart velocity and angular velocity, respectively.

Define

$$A_1(\theta) = -\frac{l}{\cos\theta}\left(\frac{4}{3} - \frac{m\cos^2\theta}{m_c + m}\right)$$

$$A_2(\theta) = -\frac{g\sin\theta}{\cos\theta},$$

$$A_3(\theta,\dot{\theta}) = ml\dot{\theta}\sin\theta + \frac{\mu_p}{ml\cos\theta},$$

$$A_4(\dot{y}) = -\frac{\mu_c\,\mathrm{sgn}(\dot{y})}{\dot{y}},$$

$$A_5 = m_c + m,$$

$$A_6(\theta,\dot{\theta}) = ml\dot{\theta}\sin\theta,$$

$$A_7(\theta) = -ml\cos\theta.$$

With these notations, Equation 13.28 can be rewritten as

$$A_1\ddot{\theta} = F + A_2 + A_3\dot{\theta} + A_4\dot{y}$$
$$\frac{A_1A_5}{A_1 + A_7}\ddot{y} = F + \frac{A_2A_7}{A_1 + A_7} + \frac{A_1A_6 + A_3A_7}{A_1 + A_7}\dot{\theta} + \frac{A_1A_4 + A_4A_7}{A_1 + A_7}\dot{y} \quad (13.31)$$

By choosing

$$
D = \begin{bmatrix} A_1 & 0 \\ 0 & \dfrac{A_1 A_5}{A_1 + A_7} \end{bmatrix}, \quad C = - \begin{bmatrix} A_3 & A_4 \\ A_1 A_6 + A_3 A_7 & A_1 A_4 + A_4 A_7 \\ \dfrac{}{A_1 + A_7} & \dfrac{}{A_1 + A_7} \end{bmatrix}
$$

$$
\phi = - \begin{bmatrix} A_2 \\ \dfrac{A_2 A_7}{A_1 + A_7} \end{bmatrix}, \quad q = \begin{bmatrix} \theta \\ y \end{bmatrix}, \quad u = \begin{bmatrix} F \\ F \end{bmatrix}
$$

the system (13.28) coincides with the model (13.1). Note that the input u in this situation is constrained in the set $\Omega = \{u = [u_1, u_2]^T, u_1 = u_2 \in \mathbb{R}\}$.

13.6.1.2 Experiment Setup and Results

In the simulation experiment, we set the discount factor $\gamma = 0.95$, the sliding surface parameter $k = 10$, $\delta_1 = 2$, and $\delta_2 = 24$. The feasible control action set Ω in Equation 13.7 is defined as $\Omega = \{u = [u_1, u_2]^T, u_1 \in \mathbb{R}, u_2 \in \mathbb{R}, u_1 = u_2 = \pm 10 \, \text{Newton}\}$. This definition corresponds to the widely used bang–bang control in industry. To make the output of the action model within the feasible set, the output of the action network is clamped to 10 if it is greater than or equal to zero and clamped to -10 if less than zero. The sampling period τ is set to 0.02 s. Both the critic model and the action model are linearly parameterized. The step size of the critic model, which is $l_c(n)$ and that of the action model, which is $l_a(n)$ are both set to 0.03. Both the update of the critic model weight W_c in Equation 13.25 and the update of the action model weight W_a in Equation 13.27 lasts for 30 s. For the uncontrolled cart–pole system with $F = 0$ in Equation 13.28, the pendulum will fall down. The control objective is to stabilize the pendulum to the inverted direction ($\theta = 0$). Time history of the state variables is plotted in Figure 13.3 for the system with the proposed self-learning SMC strategy. From this figure, it can be observed that θ is stabilized in a small vicinity around zero (with a small error of ± 0.1 rads), which corresponds to the inverted direction.

13.6.2 Under-Actuated Two-Link Manipulator

The underactuated two-link manipulator is a planer robot arm with 2-DoF. Different from a conventional fully actuated manipulator with two motors actuating on the elbow joint and the shoulder joint separately, the control input for the manipulator investigated here can only be applied at the elbow joint as shown in Figure 13.4. In this part, we investigate the application of the proposed strategy to control such a system.

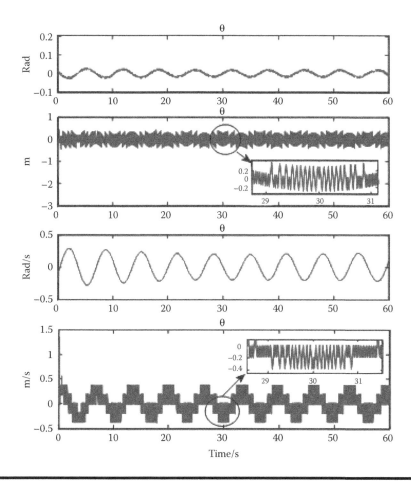

Figure 13.3   State profiles of the cart–pole system with the proposed control strategy.

13.6.2.1 The Model

The dynamics of the model can be written as follows [40]:

$$d_{11}\ddot{\theta}_1 + d_{12}\ddot{\theta}_2 + h_1\dot{\theta}_2 + \phi_1 = 0$$
$$d_{21}\ddot{\theta}_1 + d_{22}\ddot{\theta}_2 + h_2\dot{\theta}_1 + \phi_2 = T \tag{13.32}$$

Where

$$d_{11} = m_1 l_{c1}^2 + m_2(l_1^2 + l_{c2}^2 + 2l_1 l_{c2}\cos\theta_2) + I_1 + I_2$$
$$d_{22} = m_2 l_{c2}^2 + I_2 \tag{13.33}$$

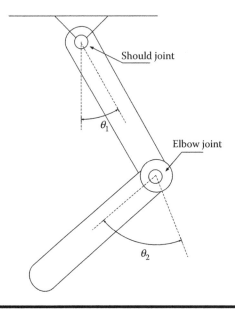

Figure 13.4 The underactuated two-link manipulator.

$$d_{12} = m_2(l_{c2}^2 + l_1 l_{c2} \cos\theta_2) + I_2$$
$$d_{21} = d_{12}$$
$$h_1 = -m_2 l_1 l_{c2} \dot\theta_2 \sin\theta_2 - 2m_2 l_1 l_{c2} \dot\theta_1 \sin\theta_2$$
$$h_2 = m_2 l_1 l_{c2} \dot\theta_1 \sin\theta_2$$
$$\phi_1 = (m_1 l_{c1} + m_2 l_1) g \cos\theta_1 + m_2 l_{c2} g \cos(\theta_1 + \theta_2)$$
$$\phi_2 = m_2 l_{c2} g \cos(\theta_1 + \theta_2)$$

with the following values of the parameters,

g: 9.8 m/s², acceleration due to gravity;
m_1: 1.0 kg, mass of the link 1;
m_2: 1.0 kg, mass of the link 2;
l_1: 0.5 m, length of link 1;
l_{c1}: 0.5 m, length from the shoulder to the mass of center of link 1;
l_{c2}: 0.5 m, length from the elbow to the mass of center of link 2;
I_1: 1 kg m², inertia moment of link 1;
I_2: 1 kg m², inertia moment of link 2; and
T: ±10 N m, torque applied to the elbow joint.

This system has four state variables: θ_1 is the angular position of the first link, θ_2 is the relative angular position between the first link and the second one, $\dot{\theta}_1$ is the angular velocity of the first link, and $\dot{\theta}_2$ is the relative angular velocity of the second link.

By choosing

$$D = \begin{bmatrix} d_{11} & d_{12} \\ d_{21} & d_{22} \end{bmatrix}, C = \begin{bmatrix} 0 & h_1 \\ h_2 & 0 \end{bmatrix}, \phi = \begin{bmatrix} \phi_1 \\ \phi_2 \end{bmatrix}, q = \begin{bmatrix} \theta_1 \\ \theta_2 \end{bmatrix}, u = \begin{bmatrix} 0 \\ T \end{bmatrix}$$

the system (13.28) falls into the same model as described by Equation 13.1. In this case, the input u is constrained in the set $\Omega = \{u = [u_1, u_2]^T, u_1 = 0, u_2 \in \mathbb{R}\}$.

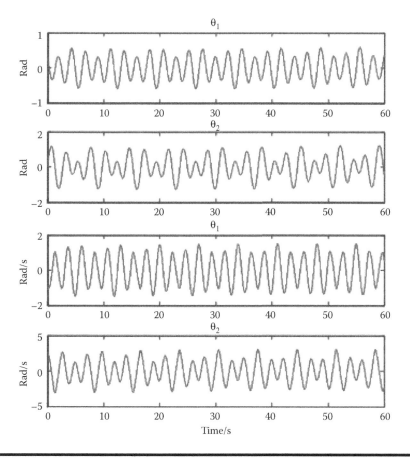

Figure 13.5 State profiles of the underactuated two-link manipulator system with the proposed control strategy.

13.6.2.2 Experiment Setup and Results

The underactuated two-link manipulator model (13.32) demonstrates undamped oscillations as shown in Figure 13.5. The goal for this domain is to stabilize the manipulator at $q = [0,0]^T$ by only applying control force at the elbow joint.

In the simulation experiment, we set the discount factor $\gamma = 0.95$, the sliding surface parameter $k = 10$, $\delta_1 = 2$, and $\delta_2 = 24$. The feasible control action set Ω in Equation 13.7 is defined as $\Omega = \{u = [u_1, u_2]^T, u_1 = 0, u_2 \in \mathbb{R}, u_2 = \pm 10 \text{Newton}\}$. The sampling period τ is set to 0.02 s. Both the critic model and the action model are parameterized with two layer nonlinear neural network models. For both of them, there are six hidden neurons in the neural networks. The first layer of the neural network is a linear map and the second layer is a sigmoid saturation mapping. The

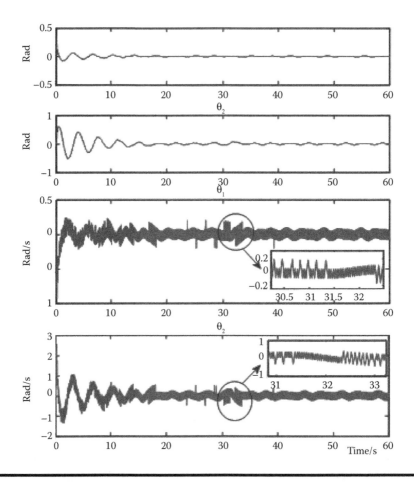

Figure 13.6 State profiles of the underactuated two-link manipulator system with the proposed control strategy.

step size of the critic model, which is $l_c(n)$ and that of the action model, which is $l_a(n)$ are both set to 0.03. Both the update of the critic model weight W_c in Equation 13.25 and the update of the action model weight W_a in Equation 13.27 last for 30 s. As shown in Figure 13.6 with the proposed system, the system is successfully stabilized to the desired angles.

13.7 Conclusions

In this chapter, the self-learning SMC is considered to solve a class of sensor–actuator systems. The control problem is formulated from the optimal control perspective and solved via iterative methods. In contrast to existing models, this method does not need preknowledge on the accurate mathematic model. The convergence of the iteration and the stability of the obtained control law are both proven in theory. The critic model and the action model are introduced to make the method more practical. Simulations show that the control law obtained by the proposed method indeed achieves the control objective.

Acknowledgment

Shuai Li would like to share the words by the entrepreneur Jack Ma that we should warm up our own left hand with the right one when striving, without expecting someone to recognize us. The worst failure is to give up.

References

1. R. Isermann. Modeling and design methodology for mechatronic systems. *IEEE/ASME Transactions on Mechatronics*, 1(1):16–28, 1996.
2. M. van de Panne and E. Fiume. Sensor-actuator networks. In *Proceedings of the 20th Annual Conference on Computer Graphics and Interactive Techniques*, SIGGRAPH '93, pp. 335–342, New York, NY, USA, 1993. ACM.
3. C. De Silva. *Sensors and Actuators: Control System Instrumentation*. Taylor & Francis, CRC Press, Oxford, UK, 2007.
4. S. Li, B. Liu, B. Chen, and Y. Lou. Neural network based mobile phone localization using bluetooth connectivity. *Neural Computing and Applications*, 23:667–675, 2013.
5. S. Li and F. Qin. A dynamic neural network approach for solving nonlinear inequalities defined on a graph and its application to distributed, routing-free, range-free localization of WSNs, *Neurocomputing*, 117:72–80, 2013.
6. S. Li, Z. Wang, and Y. Li. Using Laplacian eigenmap as heuristic information to solve nonlinear constraints defined on a graph and its application in distributed range-free localization of wireless sensor networks. *Neural Processing Letters*, 37:411–424, 2013.

7. S. Li, S. Chen, B. Liu, Y. Li, and Y. Liang. Decentralized kinematic control of a class of collaborative redundant manipulators via recurrent neural networks. *Neurocomputing*, 91:1–10, 2012.

8. S. Li, H. Cui, Y. Li, B. Liu, and Y. Lou. Decentralized control of collaborative redundant manipulators with partial command coverage via locally connected recurrent neural networks. *Neural Computing & Applications*, 23(3):1051–1069, 2013.

9. F. P. Beer. *Vector Mechanics for Engineers: Statics and Dynamics*. McGraw Hill Science, Engineering, Math, 2003.

10. A. M. Bloch. *Nonholonomic Mechanics and Control*, Volume 24. Springer, New York, 2003.

11. Frank L Lewis, Darren M Dawson, and Chaouki T Abdallah. *Manipulator Control Theory and Practice*, volume 15. Marcel Dekker, New York, 2004.

12. S. Li, M. Q. H. Meng, and W. Chen. Sp-nn: A novel neural network approach for path planning. *IEEE International Conference on Robotics and Biomimetics*, Sanya, China, pp. 1355–1360, 2007.

13. H. Yu, Y. Liu, and T. Yang. Tracking control of a pendulum-driven cart-pole underactuated system. *IEEE International Conference on Systems, Man and Cybernetics, ISIC,* Montreal, Canada, pp. 2425–2430, Oct. 2007.

14. R. Seifried. Two approaches for feedforward control and optimal design of underactuated multibody systems. *Multibody System Dynamics*, 27:75–93, 2012.

15. A. Isidori. *Nonlinear Control Systems II*, Volume 2. Springer-Verlag, 1999.

16. J. A. Primbs, V. Nevistic, and J. C. Doyle. Nonlinear optimal control: A control Lyapunov function and receding horizon perspective. *Asian Journal of Control*, 1:14–24, 1999.

17. R. Ortega, A. Loria, P. J. Nicklasson, and H. Sira-Ramirez. *Passivity-based Control of Euler-Lagrange Systems*. Springer, New York, USA, 1st edition, 1998.

18. V. Azhmyakov. Optimal control of mechanical systems. *Differential equations and Nonlinear Mechanics*, 12(3):16, 2007.

19. W. Huo. Predictive variable structure control of nonholonomic chained systems. *International Journal of Computer Mathematics*, 85(6):949–960, 2008.

20. B. Dumitrascu, A. Filipescu, V. Minzu, and A. Filipescu. Backstepping control of wheeled mobile robots. *15th International Conference on System Theory, Control, and Computing (ICSTCC)*, Sinaia, Romania, pp. 1–6, Oct. 2011.

21. I. I. Hussein and A. M. Bloch. Optimal control of underactuated nonholonomic mechanical systems. *IEEE Transactions on Automatic Control*, 53(3):8, 2005.

22. D. Pazderski, K. Kozowski, and B. Krysiak. Nonsmooth stabilizer for three link nonholonomic manipulator using polar-like coordinate representation. In Krzysztof Kozlowski, editor, *Robot Motion and Control 2009, volume 396 of Lecture Notes in Control and Information Sciences*, pp. 35–44. Springer Berlin/Heidelberg, 2009.

23. F. Cuesta, A. Ollero, B. C. Arrue, and R. Braunstingl. Intelligent control of nonholonomic mobile robots with fuzzy perception. *Fuzzy Sets and Systems*, 134(1):47–64, 2003.

24. R-J. Wai and C-M. Liu. Design of dynamic petri recurrent fuzzy neural network and its application to path-tracking control of nonholonomic mobile robot. *IEEE Transactions on Industrial Electronics*, 56(7):2667–2683, 2009.

25. H. Kinjo, E. Uezato, S. Duong, and T. Yamamoto. Neurocontroller with a genetic algorithm for nonholonomic systems: Flying robot and four-wheel vehicle examples. *Artificial Life and Robotics*, 13:464–469, 2009.

26. F. L. Lewis and D. Vrabie. Reinforcement learning and adaptive dynamic programming for feedback control. *IEEE Circuits and Systems Magazine*, 9:32–50, 2009.

27. J. Si, A. Barto, W. Powell, and D. Wunsch. *Handbook of Learning and Approximate Dynamic Programming*. Wiley-IEEE Press, Hoboken, USA, 2004.

28. D. P. Bertsekas. *Dynamic Programming and Optimal Control*. Athena Scientific, Nashua, USA, 3rd ed., Nov 2005.

29. J. J. Murray, C. J. Cox, G. G. Lendaris, and R. Saeks. Adaptive dynamic programming. *IEEE Transactions on Systems, Man, and Cybernetics*, 32:140–153, 2002.

30. M. Morari and J. H. Lee. Model predictive control: Past, present and future. *Computers and Chemical Engineering*, 23:667–682, 1999.

31. F. Allgower, R. Findeisen, and Z. K Nagy. Nonlinear model predictive control: From theory to application. *Journal of the Chinese Institute of Chemical Engineers*, 35(3):299–315, 2004.

32. J. Skaf, S. Boyd, and A. Zeevi. Shrinking-horizon dynamic programming. *International Journal of Robust and Nonlinear Control*, 20(17):1993–2002, 2010.

33. S. J. Qin and T. A. Badgwell. A survey of industrial model predictive control technology. *Control Engineering Practice*, 11(7):733–764, 2003.

34. S. Chen, S. Li, B. Liu, Y. Lou, and Y. Liang. Self-learning variable structure control for a class of sensor-actuator systems. *Sensors*, 12(5):6117–6128, 2012.

35. D. Q. Mayne and H. Michalska. Receding horizon control of nonlinear systems. *IEEE Transactions on Automatic Control*, 35(7):814–824, 1990.

36. A. Kirk and B. Sims. Handbook of metric fixed point theory. 2001.

37. H. Khalil. *Nonlinear Systems*. Prentice Hall, Upper Saddle River, New Jersey, January 2002.

38. C. M. Bishop. *Pattern Recognition and Machine Learning*, Volume 4. Springer, New York 2006.

39. J. Si and Yu-Tsung Wang. Online learning control by association and reinforcement. *IEEE Transactions on Neural Networks*, 12(2):264–276, 2001.

40. M. W. Spong. The swing up control problem for the acrobot. *Control System IEEE*, 15(1):49–55, 1995.

Chapter 14

Defenses against Packet-Dropping Attacks in Wireless Multihop Ad Hoc Networks

L. Sánchez-Casado, R. Magán-Carrión,
P. García-Teodoro, and J.E. Díaz-Verdejo

Contents

14.1 Introduction

Ad hoc networks constitute a technology of increasing use in certain areas, such as environmental and military applications, disaster management, and so on. This fact is mainly motivated by some particular characteristics of these networks, which are among others: geographical distribution without a fixed infrastructure, self-configuration capability, and wireless-based communications. It is also remarkable for these environments that the nodes in the network with no direct communication among them can communicate each other through other nodes. This is the so-called *multihop* transmission strategy.

As wireless multihop ad hoc networks proliferate, many security issues associated with this communication paradigm become more relevant and thus need to be conveniently addressed. In this line, Table 14.1 shows some principal security threats reported for this kind of environment [1–3]. Among them, there are several attacks where a malicious node, after introducing itself in some way in the origin–destination routes (multihop path), controls communications and alters transmissions by discarding packets. This kind of well-known attack generally includes *sinkhole, blackhole, grayhole,* and *selfish*.

Sinkhole attacks are usually referred to misbehaving nodes that try to introduce themselves in the routing/forwarding path to seize communications. To do so, a malicious node modifies routing messages either by publishing that it has the shortest path to the destination or by spoofing the destination address to guarantee that the sender chooses it as an intermediate hop. Blackhole and grayhole attacks are two of the most popular attacks in multihop ad hoc networks. Both are related with the packet-forwarding process carried out by intermediate nodes. When the node completely drops all the received packets, the attack is considered a blackhole attack. On the other hand, the grayhole attack is caused by a node dropping packets in a selective way, *for example,* one out of N packets received, one packet every certain time, only packets corresponding to specific flows, and so on. Selfish nodes evade their responsibility on forwarding packets in the network with the principal aim of preserving or economizing its energetic resources.

All of the abovementioned attacks can be grouped into a generic type named *packet-dropping attack*, which constitutes a major security concern in current wireless multihop ad hoc networks [4,5]. As mentioned, nodes exhibiting this behavior maliciously drop received data or routing messages instead of forwarding them, which in fact disrupt the normal operation of the network [6]. Though the specific damage caused by packet-dropping attacks depends on the discarding level implemented in each case (e.g., indiscriminate vs. selective dropping, or actual malicious behavior vs. "just" saving resources-related selfish behavior),

Table 14.1   Some Principal Attacks Reported in the Literature for Wireless Multihop Ad Hoc Networks

Attack	*Description*
Physical layer	
Eavesdropping	Listening to private communications, that is, intercepting data
Jamming (random, periodic, etc.)	Generating signal interferences, which provokes communication disruption
Link layer	
Collision	Generating selective interferences to disrupt MAC mechanisms, which affects the capture of a channel for legitimate transmissions
Exhaustion	Repeated collisions and/or continuous retransmissions to occupy the channel
Sleep deprivation, *a.k.a.* resource consumption	Repeated collisions that induce the node to continuous retransmissions, thus causing its death
Network layer	
Blackhole	Sending fake routing information that claims an optimum route to make other nodes relay data packets through the malicious node. In a second step, this node could drop or discard traffic
Delay	Introducing time delays in the retransmission of control packets, thus disrupting the normal routing operations
Grayhole, *a.k.a.* selective forwarding	Blackhole attack where the node drops packets selectively, for example, with a certain probability, one packet every certain time, or only packets corresponding to specific flows
HELLO flooding	Massive sending of HELLO packets to overwhelm neighbors
Link spoofing	Advertising fake links with nonneighbors, thus disrupting routing operations
Link withholding	Ignoring a link advertisement, which can result in node isolation

continued

Table 14.1 (continued) Some Principal Attacks Reported in the Literature for Wireless Multihop Ad Hoc Networks

Link-broken error	Sending fake control messages, which gives rise to connectivity loss
Routing cache poisoning	Faking routing table information, thus disrupting the routing function
Routing table overflow	Advertising an excessive number of routes to nonexisting nodes, which prevents neighbors from creating new legitimate routes
Rushing	Artificial quick retransmission of routing packets, which can result in building fake routes
Selfish	Bypassing certain protocols rules to save resources (e.g., battery), which decreases network performance
Sinkhole	Sending fake routing information that claims an optimum route to make other nodes route data packets through the malicious node to inspect and filter the traffic in some way
Wormhole	Two colluding attackers record packets at one location and replay them at another using a private high speed link
Other layers or any of them	
Jellyfish	Introducing time delays to TCP retransmissions, which decreases end-to-end performance
Sybil	Adopting multiple identities, for example, becoming a legitimate part of the network
Tampering	Physically manipulating a node to affect some functionality or compromise it

its potential impact and relevance in communications is unquestionable. Huge efforts are carried out by the research community to address this problem; the number of proposals in literature specialized in this line are continuously increasing.

However, some limitations can be checked for almost all of the contributions in the field. First, most of the published papers deal with only a partial aspect of the problem. On the one hand, some of them are limited to a particular type of data discarding (i.e., blackhole or sinkhole or grayhole or selfish) instead of studying all of them as a global typology. On the other hand, the majority of the papers are only focused either on preventing or detecting or reacting against these malicious behaviors, while they avoid the rest of possible defense lines.

Another relevant shortcoming that usually affects works on multihop ad hoc security is the existence of bit confusion in specifications. Thus, both the definition and scope of each particular attack and the type of defense lines specifically developed in each case are wrong or at least mixed up, in occasions.

In summary, the great majority of current proposals are interesting but partial and, conversely, incomplete, if not slightly confusing. This chapter tries to solve these limitations by presenting a general survey on defenses against packet-dropping attacks with the following main characteristics:

1. As previously stated in the previous paragraphs and through Table 14.1, each specific attack (sinkhole/blackhole/grayhole/selfish) is clearly defined and differentiated from the rest.
2. Despite this difference, especially regarding the research works we find in the literature, we propose to group all of them as belonging to a common class, *packet dropping*, with similar final consequences on network performance.
3. A detailed state-of-the-art covering the most relevant proposals existent in the literature to fight against this kind of attack is presented afterwards. Aimed at providing an organized vision of the subject, the study is carried out based on the specific defense line considered in each case. Despite some other organizations that can perform for that (e.g., network layer or protocol affected), we consider the traditional defense in-depth approach a good option (see Figure 14.1).
4. Finally, new trends and open challenges in the field of wireless multihop ad hoc security are discussed with the aim of completing as much as possible the information about the topic.

According to these principal contributions, along with this first section, the rest of the chapter is structured as follows. Section 14.2 presents the main works related with the development of prevention techniques against packet-dropping attacks. Among others, cryptographic and credit-based algorithms are the most widely used prevention schemes at present [7]. In Section 14.3, current detection proposals are discussed, which usually rely on observing the occurrence of misbehaviors and intrusion events in the monitored environment. Section 14.4 is devoted to analyze response/reaction-related security approaches, which are basically oriented to

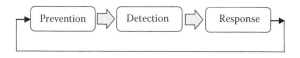

Figure 14.1   Defense lines in a traditional defense in-depth approach.

isolate the malicious nodes to preserve the operation of the network and services. In some cases, the reaction schemes are aimed at serving as a feedback mechanism to strengthen the network by adapting the considered security mechanisms to the particular conditions observed, as shown in Figure 14.1. After that, Section 14.5 brings new trends and open challenges in the topic of multihop ad hoc security in general, and for packet-dropping attacks in particular. Finally, the chapter concludes with a summary of the main aspects contributed.

14.2 Preventing Packet-Dropping-Related Attacks

A first concern quickly arises regarding the prevention of packet-dropping attacks when reviewing the literature. It is directly related to the concept of prevention itself. In a strict sense, only mechanisms or methods avoiding the potential attacks should be considered preventive. But there are many authors that label their approach as preventive, despite being based on a detection phase. In this sense, these algorithms should be included either in the detection or the reaction category. All these methods will be described later in Sections 14.3 and 14.4. On the other hand, there exists approaches which explicitly encourage the nodes not to misbehave, although they do not neglect the possibility of dropping attacks. We will consider these approaches as preventive, as there is no detection or reaction mechanism involved. It is a difficult task to label many of the approaches that will be described later as belonging to a single category because they can mix various techniques from different categories.

According to [8], a wireless multihop ad hoc network should "(1) *provide an effective security mechanism to deal with misbehaving nodes in the network, (2) encourage cooperation among nodes in the network* ...." Therefore, for the purposes of this chapter, we will consider as preventive mechanisms, those not allowing or discouraging misbehaving nodes. In this context, four main categories of mechanisms can be identified according to the main method used for the prevention scheme: authentication-based, based on changes in the routing protocol, reputation-based, and credit-based systems. They are described in the following sections.

14.2.1 Authentication-Based Prevention Schemes

The methods in this category usually provide for preventive schemes to protect the routing procedure, that is, to guarantee the correctness of the announced routes, mainly based on key management or encryption techniques, by checking the identity of the nodes involved in communications. Therefore, their primary purpose is to prevent unauthorized nodes from joining the network, which constitutes a defense against external attacks.

Most of the techniques in this category are based on authenticated routing. Some examples of this kind of method are Ariadne [9] and ARAN [10]. Ariadne uses an end-to-end authentication based on shared key pairs, while ARAN uses a hop-by-hop authentication.

These techniques are suitable to prevent foreigner nodes from being able to disrupt the network operation through fake route announcements and, consequently, are able to avoid subsequent packet-dropping attacks. However, most of them fail in its preventive behavior if the attacker is an insider. On the other hand, the use of ciphering constitutes a drawback from the viewpoint of energy saving and performance, even if a trusted authority is not required.

14.2.2 Prevention Based on Routing Protocol Modification

Most of the routing protocols for multihop ad hoc networks, especially in the case of mobile ad hoc networks (MANETs), have not been designed with security requirements in mind. Therefore, a method for the prevention of dropping attacks is the introduction of changes in the used protocol to fix the vulnerabilities that make the attacks possible. In this sense, multipath routing can be considered as a first kind of prevention, as the objective is to support a secure and reliable communication in case a route is compromised. Nevertheless, other approaches that use additional exchanges of information among nodes are described in the literature see, for example, [11–14].

Although in some cases the changes include the use of ciphering, it is worth mentioning that the primary goal is not to guarantee the identity of the nodes involved in the communication, but to guarantee the freshness and correctness of the communication routes between nodes. This way, there exist many proposals in literature based on the cross-checking of the routes by comparison with the neighbors' ones or by explicitly requesting the final or intermediate nodes to confirm them or to send additional information about the routes. An example of the latter is the case of secure ad hoc on-demand distance vector (SAODV) [11,12], which is an extension of ad hoc on-demand distance vector (AODV) to counter for dropping attacks. It is based on the use of new secure route request (SRREQ) and secure route reply (SRREP) packets with a secret code each time a route reply (RREP) packet is received. When the source node receives at least two SRREP packets, it chooses the shortest path as a secure path to the destination. Another similar procedure is that introduced in [13], which verifies the security of the path after receiving an RREP packet by sending back the next hop information within the RREP. Alternatively, the proposal in [14], also aimed at preventing the blackhole attack, uses route confirmation request (CREQ) and route confirmation reply (CREP) packets to confirm the route validity by explicit comparison with neighbors' routes.

An alternative approach is based on somehow enforcing the cooperation of the nodes through the introduction of new mechanisms in the protocol. This way, one more hop (OMH) [15] uses asymmetric keys to cipher the packets in such a way that only the next node to a node in the route knows whether the destination of the packet is the previous node or not. Thus, every node receiving a packet needs to forward it in order to be informed by the next node about the keys required to decipher the content, in case it is the final node.

Some other secure protocols designed to address dropping attacks are SRPM [16] and E-SRPM [17], which keep the information of two hop neighbors to increase the security level and look for the available data rates in order to search such a route which is not only secure but also have better data rates. Authors in [18] propose two flexible protocols to secure resource-limited spontaneous ad hoc networks by providing authenticity, integrity checking, and verification distribution. CSROR [19] is a secure and resource-aware protocol based on a cross-layer information exchange scheme. In CSROR, the route selection is done at destination node on the basis of path security, bandwidth, and battery life.

The main disadvantage of this family of solutions is the increased cost in terms of the number of packets required and the higher number of nodes involved in the communication. On the other hand, some of them use encryption, which is an additional drawback.

14.2.3 Reputation-Based Prevention Schemes

Reputation-based prevention methods monitor the nodes' behavior during the operation of the network in order to assign them a reputation or trust level. Only nodes with an adequate level of reputation will be considered during the routing of packets. According to [20], "*Reputation of an agent is a perception regarding its behavior norms, which is held by other agents, based on experiences and observation of its past actions.*" Therefore, in order to assign the reputation level, two main components are required: a model for the normal or proper behavior of the nodes when forwarding packets, and a way to observe, measure, and store the reputation value.

An additional question arises for these kinds of procedures. Although in many of them there is no explicit detection of misbehaving nodes, the reputation level of a node can be used as an indicator of its proper behavior and, therefore, it would be correct to label them as attackers. And, having taken into account that the untrusted nodes will be avoided during the forwarding process of the packets, the procedure can also be somewhat considered as a response scheme. Nevertheless, we have included this category in the prevention phase, as some of the proposed systems limit their operation to only include proven trustable nodes in route selection algorithms, which does not imply the other nodes to be misbehaving.

Thus, one of the first proposals using reputation was the Pathrater algorithm [21], which encourages nodes to forward packets to increase its rating. Nevertheless, it is clearly based upon detecting misbehaving nodes by using the watchdog method. Therefore, it will be detailed in the following sections.

Some solutions make use of a trusted authority for storing the reputation of each other, while others adopt a decentralized management approach and introduce a recommendation protocol to exchange trust-related information. As an example, in CORE [22] each node keeps track of a reputation table through the observation of neighbors' behavior and the exchange of information with the nodes involved in each operation. When a request to relay a packet arrives at a node, it is forwarded only if the requester has a positive reputation value. A similar approach is the so-called friend and foes proposed in [23], which is based on the society principle stating that people agree to cooperate in a duty as long as they notice there is a fair tasks distribution in the group. To build an opinion for a node, each participating node advertises its set of friends and foes, that is, the set of nodes to whom it is/is not willing to forward packets. Another decentralized approach is secure and objective reputation-based incentive (SORI) [24], where the nodes exchange reputation information only with their neighbors.

A hybrid approach using reputation is that in [25], where a routing protocol is proposed to combat the blackhole attack, includes a trust-based method where the sender takes opinions of the neighbors which replied with an RREP packet.

14.2.4 Credit-Based Prevention Schemes

In credit-based approaches each node receives a micropayment for its cooperation in forwarding network messages, while it also pays those nodes retransmitting its messages [26]. Two models can be applied: the message purse model and the message trade model [27]. In the message purse case, it is the source node that pays the intermediate nodes for their service in forwarding packets. Therefore, a node should have enough credits to start a new transmission. On the contrary, in the message trade case the messages are considered as merchandise and, consequently, it is the receiver who pays the sender and the intermediate nodes. Both approaches suffer from the same problem, which is related with handling credits. Thus, to avoid cheating, secure payments have to be deployed and proof of the successful forwarding of the packets is required. This usually involves a central trusted authority, which is usually impractical in multihop ad hoc networks.

The simplest credit-based method is tit-for-tat (TFT) [28], in which two neighbor nodes exchange the same amount of messages. This method distinguishes between two types of messages: primary and secondary. Primary messages are those in which the node is directly interested, that is, the node is the origin or the destination of the message. Secondary messages are those in which the node is not interested. The key idea is to involve the nodes in forwarding the secondary messages in order to earn credits for the transmission or reception of primary messages. This method does not require any credit accounting or trusted authority, but it is only valid in delay tolerant networks, as the messages should wait in the queue till the nodes have enough credits.

Sprite [29] is a cheat-proof credit-based proposal that uses digital signatures for any single transaction. There exists a central trusted authority, the credit clearance service, which is responsible for keeping account of the credits. Apart from the need of this central authority, the main limitation is the use of signatures, as they are costly operations and have to be done by every forwarding node.

An improvement over Sprite, named Express, is described in [26]. It is based on the substitution of the signatures by hash chains, which reduces the processing costs for the nodes.

Mobicent [30] is a more recent typical credit-based solution in which a virtual bank performs the charging and rewarding processes. The credit charged to a node for sending a packet is equally distributed to the intermediate nodes.

The credit-based systems must have an incentive for the cooperation of nodes in the forwarding process in order to earn credits for their own transmissions, which is especially relevant to avoid selfish behaviors. But they require a trusted party to avoid cheating on the credits and can be unfair if not all nodes send a similar amount of information.

14.3 Detection of Dropping Attacks in Wireless Multihop Ad Hoc Networks

Despite great efforts carried out by the research community to propose preventive solutions for the dropping problem, it is still necessary to perform a subsequent detection procedure, as shown in Figure 14.1. Thus, a big number of approaches have been proposed in literature to handle packet dropping in wireless multihop ad hoc networks [4]. In the next section, we classify them into two main categories according to their basic operation: ACK-based and intrusion detection related.

14.3.1 ACK-Based Schemes

In this category, nodes request an explicit acknowledgment from their neighbors to confirm the success on the reception of the packets they send.

A two-hop ACK-based scheme is proposed in [31], where each node asks its two-hop neighbors for an ACK packet to detect misbehaving nodes. As the next hop is able to send a forged ACK packet back on behalf of the intended two-hop neighbor, an authentication mechanism is used. In order to reduce the overhead involved, the authors propose in [32] each node to ask its two-hop neighbors randomly instead of continuously. However, these two schemes fail when any two-hop neighbor refuses to send back an ACK. In such a situation, the requester node is unable to determine who the malicious node is.

To overcome the previous ambiguity in detecting malicious nodes, Liu et al. [33] proposed TWOACK to detect malicious links instead. The main idea is to send two-hop acknowledgment packets in the opposite direction of the routing path. In this scheme, each sender maintains a list of data packets sent out but not yet acknowledged, a counter of the forwarded data packets, and a counter of the missed packets. Also, to reduce the incurred routing overhead, authors in [34] present an improvement of their scheme by proposing 2ACK, where only a fraction of the packets are acknowledged according to the value of an *acknowledgment ratio*.

In [35] a modification of the AODV protocol is introduced to detect multiple blackholes in the group. The scheme uses a table which provides a given fidelity level to every participating node. When the destination correctly receives a data packet, it will send an acknowledgment to the source and, therefore, the fidelity level of the intermediate nodes will be incremented. If no acknowledgment is received, the intermediate node's level will be decremented. If the fidelity value of a given node reaches zero value, it will be labeled as malicious. The main drawback of this solution is the processing delay introduced in the network.

The authors in [36] complete their previous works in [31] and [32] by suggesting a modular solution that employs two-hop cryptographic acknowledgments for unicast packets, while a passive feedback mechanism to monitor broadcast packets is also employed. The gathered information is afterwards used as the basis for an accusation-based collaborative mechanism to detect dropping attacks.

The main idea of [37] is using a Merkle tree, a binary tree in which each leaf carries a given value and the value of an interior leaf (including the root) is a one-way hash function of the leaf's children values. For detecting single and cooperative blackhole attacks, each node contains a hash, which is a combination of its own identity and a secret value that only the node knows. Then, each node in the path acknowledges the reception of the message to the source, which constructs a Merkle tree whose leaves are the acknowledgments and calculates the root value y^r. Thereafter, y^r is compared with a precomputed value y^p, obtained during an initialization process generally corresponding to the route discovery mechanism. If they are equal, the path is secure against droppers. Because of the huge overhead implied, authors propose two versions: total acknowledgment (TA), acceptable if we are dealing with important data; and random acknowledgment (RA), which generates a relatively small overhead but provides no guarantee, and which may be used with less important data.

14.3.2 Intrusion Detection-Related Schemes

Intrusion detection techniques have been recurrently used in the literature to deal with the potential occurrence of nonlegitimate events in a communication environment (either host or network related) [38]. Consequently, several intrusion detection systems (IDSs) have been proposed to determine the potential existence of droppers in wireless multihop ad hoc environments. Based on the approach followed to perform the intrusion detection process, the next works are grouped into different classes.

Some techniques simply *monitor* the target environment, comparing the value of the collected features with a given threshold, which could be adaptive or not. As previously discussed, Marti et al. [21] presented in their pioneering work watchdog and Pathrather. Watchdog uses a monitor node that saves the recently sent packets by itself and compares them with the overheard packets forwarded by the next hop. If a sent packet does not match longer than a timeout, a failure tally is incremented for the next hop. If the tally exceeds a given threshold, the node is determined to be malicious. In [39], Kurosawa et al. deal with blackhole attacks in MANETs by introducing an anomaly-detection scheme which makes use of a dynamic training method. They consider the number of route request (RREQ) packets sent and

RREP packets received, as well as the average of the differences between the destination sequence numbers sent in RREQ packets and the ones received in RREP packets, to express the state of the network. Thus, this training set of features is employed to calculate the detection threshold based on the normal state of the network, which is dynamically updated at regular time intervals to improve the detection accuracy. For the detection process, every sample in the data set is compared with the threshold to detect deviations from the normal network state. In [40], the authors propose a solution called DPRAODV to counter blackhole attacks, in which the node receiving an RREP message from an intermediate node checks whether the sequence number value exceeds a given threshold. To reduce inaccuracies which can lead to false alarms, this threshold value is dynamically updated at every time interval. If the sequence number is higher than the threshold, the intermediate node is suspected to be malicious and is added to a blacklist.

Other approaches carry out some sort of *matching* techniques. For instance, intrusion detection-based on anomaly detection (IDAD) [41] is a host-based IDS solution to detect both single and multiple blackholes. This scheme compares every activity of a host with a precollected set of anomaly and attack activities, called *audit data*. The parameters used as audit data are a set of entries obtained from each anomaly RREP packet: destination sequence number, hop count, route lifetime, destination IP address, and timestamp. This way, the IDAD system is able to differentiate normal from abnormal RREP packets just by checking if the received RREP resembles one of those listed in the audit data. In such a case, the given node will be concluded to be malicious.

Supervised/unsupervised *machine learning* approaches are applied in many proposals to perform the detection process. Zhang et al., in [42], introduced a local and cooperative scheme in which each mobile node runs a support vector machine (SVM)-based IDS agent that monitors local traces, collecting data like user and system activities or communications within the radio range. Also, each agent is responsible for detecting, locally and independently, signs of intrusions. However, if an anomaly is detected among the local data, or if the evidence is inconclusive and needs further investigation, neighboring IDS agents will collaboratively investigate in a broader range, participating in the cooperative and global detection procedure that is launched. A cross-feature method is described in [43], where a total of 141 traffic and topology-related features are defined. This method also executes a data mining analysis to extract correlations and interrelations between features, in order to reduce this space of features. Then, a classifier like C4.5, RIPPER, or Naïve-Bayes is used to carry out the anomaly-detection procedure. The authors in [44] introduce a multilayer approach composed of three different subsystems that use a Bayesian classifier, Markov chains, and an association rule algorithm for intrusion detection in medium access control (MAC) , routing, and application layers, respectively. The results from the three layers are integrated into a local module, and the final result is sent to a global module. CRADS [45] combines the use of a nonlinear SVM-based detector and some data-reduction techniques to decrease the size of the feature set, thus minimizing the learning overhead. In a similar line, the authors in [46] use a linear classification algorithm, namely Fisher discriminant analysis (FDA), to remove data with low-information content, making the SVM classifier feasible for ad hoc nodes.

Additionally, some schemes make use of *reputation* methods to establish, in a cooperative way, a confidence level for each node that allows the detection process. In [47], the CONFIDANT protocol tries to detect malicious nodes. A *monitor module* supervises, through a passive-feedback technique, the behavior of its first-hop neighbors. If a suspicious event is detected, details are passed to a *reputation module*, which manages a table containing the rating for all the known nodes. Depending on how significant and how frequent the event is, the rating can be updated and the node labeled as malicious. The use of a *trust manager* and a *path manager* module will be discussed later. In the aforementioned friend and foes approach [23], each node performs the detection through a passive-feedback technique and by maintaining *credits* for each other, indicating the number of packets forwarded by other nodes. Then, the node classifies the rest in three categories that are periodically updated: friends, for which the node accepts to relay packets; foes, for which no service is provided; and selfish, corresponding to those that consider the node as a foe. The concept of *inner-circle consistence* was adopted in [48] to identify and detect forged route replies. The idea is to let each node discover its *k*-hop neighborhood. All its neighbors form the inner-circle, responsible for voting malicious outgoing data from the node. Specifically, route replies need to get approval from its inner-circle, which verifies the validity of the messages. If a reply contains false routing information to attract packets, an attack is detected through a voting process performed by each inner-circle node.

Finally, some works extract an *analytical model* for representing the dynamics of a given protocol, detecting inconsistencies during its operation. In [6], the authors obtain the extended finite state automaton (EFSA) for the AODV routing protocol, modeling its normal state and proposing both specification-based and statistical-based detection. The first approach detects anomalous events that are direct violations of the specifications defined by EFSA. Thus, the attackers can be detected by monitoring some particular transitions. In anomaly detection, a set of statistical features based on anomalous events associated with different attacks is defined, as well as another set that defines the normal state. Then, a rule-based classifier (RIPPER) is used to process these sets and to generate a collection of rules useful to detect these attacks. The authors in [49] propose a theoretical model for the different causes of packet loss, detecting dropping attacks in dynamic source routing (DSR)-based networks, and distinguishing these attacks from other legitimate circumstances, like collisions or channel errors. However, a very limited topology is studied there, and no mobility aspects are considered. This needs more investigation. In [5], a heuristic is proposed to complete the model in [49] to properly deal with mobility scenarios, which cause legitimate packet drops when a node moves out of the communication range. These reasons can cause a large number of false positives if not properly treated. For that, some features from MAC and routing layers are considered. As a result of such multilayer approaches, a much better detection efficiency is obtained than that raised in the referred paper [49].

14.4 Response Schemes against Packet-Dropping Attacks

As commented before, a large amount of proposals exist in the literature that deal with packet-dropping attacks. They are mainly related to security prevention and detection defense lines. Although these security lines are needed, they are not sufficient to avoid the consequences due to the potential apparition of attack events. Therefore, a reaction defense line is recommended to mitigate such undesired consequences (Figure 14.1). Prevention (resistance), detection (recognition), and response (recovery) defense lines strengthen the target system and contribute to its *survivability*, which is defined as *"the ability of a system to fulfill its mission, in a timely manner, in the presence of attacks, failures, or accidents"* [50].

Most of the ad hoc networks security solutions are focused on prevention and detection techniques, while others response-based mechanisms are mentioned less often. The current reaction-related approaches are generally intended to isolate or elude misbehaving nodes in order to preserve the network operation and performance [51].

Even though it is not an easy task to provide a definitive classification for response mechanisms, we tentatively propose the following types, groups, or classes: node exclusion, node exclusion and announcement, and node isolation.

14.4.1 Node Exclusion

These types of reaction techniques are aimed at eluding the misbehaving node in such a way that it is avoided as an intermediate node in multihop origin–destination routes. These reaction schemes present two main features. First, only the nodes that belong to the malicious node neighborhood are aware of the misbehavior occurrence. Second, and as a subsequent action of the previous fact, the neighbors will try to elude those routes to which the malicious node belongs.

Although trust-based management systems are widely used as prevention and detection techniques, they are also considered as response mechanisms. Thus, a security extension of DSR routing protocol is introduced in [52]. A hidden Markov model (HMM) for each node is in charge of computing the node's trustworthiness, which can be considered *"a quantitative value of trust that indicates the probability that a node will behave as expected"* [53]. Trust-based secure MANET routing using HMMs (TSR) acts against selfish behaviors by the selection of the route whose nodes have higher trustworthiness values. This way, the misbehaving node will be eluded. In [54], the DSR MANET routing protocol is modified by attaching two agents to each network node: a MOnitoring Agent (MOA) and a ROuting Agent (ROA). The first one monitors the network node behavior to assign a trust value.

When a malicious node is detected its assigned trust value is decreased. Afterwards, the ROA agent selects a trustworthy route, discarding nodes with less trust level.

Other existing mechanisms use the information provided by a reputation system to trigger the response. The authors in [24] introduce SORI, a secure and objective reputation-based incentive scheme. SORI discourages selfish behaviors by discarding, with a certain probability, the packets generated by a selfish node. This way, a smaller reputation value causes a higher discarding probability, which over time, limits the transmission capability of the selfish node. More recently, in [55] a reputation-based routing algorithm is proposed as a response mechanism. In this case, the future trustworthiness of each node is evaluated by means of a dynamic prediction algorithm that takes into account its historical behavior. Similarly to [52], when a node begins its malicious activities the system reduces the associated trustworthiness so that no more packets are sent to or from this node when a fixed threshold is surpassed. This proposal has a particularity: the malicious node can be recovered as a benign node. Hence, the system is now acting as a tolerance method as it tries to maintain the node in the network to avoid disconnections that would negatively affect the network performance.

Other systems that make use of simpler thresholds to execute the response procedure are also proposed in the literature. For example, in [56] the malicious node is blocked at the source node routing table when the RREP sequence number has a high value. In [57], during the route-establishment phase, each node creates a legitimacy table whose entries—one per node in the network—are calculated using two factors: the number of times that the entry node has been chosen as an intermediate node, and the number of times the destination node has been reached through such intermediate nodes. When a malicious behavior is detected the legitimacy value for the corresponding node is decreased. Afterwards, the nodes with higher legitimacy value will be chosen as intermediate nodes in the route, which in fact results in confining the maligned node.

14.4.2 Node Exclusion and Announcement

These response mechanisms improve the previous ones by notifying the existence of the malicious node/s to the rest of the network by means of different messages. Then, any node is able to discard the misbehaving node as a routing intermediary, making the response action more global than in the previous case.

A reputation-based trust management is described in [58] (an ulterior work to [47]). Here, the response action is carried out cooperatively between a reputation manager module, a trust manager module, and a path manager module. Once a suspected event is detected for a node, it is passed to the reputation manager in order to evaluate the historical behavior of the node. If a threshold is exceeded, a notification is passed to the path manager, which will remove this node from the route. Additionally, an ALARM message is sent by the trust manager to the neighborhood. Every ALARM message received in a node is passed to the trust manager module to determine if the associated node has been evaluated in the same way by other trusted nodes. If there exist sufficient evidences about the malignity of the node, it is notified to the reputation module for the malicious node to be discarded as a routing alternative.

In [59], a response mechanism is taken in one of two ways: directly or indirectly. In the first case, each node is in charge of removing a detected malicious node from its routing table. In the second case, a monitor node will send an alarm message to the neighborhood. Depending on the amount of alarm messages received at a given node, it will result in the removal the misbehaving node from its routing table.

The authors in [60] introduced a blocking-related response mechanism. There exists a set of agents monitoring the network with communication capabilities between each other. This way, when a malicious node is detected, the subsequent response action is launched. First, a blocking message in sent to the associated nodes of the agent who discovered the attack. Moreover, this blocking message for the misbehaving node is disseminated among the rest of the agents for being excluded from the network.

In [61], a modified DSR routing protocol is presented. A blackhole node is eluded by means of the creation a blacklist of nodes and its dissemination throughout the network. Therefore, all nodes know who is a blackhole and they will not process any packet from it. In the friend and foes algorithm introduced in [23] two lists of nodes are broadcasted by each node. The first one is the set of nodes to whom we are willing to forward packets and the second one is the set of nodes that we are not willing to forward packets. Thus, a benign node will refuse a control packet from a selfish node, which will force to the establishment of an alternative path.

In [62], the intermediate nodes react by discarding RREP packets from a given node if the sequence number exceeds a fixed threshold. This value is calculated with the sequence number stored in the routing table of the intermediate node, the sequence number of the incoming RREP packet, and the number of RREPs received. Also, the malicious node identification is disseminated to the others nodes by adding the malicious node information into the RREP message.

14.4.3 *Node Isolation*

Schemes intended to actively isolate the misbehaving node are introduced here. In this class of response mechanisms, a misbehaving node is besieged or surrounded by others that are in charge of blocking the incoming and outgoing communications of the former. Therefore, not a mere "passive" exclusion of the node out of the routing paths is performed.

A cross-layer mechanism is provided in [63] for that purpose. Although the attack occurs at the routing layer, the reaction is performed at the physical layer by means of creating a radio quarantine zone around the attacker. Nodes into the quarantine zone will not be able to send or receive packets. This is aided by a positioning system, which provides the locations of the nodes over the time.

Reaction techniques based on the inclusion of autonomous agents, are now described. In [64], the authors introduced a scheme imitating the human immune system. There exists an immune agent (IA) distributed along the network. The IA is in charge of detecting, classifying, isolating, and recovering the system from the attack (the last action is only performed if needed). A given node is isolated from the rest of the network when it has carried out a certain number of attacks. Moreover, the isolated node can be recovered as a benign node when it is no longer a threat for the environment. A similar scheme is proposed in [65], where there exist two types of agents in the system: detection and counterattack. When a threat is detected, an activation message is first broadcasted to the counterattack agents. Only the counterattack agents belonging to the neighborhood of the attacker node are going to be activated. Afterwards, they will block any packet going to or from the misbehaving node. In [66] an ad hoc network is partitioned in clusters, where a cluster head (CH) supervises the corresponding nodes in each cluster. When the CH detects a malicious node, an action agent (AA) is created, cloned, and positioned in each neighbor. Afterwards, each AA checks if the malicious node is one-hop located. If it is so, the AA remains in the neighbor node; otherwise, it is autocloned and positioned in the neighborhood. This operation is repeated until the misbehaving node is surrounded. The next step can be diverse: to isolate the malicious node from the network, to remove the node from the routing tables, to block traffic to and from the malicious node, to reduce the trust level of this node to avoid its incorporation in valid routes, and so on.

14.5 New Trends and Open Challenges in Wireless Multihop Ad Hoc Networks Security

We have shown the existence of a vast literature and a number of associated proposals on wireless multihop ad hoc networks security. However, despite such big efforts, it is still necessary to empower current technologies and boost new approaches if we want to improve system performance and users' confidence in this kind of environments. New trends and several challenges in this line are discussed. They all are beyond the trivial recommendation of improving both current prevention, detection, and response schemes.

Some authors defend the necessity of designing *new protocols and procedures* to reinforce traditional security aspects such as that of authentication. Thus, more robust routing protocols and collaborative procedures are being developed to strengthen reliability, for example, [67–69]. Although these mechanisms can be used in a dynamic way in a number of tasks (access control, trust and reputation, and so on), they all are usually related with a perspective of prevention security. In other words, the continuous apparition of new attacks concludes the necessity of improving the initial security conditions considered for a target network.

Moreover, as new types of attacks and variants appear, more powerful and reliable detection schemes are required at our disposal. The usual response given by the community is for the development of more specialized detection approaches. This diversification or specialization in detection gives way to two main consequences. On the one hand, it provides a better performance in terms of detection figures. On the other hand, however, this leads to a significant increase in detection cost, as the number of attacks and variants we want to be able to detect is broadened. To avoid this inconvenience while not affecting the detection accuracy, we defend the convenience of developing *holistic detection schemes*. This way, the construction of semantic models will help in implementing novel detection paradigms that surpass attack particularities and provide more global detection capabilities.

Also, as new attacks and variants appear, it is also recommended to devise new reaction schemes for guaranteeing the continuity and survivability of the monitored communications system. Opposite to current response schemes, which are generally performed locally, novel *global reaction schemes* based on the collaboration of the whole network are desirable. Otherwise, the response could be useless. For example, if a packet dropper is isolated and prohibited to participate in communications by a group of neighbor nodes, the malicious node could avoid the restriction by simply moving to a different area of the network.

Another challenge and recommendation from our point of view is to design and implement *integral defense mechanisms*. That is, to mix together prevention, detection, and reaction mechanisms in such a dynamic way that the security system acts as a whole instead of the mere sum of the parts. In other words, it is a desirable dynamic unsupervised adaptation of the system. This global adaptation must converge to stable and optimal solutions, which in fact have to be carefully controlled by the defense system itself. In other words, every functional element must be conveniently interrelated with the rest to provide with global solutions. For example, as a new attack instance is detected, it is evaluated in terms of its risk before triggering the adequate response/s and, if necessary, new prevention schemes may be carried out to protect our environment. Additionally, the model used in the detection process can be dynamically reestimated and thus, adapted to the conditions of the network over time.

One of the main consequences of the abovementioned research lines is the necessity of intra- as well as internode collaboration. However, this implies a new level of complexity and, as a consequence, a higher consumption of physical and logical resources. Since the disposal of such resources (i.e., battery and disk space) is restricted in some new devices, environments, and applications, a trade-off between security and cost is mandatory. In this line, the deployment of holistic approaches for *resource consumption saving* when dealing with the different security mechanisms is highly recommended. This trade-off between security and cost is also relevant from the viewpoint of the impact on the *quality of service* (QoS) of the communications. As a consequence, some of the current proposals existing in the literature are not valid from a practical perspective because they obviate resource consumption and/or their real impact on network performance. This, in fact, implies developing alternative schemes and methods.

14.6 Conclusion

This work constitutes a global survey on packet-dropping security threats for wireless multihop ad hoc networks. Beginning with the existence of several attacks reported in the literature with a similar objective of dropping packets, and sometimes a bit confusing in their final purposes, the chapter provides detailed state-of-the-art approaches based on the defense lines deployed to fight against this kind of attacks. Thus, prevention, detection, and reaction schemes developed in the literature during

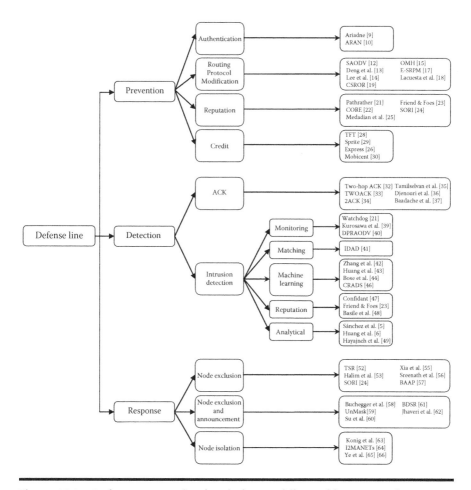

Figure 14.2   Defense approaches in wireless multihop ad hoc networks.

the past years are subsequently described here to organize the knowledge existing in the field. A summary of the different approaches can be found in Figure 14.2. Moreover, new trends and open challenges are also highlighted in order to point out what would constitute the near future in the target topic.

In summary, this chapter contributes a complete study of the packet discarding behavior problem in wireless multihop ad hoc networks, which is of high interest in the research community to improve security and, thus, service providing in this kind of (more and more accepted) environments. This work will actively help researchers to better understand packet dropping-related security attacks in multihop ad hoc networks.

Acknowledgment

This work has been partially supported by the Spanish MICINN through project TEC2011-22579.

References

1. D. Martins, H. Guyenne, Wireless sensor network attacks and security mechanisms: A short survey, *Proc. 13th International Conference on Network-Based Information Systems (NBiS)*, Takayama, Gifu, Japan, pp. 313–320, 2010.
2. B. Kannhavong, H. Nakayama, Y. Nemoto, N. Kato, A survey of routing attacks in mobile ad hoc networks, *IEEE Wireless Communications*, 14(5), 85–91, 2007.
3. B. Wu, J. Chen, J. Wu, M. Cardei, A survey on attacks and countermeasures in mobile ad hoc networks, in *Wireless Network Security on Signals and Communication Technology*, Y. Xiao, X. Shen, D.-Z. Du (Eds.), Springer, New York, pp. 103–135, 2007.
4. S. Djahel, F. Nait-Abdesselam, Z. Zhang, Mitigating packet dropping problem in mobile ad hoc networks: Proposals and challenges, *IEEE Communications Surveys & Tutorials*, 13(4)658–672, 2011.
5. L. Sánchez-Casado, G. Maciá-Fernández, P. García-Teodoro, An efficient cross-layer approach for malicious packet dropping detection in MANETs, *Proc. 11th IEEE International Conference on Trust, Security and Privacy in Computing and Communications (TrustCom)*, Liverpool, UK, pp. 231–238, 2012.
6. Y. Huang, W. Lee, Attack analysis and detection for ad hoc routing protocols, *Proc. 7th International Symposium on Recent Advances in Intrusion Detection 2004 (RAID)*, Sophia Antipolis, French Riviera, France, pp. 125–145, 2004.
7. A. El-Mousa, A. Suyyagh, Ad hoc networks security challenges, *Proc. 7th International Multi-Conference on Systems, Signals and Devices (SSD)*, Amman, Jordan, pp. 1–6, 2010.
8. R. Raghuvanshi, R. Kaushik, J. Singhai, A review of misbehaviour detection and avoidance scheme in adhoc network, *Proc. International Conference on Emerging Trends in Networks and Computer Communications (ETNCC)*, Udaipur, India, pp. 301–306, 2011.

9. Y. Hu, A. Perrig, B. Johnson, Ariadne: A secure on-demand routing protocol for ad hoc networks, *Proc. 8th Annual International Conference on Mobile Computing and Networking (MOBICOM)*, Atlanta, Georgia, USA, pp. 12–23, 2002.

10. K. Sanzgiri, B. Dahill, B.N. Levine, C. Shields, E.M. Belding-Royer, A secure routing protocol for ad hoc networks, *Proc. 10th IEEE International Conference on Network Protocol (ICNP)*, Paris, France, pp. 1–10, 2002.

11. S. Lu, L. Li, K. Lam, L. Jia, SAODV: A MANET routing protocol that can withstand black hole attack, *Proc. 5th International Conference on Computational Intelligence and Security (CIS)*, Beijing, China, pp. 421–425, 2009.

12. D. Cerri, A. Ghioni, Securing AODV: The A-SAODV secure routing protocol, *IEEE Communications Magazine*, 46(2), 120–125, 2008.

13. H. Deng, W. Li, D. P. Agrawal, Routing security in wireless ad hoc networks, *IEEE Communication Magazine*, 40(10), 70–75, 2002.

14. S. Lee, B. Han, M. Shin, Robust routing in wireless ad-hoc networks, *Proc. 31st International Conference on Parallel Processing Workshops (ICPPW)*, Vancouver, British Columbia, Canada, pp. 73–78, 2002.

15. C. Song, Q. Zhang, OMH-suppressing selfish behavior in Ad hoc networks with one more hop, *Mobile Network Applications*, 14(2), 178–187, 2009.

16. S. Khan, K.-K. Loo, N. Mast, T. Naeem, SRPM: Secure routing protocol for IEEE 802.11 infrastructure based wireless mesh networks, *Journal of Network and Systems Management*, 18(2), 190–209, 2010.

17. S. Khan, N.A. Alrajeh, K.-K. Loo, Secure route selection in wireless mesh networks, *Computer Networks*, 56(2), 491–503, 2012.

18. R. Lacuesta, J. Lloret, M. Garcia, L. Peñalver, Two secure and energy-saving spontaneous ad-hoc protocol for wireless mesh client networks, *Journal of Network and Computer Applications*, 34(2), 492–505, 2011.

19. S. Khan, J. Loo, Cross layer secure and resource aware on demand routing protocol for hybrid wireless mesh networks, *Journal of Wireless Personal Communications*, 62(1), 201–214, 2012.

20. J. Liu, V. Issarny, Enhanced reputation mechanism for mobile Ad Hoc networks, *Trust Management, Lecture Notes in Computer Science*, 2995, 48–62, 2004.

21. S. Marti, T.J. Giuli, K. Lai, M. Baker, Mitigating routing misbehavior in mobile ad hoc networks, *Proc. 6th Annual International Conference on Mobile Computing and Networking (MOBICOM)*, Boston, Massachusetts, USA, pp. 255–265, 2000.

22. P. Michiardi, R. Molva, CORE: A collaborative reputation mechanism to enforce node cooperation in mobile ad-hoc networks, *Proc. 6th IFIP Conference on Communication and Multimedia Security (CMS)*, pp. 107–121, 2002.

23. H. Miranda, L. Rodrigues, Friends and foes: Preventing selfishness in open mobile ad hoc networks, *Proc. 23rd International Conference on Distributed Computing Systems Workshops (ICDCSW)*, Providence, Rhode Island, USA, pp. 440–445, 2003.

24. Q. He, D. Wu, P. Khosla, SORI: A secure and objective reputation-based incentive scheme for ad-hoc networks, *Proc. IEEE Conference on Wireless Communications and Networking (WCNC)*, Atlanta, Georgia, USA, 2, 825–830, 2004.

25. M. Medadian, M. H. Yektaie, A. M. Rahmani, Combat with black hole attack in AODV routing protocol in MANET, *Proc. 1st Asian Himalayas International Conference on Internet (AH-ICI)*, Kathmandu, Nepal, pp. 1–5, 2009.

26. H. Janzadeh, K. Fayazbakhsh, M. Dehghan, M. S. Fallah, A secure credit-based cooperation stimulating mechanism for MANETs using hash chains, *Future Generation Computer Systems*, 25, 926–934, 2009.

27. J. Miao, O. Hasan, S. Ben Mokhtar, L. Brunie, K. Yim, An analysis of strategies for preventing selfish behavior in mobile delay tolerant networks, *Proc. 6th International Conference on Innovative Mobile and Internet Services in Ubiquitous Computing (IMIS)*, Palermo, Italy, pp. 208–215, 2012.

28. L. Buttyan, L. Dora, M. Felegyhazi, I. Vajda, Barter trade improves message delivery in opportunistic networks, *Ad Hoc Networks*, 8(1), 1–14, 2010.

29. S. Zhong, J. Chen, Y.R. Yang, Sprite: A simple, cheatproof credit based system for mobile ad hoc networks, *Proc. 22nd IEEE Conference on Computer Communications (INFOCOM)*, San Francisco, California, USA, pp. 1987–1997, 2003.

30. B.B. Chen, M.C. Chan, Mobicent: A credit-based incentive system for disruption tolerant network, *Proc. 29nd IEEE Conference on Computer Communications (INFOCOM)*, San Diego, California, USA, pp. 1–9, 2010.

31. D. Djenouri, N. Badache, New approach for selfish nodes detection in mobile ad hoc networks, *Proc. Workshop of the 1st International Conference on Security and Privacy for Emerging Areas in Communication Networks (SecurComm)*, Athens, Greece, pp. 288–294, 2005.

32. D. Djenouri, N. Ouali, A. Mahmoudi, N. Badache, Random feedbacks for selfish nodes detection in mobile ad hoc networks, *Proc. 5th IEEE International Workshop on IP Operations and Management (IPOM)*, Barcelona, Spain, pp. 68–75, 2005.

33. K. Balakrishnan, J. Deng, P.K. Varshney, TWOACK: Preventing selfishness in mobile ad hoc networks, *Proc. IEEE Wireless Communications and Networking Conference (WCNC)*, New Orleans, Louisiana, USA, pp. 2137–2142, 2005.

34. K. Liu, J. Deng, P.K. Varshney, K. Balakrishnan, An acknowledgment-based approach for the detection of routing misbehavior in MANETs, *IEEE Trans. Mobile Computing*, 6(5), 536–550, 2007.

35. L. Tamilselvan, V. Sankaranarayanan, Prevention of co-operative black hole attack in MANET, *Journal of Networks*, 3(5), 13–20, 2008.

36. D. Djenouri, N. Badache, On eliminating packet droppers in MANET: A modular solution, *Ad Hoc Networks*, 7(6), 1243–1258, 2009.

37. A. Baadache, A. Belmehdi, Fighting against packet dropping misbehavior in multihop wireless ad hoc networks, *Journal of Network and Computer Applications*, 35(3), 1130–1139, 2012.

38. P. García-Teodoro, J.E. Díaz-Verdejo, G. Maciá-Fernández, E. Vázquez, Anomalybased network intrusion detection: Techniques, systems and challenges, *Computer & Security*, 28, 18–28, 2009.

39. S. Kurosawa, H. Nakayama, N. Kato, A. Jamalipour, Y. Nemoto, Detecting blackhole attack on AODV-based mobile ad hoc networks by dynamic learning method, *International Journal of Network Security*, 5(3), 338–346, 2007.

40. P.N. Raj, P.B. Swadas, DPRAODV, A dynamic learning system against blackhole attack in AODV based MANET, *International Journal of Computer Science Issues*, 2, 54–59, 2009.

41. Y.F. Alem, Z.C. Xuan, Preventing black hole attack in mobile ad-hoc networks using anomaly detection, *Proc. 2nd International Conference on Future Computer and Communication (ICFCC)*, Wuhan, China, 3, 672–676, 2010.

42. Y. Zhang, W. Lee, Y.A. Huang, Intrusion detection techniques for mobile wireless networks, *Wireless Networks*, 9(5), 545–556, 2003.

43. Y. Huang, W. Fan, W. Lee, P.S. Yu, Cross-feature analysis for detecting ad-hoc routing anomalies, *Proc. 23rd IEEE International Conference on Distributed Computing Systems (ICDCS)*, Providence, Rhode Island, USA, pp. 478–487, 2003.

44. S. Bose, S. Bharathimurugan, A. Kannan, Multi-layer integrated anomaly intrusion detection system for mobile Adhoc networks, *Proc. IEEE International Conference on Signal Processing and Networking (ICSCN)*, Chennai, India, pp. 360–365, 2007.

45. J.F.C. Joseph, A. Das, B.C. Seet, B.S. Lee, CRADS: Integrated cross layer approach for detecting routing attacks in MANETs, *Proc. IEEE Wireless Communications and Networking Conference (WCNC)*, Las Vegas, Nevada, USA, 1525–30, 2008.

46. J.F.C. Joseph, B.S. Lee, A. Das, B.C. Seet, Cross-layer detection of sinking behavior in wireless ad hoc networks using SVM and FDA, *IEEE Trans. on Dependable and Secure Computing*, 8(2), 233–245, 2011.

47. A.S. Buchegger, J.Y. Le Boudec, Performance Analysis of the CONFIDANT protocol, *Proc. 3rd ACM International Symposium on Mobile Ad Hoc Networking & computing (MOBIHOC)*, Lausanne, Switzerland, pp. 226–236, 2002.

48. C. Basile, Z. Kalbarczyk, R.K. Iyer, Inner-circle consistency for wireless ad hoc networks, *IEEE Trans. Mobile Computing*, 6(1), 39–55, 2007.

49. T. Hayajneh, P. Krishnamurthy, D. Tipper, K. Taehoon, Detecting malicious packet dropping in the presence of collisions and channel errors in wireless ad hoc networks, *Proc. IEEE International Conference on Communications (ICC)*, Dresden, Germany, pp. 1–6, 2009.

50. M. Lima, A. dos Santos, G. Pujolle, A survey of survivability in mobile ad-hoc networks, *IEEE Communications Surveys & Tutorials*, 11, 66–77, 2009.

51. R. H. Jhaveri, S. J. Patel, D. C. Jinwala, DoS attacks in mobile ad hoc networks: A survey, *Proc. 2nd International Conference on Advanced Computing Communication Technologies (ACCT)*, Rohtak, Haryana, India, pp. 535–541, 2012.

52. M.E.G. Moe, B.E. Helvik, S.J. Knapskog, TSR: Trust-based secure MANET routing using HMMs, *Proc. 4th ACM Symposium on QoS and Security for Wireless and Mobile Networks (Q2SWinet)*, Vancouver, British Columbia, Canada, pp. 83–90, 2008.

53. J.H. Cho, A. Swami, I.R. Chen, A survey on trust management for mobile ad hoc networks, *IEEE Communications Surveys & Tutorials,* 13(4), 562–583, 2011.

54. I.T.A. Halim, H.M.A. Fahmy, A.M. Bahaa El-Din, M.H. El-Shafey, Agent-based trusted on-demand routing protocol for mobile ad hoc networks, *Proc. 4th International Conference on Network and System Security (NSS)*, Melbourne, Australia, pp. 255–262, 2010.

55. H. Xia, Z. Jia, X. Li, L. Ju, E.H.M. Sha, Trust prediction and trust based source routing in mobile ad hoc networks, *Ad Hoc Networks*, 11(7), 2096–2114, 2013.

56. N. Sreenath, A. Amuthan, P. Selvigirija, Countermeasures against multicast attacks on enhanced-on demand multicast routing protocol in MANETs, *Proc. 2nd International Conference on Computer Communication and Informatics (ICCCI)*, Coimbatore, Tamil Nadu, India, pp. 1–7, 2012.

57. S. Gupta, S. Kar, S. Dharmaraja, BAAP: Blackhole attack avoidance protocol for wireless network, *Proc. 2nd International Conference on Computer and Communication Technology (ICCCT)*, pp. 468–473, 2011.

58. S. Buchegger, J.Y. Le Boudec, Nodes bearing grudges: Towards routing security, fairness, and robustness in mobile ad hoc networks, *Proc. 10th Euromicro Workshop*

on Parallel, Distributed and Network-based Processing (PDP), Canary Islands, Spain, pp. 403–410, 2002.

59. I. Khalil, S. Bagchi, C.N. Rotaru, N.B. Shroff, UnMask: Utilizing neighbor monitoring for attack mitigation in multihop wireless sensor networks, *Ad Hoc Networks*, 8(2), 148–164, 2010.

60. M.-Y. Su, K.-L. Chiang, W.-C. Liao, Mitigation of black-hole nodes in mobile ad hoc networks, *Proc. 8th International Symposium on Parallel and Distributed Processing with Applications (ISPA)*, Taipei, Taiwan, pp. 162–167, 2010.

61. P.C. Tsou, J.M. Chang, Y.H. Lin, H.C. Chao, J.L. Chen, Developing a BDSR scheme to avoid black hole attack based on proactive and reactive architecture in MANETs, *Proc. 13th International Conference on Advanced Communication Technology (ICACT)*, Gangwon-Do, South Korea, pp. 755–760, 2011.

62. R.H. Jhaveri, S.J. Patel, D.C. Jinwala, A novel approach for GrayHole and blackHole attacks in mobile ad hoc networks, *Proc. 2nd International Conference on Advanced Computing Communication Technologies (ACCT)*, Rohtak, Haryana, India, pp. 556–560, 2012.

63. A. Konig, M. Hollick, R. Steinmetz, On the implications of adaptive transmission power for assisting MANET security, *Proc. 29th IEEE International Conference on Distributed Computing Systems Workshops (ICDCS)*, Montreal, Quebec, Canada, pp. 537–544, 2009.

64. Y.A. Mohamed, A.B. Abdullah, Immune-inspired framework for securing hybrid MANET, *Proc. IEEE Symposium on Industrial Electronics Applications (ISIEA)*, 1, 301–306, 2009.

65. X. Ye, J. Li, A security architecture based on immune agents for MANET, *Proc. International Conference on Wireless Communication and Sensor Computing (ICWCSC)*, Chennai, India, pp. 1–5, 2010.

66. X. Ye, J. Li, R. Luo, Hide Markov model based intrusion detection and response for MANETs, *Proc. 2nd International Conference on Information Technology and Computer Science (ITCS)*, Kiev, Ukraine, pp. 142–145, 2010.

67. P.L.R. Chze, W.K.W. Yan, Kan Siew Leong, A user-controllable multi-layer secure algorithm for MANET, *Proc. 8th International Wireless Communications and Mobile Computing Conference (IWCMC)*, Limassol, Cyprus, pp. 1080–1084, 2012.

68. A. Nabet, R. Khatoun, L. Khoukhi, J. Dromard, D. Gaiti, Towards secure route discovery protocol in MANET, *Proc. 3rd Global Information Infrastructure Symposium (GIIS)*, Da Nang, Vietnam, pp. 1–8, 2011.

69. C.A. Melchor, B.A. Salem, P. Gaborit, K. Tamine, AntTrust: A novel ant routing protocol for wireless ad-hoc network based on trust between nodes, *Proc. 3rd International Conference on Availability, Reliability and Security (AReS)*, Barcelona, Spain, pp. 1052–59, 2008.

Chapter 15

Security Issues in Machine-to-Machine Communication

Shuo Chen and Maode Ma

Contents

15.1 Introduction

The Internet has made the world seem smaller. Using the Internet leads to rapid sharing of huge amounts of information among people around the world. However, there is still a serious gap existing between the cyber world and the physical world [1]. The emerging cyber–physical systems (CPS) are going to fill this gap and connect all the objects of the physical and cyber world. Over the CPSs, the connected objects and items which are capable to report their locations and states will be able to exchange information among each other automatically without human operation. Since our lives are becoming increasingly interlinked by mobile phones, networked appliances, and other intelligent devices [2], the CPSs could make our lives more convenient and comfortable. A general architecture of a CPS can be shown in Figure 15.1.

There will be three major types of the components to form three tiers in a CPS. One type of component is a group of sensors to form an environmental tier. The second type of the component is the actuator, which can form a service tier. And the last type is the controller forming the control tier. The sensors collect information from physical systems and send the information to the network, which is handled by the distributed controllers in the cyber world. After processing the information, the controllers communicate with the actuators to issue appropriate operation commands. Then, the actuators will act to impose the physical world through activating the related operations and generate feedback. Based on the closed process of sensing, decision, execution, and feedback, the CPS can achieve self-awareness, self-judgment, and self-adjustment [4].

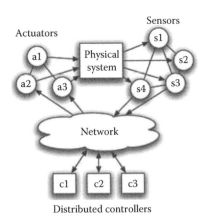

Figure 15.1 General architecture of a CPS. (From A. A. Cardenas, S. Amin, and S. Sastry, Secure control: Towards survivable cyber-physical systems, *Proceedings of IEEE 28th International Conference on Distributed Computing Systems Workshops (ICDCS)*, Beijing, China, June 2008, pp. 495–500.)

The major function of the environmental tier in a CPS is to collect and transmit the environmental information over the communication network, which connects the three tiers of the CPS, without human intervention. The fundamental feature of the environmental tier in a CPS is communication without human operations, which is called machine-to-machine (M2M) communication or machine-type-communication (MTC), where intelligent devices, including sensors, will communicate to each other end-to-end (E2E). The service-providing, decision-making, and autonomous control components and technologies consist of the service tier and the control tier. The M2M communication in a CPS integrates wireless sensor networks (WSNs) with other communication systems such as a cellular network or an optical network. By utilizing both wireless and wired technologies, the M2M could monitor the physical or environmental conditions and exchange the information among the components in different tiers.

M2M communication system consists of three interlinked domains: (1) An M2M area domain including an M2M area network with M2M gateways, (2) a communication network domain including wired/wireless networks such as xDSL and 3G, and (3) an application services domain [5] consisting of the end users and applications required in the CPS. The architecture of an M2M system is shown in Figure 15.2.

The collected information from the environment will be delivered from the M2M area domain to the network domain. The first destination of the information in the delivery is usually the M2M gateway, which decides the communication protocols used and transforms the received information into the formats required by the corresponding communication systems. A middleware layer with routing and converting functions could exist in the network domain. The layer may perform

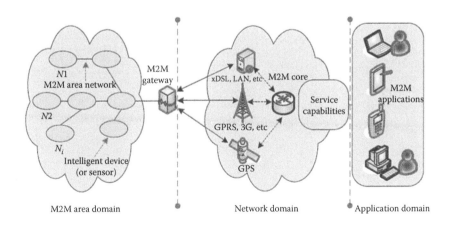

Figure 15.2   Architecture of an M2M system. (From M. Chen, J. Wan, and F. Li, Machine-to-machine communications: Architectures, standards, and applications, *KSII Transactions on Internet and Information Systems,* **6(2), 2012.)**

network management roles such as auto configurations, logging, notification, and so on. The communication systems in the network domain can be of any type such as wireless local area networks (WLANs), telephone lines, ethernets, satellite, or cellular networks, which will exchange information over a long distance. At the end, the information will be integrated into various applications in the application domain such as smart metering and smart grid, and so on [6].

The following is a specific application of M2M communication—home M2M network, see Figure 15.3.

In the application, the 802.15.4 (ZigBee/6LoWPAN) protocol is well suited for low-power/low-data rate applications such as HVAC control and appliances. The 802.11 (Wi-Fi) protocol works well for higher data rate applications such as audio and video streaming. Cellular is the best fit for applications that need to roam into and out of the home network. The Bluetooth protocol is well suited for low-data rate communications such as audio connections and file transfer [7].

The reminder of this chapter is organized as follows. In Section 15.2, the security challenges, threats, and design requirements are introduced. In Section 15.3, the current security solutions are reviewed and key technologies which would be applied in CPS and M2M communication are introduced. The direction of future research is discussed in Section 15.4. Finally, the chapter is concluded in Section 15.5.

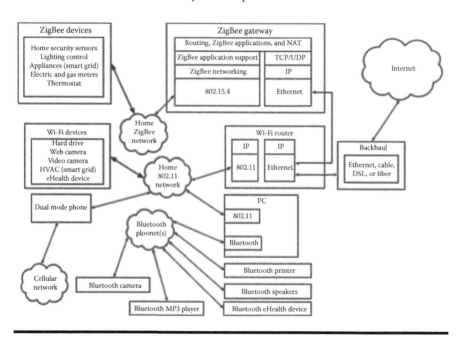

Figure 15.3 Home M2M network. (From M. Starsinic, System architecture challenges in the home M2M network, *Proceedings of the Long Island Systems Application and Technology Conference (LISAT)*, Farmingdale, NY, USA, May 2010, pp. 1–7.)

15.2 Security of M2M Communication

Since CPS is a distributed, complex, and hybrid real-time dynamic system with many different types of applications criticality operating at different time and space scales, it is easy to suffer various challenges to lose its functionality, partially or entirely. Correspondingly, CPS may face lots of threats due to the challenges from system attackers. In order to ensure the security of CPS, there are many security requirements that should be fulfilled by the designers of CPS.

15.2.1 Challenges

Compared to the decision-making and autonomous control functionalities that are mainly working in the cyber world or the physical world, the M2M communication system, which bridges the physical world and cyber world, is more fragile in the CPS.

The M2M communication network in a CPS has a few weakness, which makes the M2M unsecure [8]. First of all, in the M2M communication system the major communication medium is the radio waveform, which is easily eavesdropped. Second, the sensor nodes, which are normally unattended, in the M2M communication have limited capabilities in terms of both energy and computing power. Thus, they are easily attacked and complex security schemes will not be feasibly used to protect them. Third, it is unfeasible to have the security protection by authentication and data integrity because there is no infrastructure or server in the domain of the M2M communication systems. Finally, the network domain of an M2M system could integrate wireless and wired mediums for communication to the core network with different security schemes. This generates a protocol gap, which could be a potential threat to the M2M communication system, between different communication protocols.

The above-mentioned features of the M2M communication have left the opportunity for the various malicious attacks to impair the system. To explore solutions to effectively protect the M2M communication will be a great challenge to the research field of M2M security.

15.2.2 Threats

The categories of possible attacks in M2M communication have been explored and specified by the third-generation partnership project (3GPP) security workgroup (SA3) as follows:

■ *Physical attacks:* Physical attacks including the insertion of valid authentication tokens into a manipulated device, inserting and/or booting with fraudulent or modified software (reflashing), and environmental/side-channel attacks, both before and after in-field deployment.

- *Compromise of credentials:* Compromise of credentials involves comprising brute force attacks on tokens and (weak) authentication algorithms, physical intrusion, or side-channel attacks, as well as malicious cloning of authentication tokens residing on the machine communication identity module (MCIM)
- *Configuration attacks:* Configuration attacks such as fraudulent software update/configuration changes, mis-configuration by the owner, subscriber, or user; and mis-configuration or compromise of the access control lists.
- *Protocol attacks on the device:* Protocol attacks directed against the device, which include man-in-the-middle attacks upon first network access, denial-of-service (DoS) attacks, compromising a device by exploiting weaknesses of active network services, and attacks on over-the-air management (OAM) and its traffic.
- *Attacks on the core network:* Attacks on the core network, the main threats to the mobile network operator (MNO), include impersonation of devices; traffic tunneling between impersonated devices; mis-configuration of the firewall in the modem, router, or gateways; DoS attacks against the core network; also changing the device's authorized physical location in an unauthorized fashion or attacks on the network, using a rogue device.
- *User data and identity privacy attacks:* User data and identity privacy attacks include eavesdropping a user's or device's data sent over the access network; masquerading as another user/subscriber's device; revealing a user's network ID, or other confidential data to unauthorized parties.

In summary, there are a lot of attacks that may occur in the M2M systems, which could endanger the operation of the M2M communication systems. If the security of an M2M system could not been ensured, there will be serious problems causing the failure of its normal operations. It is very important to construct an effective security framework against various attacks to protect the M2M communication systems.

15.2.3 Requirements

From a traditional perspective, Rongxing Lu et al. [9] have described the security requirements for M2M communication as follows:

- *Confidentiality:* Confidentiality prevents unauthorized disclosure of sensory data in transmission from passive attackers, which ensures that only authorized entities can read these data in M2M communication systems.
- *Integrity:* Integrity must be ensured so that illegal alterations of the sensory data (e.g., modifying, deleting, delaying, or replaying data) can be detected. In an M2M communications system, it is critical to meet the integrity requirements since illegal alterations may result in serious consequences,

especially in life-critical M2M application contexts such as a remote e-healthcare systems.

■ *Authentication:* Authentication is a prerequisite for secure M2M communications, allowing the base station (BS) in the application domain to corroborate the sensory data of the M2M nodes in the M2M domain.

■ *Nonrepudiation:* Nonrepudiation guarantees that M2M nodes, once sending data, cannot deny the transmission.

■ *Access control:* Access control is the ability to limit and control access to the BS in the application domain. Specifically, it allows only authorized M2M application systems to gain access to the BS.

■ *Availability:* Availability ensures that whenever M2M application systems access the BS, the BS is always available.

■ *Privacy:* Privacy is also of paramount importance in some privacy-sensitive M2M communications systems (e.g., e-healthcare systems). For example, if sensitive patient health information (PHI) is illegally disclosed or improperly used, e-healthcare systems can cause undesirable effects on patients' lives.

Since M2M communication is a developing technology and it possesses some unique weaknesses as mentioned above, the design of M2M communication should also fulfill the following requirements.

■ Devices of M2M area domain normally have limited calculation, communication, and storage resources, and cannot meet the demand of traditional security technology. Therefore, designing a lightweight key management protocol is a top security issue.

■ M2M integrates many different communication technologies in network domains. Defense capability of M2M communication varies in different subsystems. Therefore, during the design of a common security architecture, we must consider the consistency and compatibility of the proposed security protocols, and ensure smooth transitions and seamless connections among different edge networks.

■ The data management in the M2M core of network domain is automatic processing technology that can help to implement massive data extraction, classification, filtering, identification, and data mining. However, intelligent processing is not enough to detect malicious information. Therefore, designing an effective trust and repudiation management mechanism is an important requirement for data management in M2M communication. To implement the separation between information content and information source is another important requirement.

■ The data precision demands of various applications in an application domain might be significantly different. It will increase the risk of privacy exposure that we only provide the same precision as the corresponding information. To provide appropriate precision for different applications is a key requirement in the design of application domains. Therefore, the M2M must make sure that the things and persons which possess the devices involved in M2M communication will never be marked or tracked by unauthorized things (such as equipments, services, application, persons, etc.). In order to achieve this security requirement, research on identification and privacy protection technology is very critical. Meanwhile, the M2M communication also proposes a big requirement of distributed database technology, since it involves massive data real-time storage and inquiry problems.

15.3 Current Solutions

In this section, we first review the M2M security solutions related to mature technologies, that is, solutions for M2M domain. Then we introduce the key technologies which would help to make the M2M communication become a practical technology in the future and list existing security solutions for the key technologies.

15.3.1 Solutions for M2M Domain

In the literature, there are many solutions found to address the security issues of the M2M communications. Most of the solutions focused on the mature technologies that are mainly related to M2M domain. The solutions could be classified into three kinds: detection, authentication, and key management.

15.3.1.1 Detection

Rongxing Lu et al. [9] proposed a mechanism to early detect the node compromise in an M2M domain. The scheme is designed to defend a M2M node-compromised attack. Since the attacker usually requires some time to compromise the M2M nodes, it is feasible for M2M nodes to form couples to monitor each other and detect node compromise early. For example, shortly after M2M node deployment, either two neighboring M2M nodes form the H-node (husband node) and W-node (wife node), or three neighboring nodes form H–W–C nodes, where "C" represents the child node. Then both H–W nodes and H–W–C nodes can periodically monitor each other with beacon messages. Once an attacker physically compromises an M2M node, its beacon message becomes exceptional, and the couple nodes can detect a node compromise attack early. Therefore, couple detection builds the first line of defense against internal attacks in the M2M domain.

Zubair et al. [10] modeled the malicious and/or abnormal events, which may compromise the security and privacy of smart grid (SG) users, as a Gaussian process. Based on this model, a novel early warning system was proposed for anticipating malicious events in the SG network. The warning system utilizes a

Bayesian data-modeling technique called Gaussian process regression. The theory is: "Gaussian processes have a prior and a posterior. Distributions are defined over functions using the Gaussian process, which is used as a prior for Bayesian inference. This prior can be flexibly obtained from training or observation data. For instance, from the SG point of view, the data collected from smart meters can be used to form the prior beliefs of a Gaussian process, which characterizes different aspects of the SG communication. Assume that the prior belief about the considered function conforms to a Gaussian process with a prior mean and covariance matrix. Through Gaussian process regression, samples of the function at different locations in the domain are observed. Given a set of observation points and their corresponding real-valued observations, it is possible to compute the posterior distribution of a new point. The optimal parameters of the Gaussian process are obtained by maximizing the log likelihood of the training data with respect to the parameters. By computing the posterior, it is possible to make predictions for unseen test cases." With the warning system, the smart grid control center can forecast malicious events, thereby enabling smart grid to react beforehand and mitigate the possible impact of malicious activity.

15.3.1.2 Authentication

Rongxing Lu et al. [9] also proposed a bandwidth efficient cooperative authentication (BECAN) to filter false reports in M2M communication. The scheme is designed to defend the node compromise attack that would not be detected because the compromising happens when the M2M nodes switch to sleep mode. Since the compromised nodes would inject false data and send them to the application domain, the objective of the BECAN scheme is to filter false data injected by compromised M2M nodes. BECAN applies the cooperative neighbor–router (CNR)-based filtering mechanism. Specifically, in the CNR-based mechanism, when an M2M node is ready to send sensory data to the M2M gateway via an established routing path, it first resorts to its neighboring nodes to cooperatively authenticate the sensory data, and then sends the data and the cooperation authentication information to the gateway via routing path. When each M2M node is equipped with TinyECC-based public key materials, the same noninteractive shared key between neighboring nodes and path nodes can be established. Then the full bipartite key graph can also be established. Because of the existence of a full bipartite key graph, the authentication information design is reasonable. Therefore, when a compromised M2M node sends false data to the gateway, the false data can be filtered if there is at least one uncompromised neighboring node participating in the reporting.

Tien-Dung Nguyen et al. [11] proposed a simple architecture M2M service that can be applied to any hospital that considers the mobility of doctors and patients. An efficient security scheme with dynamic ID-based authentication is applied in the M2M system. The proposed scheme uses a dynamic ID-based authentication with pair-wise key predistribution to establish a pair-wise key between a mobile sink and any sensor node. Since sensors are hardware- and power-limited, they consider computationally efficient methods to prevent attacks on the network. For example, the proposed security mechanism uses a dynamic ID-based authentication and collision-resistant hash function, to authenticate the source of the beacon signal before sensor nodes are allowed to transmit their aggregated data to the trusted mobile sink. The security analysis indicates that the proposed scheme provides a higher probability for noncompromised sensors to establish a secure communication in M2M service.

Sachin et al. [12] describe a novel method for over-the-air automated authentication and verification of M2M WSNs using the existing authentication assets of a cellular telecom operator. They extend the standard generic bootstrapping architecture (GBA) provided in the 3GPP specifications to implement their solution with minimal additional hardware and software requirements. Their method is divided into two procedures: bootstrapping authentication using GBA and subsequent bootstrapping usage. In the first procedure, the coordinator node authenticates itself to the cellular operator and derives key material. This shared key material is then used to securing the subsequent communication between the coordinate node and the M2M server. Their solution verifies not only the SIM card but the WSN's coordinator state. This aspect is very important because WSNs are deployed in remote locations, often devoid of physical security, and so any malicious user can easily steal the SIM card and substitute the original coordinator with a malicious device containing the original SIM. In [12], Sachin et al. have explained their solution using Zigbee as an example; the solution is equally valid for other M2M and sensor network standards. One of the key benefits of their approach is that it eschews expensive add-ons to existing operator infrastructure and yet addresses the key concern of checking the integrity of remotely deployed WSNs.

15.3.1.3 Key Management

Yosra et al. [13] propose a novel approach for establishing session keys for highly resource-constrained sensor nodes encountered in these M2M environments with an external server. The proposed system exploits the heterogeneity of M2M systems by delegating cryptographic computational loads to less resource-constrained nodes in a collaborative scheme. They present a novel key establishment protocol in which a highly resource-constrained node obtains assistance from more powerful M2M nodes in order to make use of asymmetric cryptography primitives to establish a shared secret key with a remote server. Furthermore, the highly resource-constrained node can do so through simple exchanges with neighbor nodes, which are considerably less energy consuming than actual use of these cryptographic primitives. A security analysis is conducted to verify that the proposed solution accomplishes its objective safely and efficiently.

15.3.2 Key Technology

Although there are already many solutions existing for M2M security, there are two features in M2M communication that may bring problems which cannot be solved by the current technologies. As mentioned above, the two features are: (1) the devices are deployed in very large amounts and the quantity will be larger and larger in the future, and (2) most of the devices are resource-constrained. To overcome the challenges brought by these two features, there are two rising technologies that are applied to M2M communication—the first one is IPv6 over low-power wireless personal area networks (6LoWPAN) and the second one is constrained application protocol (CoAP). Nowadays, there are already some security solutions for 6LoWPAN that belongs to the network domain.

15.3.2.1 6LoWPAN [14]

Owing to the amount of devices, M2M will need a very large address space that could only be provided by IPv6. So applying IPv6 to M2M communication is definitely the future trend. The Internet engineering taskforce (IETF) has been developing a new standard named 6LoWPAN to enable the use of IPv6 in low-power and lossy networks (LLNs), such as those based on the IEEE 802.15.4 standard.

IEEE 802.15.4 only prescribed the standard of PHY and MAC; it did not come down to the criterion above the network layer. To realize the networking of devices and the interoperability of different equipments, it is necessary to institute a uniform standard of the network layer. 6LoWPAN facilitates IPv6 connectivity over 802.15.4 compliant devices that are throughput and battery-limited, by compressing the IPv6 packets. 6LoWPAN technology's bottom layer adopts PHY and MAC layer standards of IEEE 802.15.4, and 6LoWPAN chooses IPv6 as the networking technology. But the payload length supported by MAC in IPv6 is much bigger than the one provided by 6LoWPAN bottom layer, in order to implement the seamless connection of the MAC layer and network layer, the 6LoWPAN working group suggested adding an adaptation layer between the MAC layer and network layer to achieve the header compression, fragmentation, reassembly, and mesh route forwarding. The reference model of 6LoWPAN protocol stack is shown in the Figure 15.4.

15.3.2.2 CoAP [15]

M2M applications are short-lived and reside in battery-operated devices, which most of the time sleep and wakeup only when there is data traffic to be exchanged. In addition, such applications require a multicast and asynchronous communication compared to the unicast and synchronous approach of standard Internet applications.

In March 2010, the IETF CoRE working group has started the standardization activity on CoAP. CoAP is an application layer protocol intended to be used in very

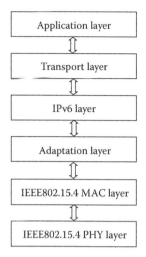

Figure 15.4   Reference model of 6LoWPAN protocol stack. (From X. Ma, and W. Luo, The analysis of 6LowPAN technology, *Proceedings of Pacific-Asia Workshop on Computational Intelligence and Industrial Application (PACIIA),* **Wuhan, Hubei, China, December 2008, pp. 963–966.)**

simple electronics devices that allows them to communicate interactively over the Internet. It is particularly targeted for small low-power sensors, switches, valves, and similar components that need to be controlled or supervised remotely, through standard Internet networks. CoAP is based on a REST architecture in which resources are identified by universal resource identifiers (URIs). The resources can be manipulated by means of the same methods as the ones used by HTTP. It consists of a subset of HTTP functionalities that have been redesigned, taking into account the low-processing power and energy consumption constraints of small embedded devices such as sensors. In addition, various mechanisms have been modified and some new functions have been included in order to make the protocol suitable to M2M applications. The HTTP and CoAP protocol stacks are illustrated in Figure 15.5.

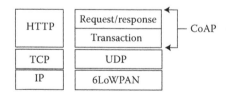

Figure 15.5   HTTP and CoAP protocol stacks. (From W. Colitti, K. Steenhaut, and N. De Caro, Integrating wireless sensor networks with the web, *Proceedings of the workshop on Extending the Internet to Low power and Lossy Networks (IP + SN 2011),* **Chicago, IL, USA, April 2011, pp. 1–5.)**

The transaction layer handles the single message exchange between end points. It also provides support for multicast and congestion control. The request/response layer is responsible for the transmission of requests and responses for the resource manipulation and transmission. This is the layer where the REST-based communication occurs. The dual layer approach allows CoAP to provide reliability mechanisms even without the use of TCP as transport protocol. In addition, it enables asynchronous communication, which is a key requirement for M2M applications.

15.3.2.3 Solutions for Network Domain

Yuanyuan Zhou et al. [16] proposed an embedded security gateway based on 6LoWPAN, which connects WSN with the IPv6 network. In order to provide security in WSN communication, their proposed gateway design adopts the security protocols for sensor networks (SPINS), which includes secure network encryption protocol (SNEP) and microtimed efficient streaming loss-tolerant authentication protocol (μTESLA). SNEP has the characteristic of low-communication overhead. In their framework, they use the gateway, which has large storage space and strong computing capability to distribute and manage keys. With the SNEP protocol, the system realizes two-party authentication, data confidentiality, integrity, as well as freshness. In their system, their concern is only in the communication between an external user (in IPv6 network) and a particular sensor node (in WSN), without broadcast queries. As a result, they introduce only the SNEP protocol, instead of the full-fledged SPINS. Besides, they build a web server in the gateway, which stores data periodically collected from sensor nodes. Users are able to query stored historical data, which allow operations such as event detection in WSN. Furthermore, they maintain a user access authority table on the web server, which implements user authentication and access control services.

Shahid Raza et al. [17] provided E2E secure communication between IP-enabled sensor networks and the traditional Internet. It is the first compressed lightweight design, implementation, and evaluation of 6LoWPAN extension for IPsec. IPsec provides authentication and privacy for IPv6. It is beneficial to use IPsec because the existing end-points on the Internet do not need to be modified to communicate securely with the WSN. Moreover, using IPsec, true E2E security is implemented and the need for a trustworthy gateway is removed. 6LoWPAN uses header compression techniques to ensure that the large IPv6 and transport-layer headers (UDP/TCP) are reduced. By supporting IPsec's authentication header (AH) and encapsulation security payload (ESP), additional IPv6 extension headers have to be included in each datagram. Independent of the achieved compression rates of AH and ESP, it is obvious that IPsec supported in 6LoWPAN will increase packet sizes as additional headers must be included. However, by using IPsec, we do not need to use the existing 802.15.4 link-layer security mechanisms that could, in turn, free some header space. Through applying IPsec, communication endpoints are able to authenticate, encrypt, and check the integrity of messages using standardized and established IPv6 mechanisms.

15.4 Future Research Direction

The classification of security issues in M2M communication and corresponding threats and defense requirements [8] can be summarized as follows. To satisfy the defense requirements is an important part of the future research.

■ *Front-end sensor security:* Front-end sensors can obtain data through the built-in sensors and transmit data through the M2M device or module, to achieve networking services of multiple sensors. This will involve the security of machines with node connectivity. Most M2M nodes are deployed in a scenario without monitoring. Then the attacker can easily access these devices, thus it could cause damage or illegal action by local operators.

An attacker could obtain confidential information of the key for M2M user or control data though eavesdropping user data, signaling data, and control data on the wireless link, or expose the signal in public places so can illegally access the data on M2M devices. Therefore, two-way authentication mechanisms of network and the corresponding encryption algorithms should be designed to prevent eavesdropping or unauthorized accessing to the data on wireless links. An attacker can cause damage of the transaction information on the M2M user by modifying, inserting, replaying, or deleting the legitimate M2M user data or signaling data transmission in wireless links. Attackers interfere user's data, signaling data, or control data in the correct transmission of a wireless link through the physical layer or protocol layer to achieve denial of service attacks of the wireless link. M2M security system should be designed to resist or mitigate the denial of service attacks, or tracking mechanism to quickly identify the location of attacker, to reduce the damage on the network.

■ *Network Security:* M2M devices will ultimately connect to core network services through a variety of means, from direct broadband or capillary wireless networks, to wired networks. Capillary networks used by M2M systems are made of a variety of links, either wireless or wired. Network's role is to provide a more comprehensive interconnection capacity, effectiveness, and economy of connection, as well as reliable quality of service. Because of the large number of nodes in M2M, it will result in denial of service when data spreads, since a large number of machines sending data leads to the congestion of network.

When an attacker enters the service network, it may eavesdrop user data, signaling data and control data, and gain unauthorized access to stored data within the network elements, or even can do passive or active flowing analysis.

An attacker, through the physical or protocol layer, interfere for transmission of user data, signaling data, or control data may use network services to impersonate legitimate users, or take advantage as posing access to legitimate users by pretending as a services network to access network services, in order to obtain unauthorized network services. To prevent unauthorized access to services, the proper validation schemes to overcome the protocol gaps are needed.

■ *Back-end security:* Back-end IT system forms on the gateway, application, or middleware, which has high security requirements, and collects and analyzes sensor data in real time or pseudo real time to increase business intelligence.

  Many researchers have begun to focus on designing techniques that would protect users' privacy such as k-anonymity. The solutions of machine and card authentication may also be able to solve the problem of security management of code resources, which use terminal numbers of international mobile equipment identity (IMEI) and international mobile subscriber identity (IMSI) of SIM modules for machine card binding. meanwhile, carrying out the interlocking management for card and machine and regularly sending the update key by M2M platform for authentication and certification to prevent the phenomenon of pirates of card or machine, ensuring the security of code resources. However, since the existence of competition among operators, the M2M users' certification information and the key may be faced with the threat of improper behavior when they are exchanged between operators, causing users' trade information to leak and economic loss. How to solve this problem should also be considered.

Moreover, as described in the previous sections, the security of M2M communication is a very challenging research topic and there are many open-research issues in this area. In the future, the research direction could focus on the following aspects.

■ *Lightweight:* The majority of the devices in M2M domain are resource-constrained. A scheme that is too complex could take time to compute and cost too much valuable energy for those deployed sensors which are unattended. So the design of a security scheme that could protect the system effectively and keep the system operating properly still needs more future research.

■ *Flexible:* In some situations, a complex scheme is not necessary for a simple attack. Finding a flexible scheme that could adjust its complexity according to the extent of threat could help save a lot of energy and extend the life of a device.

■ *Heterogeneous:* M2M communication systems integrate many different communication technologies. So we should design the security scheme to make the security-related data able to be computed, transmitted, and analyzed seamlessly under the heterogeneous communication network.

■ *Real time:* Since the quantity of the devices in the M2M domain is very large, the amount of the data transmitted in the network is massive. How to analyze the massive received data and compute the security metrics quickly is very challenging. Moreover, the ability of real time response is also very important because some applications of M2M communication may be life-vital, for example, e-health.

■ *Self restore:* The devices in an M2M domain are deployed in a very large scope. So it is impractical to send people to repair the devices that being break down under attack. Searching an approach that could make the device repair itself intelligently is a promising direction in the future.

15.5 Conclusion

In this chapter, we first introduced the architectures of CPS and M2M communi-cation and we considered M2M communication as the environmental tier of CPS. Second, we listed the challenges and threats faced by M2M communication and propose the corresponding general and special security requirements. Third, we reviewed the security solutions for the current widely applied technologies which are mainly in the M2M domain, introduce the rising technologies—6LoWPAN and CoAP—which would be the foundation of M2M communication in the future and listed the existing security solutions for 6LoWPAN. Since 6LoWPAN and CoAP are still underdeveloped, there will be more research work focused on their security solution. Finally, we summarized the classification of security issues and corresponding requirements in M2M communication and envisioned the future research directions. We believe that under the protection of the security solutions in the future, the M2M and CPS technologies could not only become significant market-changing forces but also better our lives considerably.

References

1. L. Sha, S. Gopalakrishnan, X. Liu, and Q. Wang, Cyber-physical systems: A new fron-tier, *Proceedings of IEEE International Conference on Sensor Networks, Ubiquitous and Trustworthy Computing (SUTC)*, Taichung, Taiwan, China, June 2008, pp. 1–9.
2. F. Michahelles, S. Karpischek, and A. Schmidt, What can the internet of things do for the citizen, *IEEE Pervasive Computing*, 9(4), 2010, 102–104.
3. A. A. Cardenas, S. Amin, and S. Sastry, Secure control: Towards survivable cyber-physical systems, *Proceedings of IEEE 28th International Conference on Distributed Computing Systems Workshops (ICDCS)*, Beijing, China, June 2008, pp. 495–500.
4. Y. Zhang, W. Duan, and F. Wang, Architecture and real-time characteristics analy-sis of the cyber-physical system, *Proceedings of IEEE 3rd International Conferenceon Communication Software and Networks (ICCSN)*, Xi'an, Shaanxi, China, May 2011, pp. 317–320.

5. M. Chen, J. Wan, and F. Li, Machine-to-machine communications: Architectures, standards, and applications, *KSII Transactions on Internet and Information Systems,* 6(2), 471–489, 2012.

6. S. Pandey, M-S. Kim, M-J. Choi, and J. W. Hong, Towards management of machine to machine networks, *Proceedings of the 13th Asia-Pacific Network Operations and Management Symposium (APNOMS),* Taipei, Taiwan, China, September 2011, pp. 1–7.

7. M. Starsinic, System architecture challenges in the home M2M network, *Proceedings of the Long Island Systems Application and Technology Conference (LISAT),* Farmingdale, NY, USA, May 2010, pp. 1–7.

8. C. Hongsong, F. Zhongchuan, and Z. Dongyan, Security and trust research in M2M system, *Proceedings of IEEE International Conference on Vehicular Electronics and Safety (ICVES),* Beijing, China, July 2011, pp. 286–290.

9. R. Lu, X. Li, X. Liang, X. Shen, and X. Lin, GRS: The green, reliability, and security of emerging machine to machine communications, *IEEE Communications Magazine,* 49(4), 2011, 28–35.

10. Z. M. Fadlullah, M. M. Fouda, N. Kato, X. Shen, and Y. Nozaki, An early warning system against malicious activities for smart grid communications, *IEEE Network,* 25(5), 2011, 50–55.

11. T-D. Nguyen, A. Al-Saffar, and E-N. Huh, A dynamic ID-based authentication scheme, *Proceedings of the Sixth International Conference on Networked Computing and Advanced Information Management (NCM),* Seoul, South Korea, August 2010, pp. 248–253.

12. S. Agarwal, C. Peylo, R. Borgaonkar, and J.-P. Seifert, Operator-based Over-the-air M2M wireless sensor network security, *Proceedings of the 14th International Conference on Intelligence in Next Generation Networks (ICIN),* Berlin, Germany, October 2010, pp. 1–5.

13. Y. Ben Saied, A. Olivereau, and D. Zeghlache, Energy efficiency in M2M networks: A cooperative key establishment system, *Proceedings of the 3rd International Congress on Ultra Modern Telecommunications and Control Systems and Workshops (ICUMT),* Budapest, Hungary, October 2011, pp. 1–8.

14. X. Ma, and W. Luo, The analysis of 6LowPAN technology, *Proceedings of Pacific-Asia Workshop on Computational Intelligence and Industrial Application (PACIIA),* Wuhan, Hubei, China, December 2008, pp. 963–966.

15. W. Colitti, K. Steenhaut, and N. De Caro, Integrating wireless sensor networks with the web, *Proceedings of the workshop on Extending the Internet to Low power and Lossy Networks (IP + SN 2011),* Chicago, IL, USA, April 2011, pp. 1–5.

16. Y. Zhou, Z. Jia, X. Sun, X. Li, and L. Ju, Design of embedded secure gateway based on 6LoWPAN, *Proceedings of IEEE 13th International Conference on Communication Technology (ICCT),* Jinan, Shandong, China, September 2011, pp. 732–736.

17. S. Raza, S. Duquennoy, T. Chung, D. Yazar, T. Voigt, and U. Roedig, Securing communication in 6LoWPAN with compressed IPsec, *Proceedings of International Conference on Distributed Computing in Sensor Systems and Workshops (DCOSS),* Barcelona, Spain, June 2011, pp. 1–8.

Chapter 16

Authentication in Wireless Mesh Networks

Aymen Boudguiga and Maryline Laurent

Contents

16.1 Introduction

During the past decades, we have been living a revolution in networking thanks to the definition and development of wireless technologies. First, researchers discovered one-hop wireless networks such as Wireless Local Area Networks (WLANs) and cellular networks where a client joins the network by passing through a one-hop Access Point (AP) or a Base Station (BS). Then researchers got interested in multihop wireless networks such as Mobile Ad Hoc Networks (MANETs) and Wireless Sensor Networks (WSNs). Multihop wireless networks brought flexibility and scalability, but raised many challenges concerning power, security, and mobility management.

In this context, Wireless Mesh Networks (WMNs) emerged to provide interconnection between different wireless technologies. WMNs are said to be self-organizing and self-managing networks. Indeed, WMNs adapt peer-to-peer networking to define a wireless backbone that offers connectivity to clients implementing different wireless technologies. In WMNs, wireless nodes cooperate to discover and create routing tables. As such, WMNs react quickly to any change in the network topology without the need for an organizing entity. Furthermore, WMNs are reliable as they define redundant paths between the communicating nodes thanks to the use of multihop routing protocols. In addition, WMNs reduce the cost of network deployment as there is no need for installing a wired infrastructure. Consequently, WMNs deployment is increasing as they serve advantageously to expand wired backbones (mostly MANs), especially in countryside areas where cabling only a few customers is very expensive.

WMNs are constructed as community networks. In fact, a community peer-to-peer network is built within the WMN, and it fully supports local traffic transfer among stations (STAs) of the community. In the future, every home will be equipped with a roof antenna to communicate with neighbors in the that community with no access to the Internet. A WMN creates a local network inside a community which enhances local data exchange. As a result, the load generated to access the network (and the Internet routers) decreases and the link bandwidth increases. As such, a client is no longer required to pass through the Internet in order to send data to another client in the community network. It uses the community peer-to-peer network to transfer the information to its destination. As a consequence, its local data exchange is faster and the Internet routers load is relieved. In addition, WMNs serve to enhance the home network capabilities. A home WMN provides the interconnection of all the network devices in a home without passing through a hub. This is realized thanks to the multihop routing protocols used in a WMN.

16.1.1 WMNs Architecture

There are three types of WMN architecture. The first is the *infrastructure mesh* that is composed of a backbone containing Mesh Routers (MRs). MRs are generally static, but may have some limited mobility. MRs act like APs for WLAN clients, as node B for Wimax users, or as BS for mobile phones in cellular networks. In addition, some MRs are used as gateways to external networks such as the Internet.

The second type of mesh architecture is the *client mesh,* which is composed of mobile nodes that do not have any gateway capabilities. Consequently, a client is only able to forward the traffic for one of its neighbors. In fact, the *client mesh* has the same characteristics as a MANET.

The last mesh architecture is called the *hybrid mesh* and it combines the two previous mesh architectures (Figure 16.1). The *client mesh* connects, through an MR, to the mesh backbone. As such, two mesh clients from different wireless technologies can be interconnected through the backbone. For example, an ad hoc client becomes able to communicate with a client of a cellular network.

In an effort to distinguish WMNs from other wireless architecture, we compare WMNs to WLANs, WSNs, and MANETSs. (For a detailed comparison between WMNs and other wireless networks and technologies, interested readers can refer to Zhang et al. [46].)

■ WLANs: WLANs aim is to connect clients to a wired distribution system through APs. WLANs have a fixed architecture, as they are parts of a wired network. APs are not mobile, only client nodes can be mobile. Generally, WLANs routing protocols rely on the number of hops as a metric for the route choice. As such, the packet path is known in advance. In addition, in a WLAN, if a problem occurs to an AP, all the STAs associated to that AP will be disconnected from the network. WLANs do not have the self-managing behavior, which distinguishes the ad hoc or mesh networks.

■ WSNs: WSNs contain nodes that have limited power and processing capacities. Generally, they serve to capture environment changes thanks to sensors.

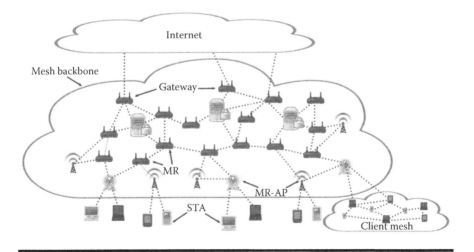

Figure 16.1 Hybrid mesh architecture.

Reports are sent to nodes that have more power, computing, and storage capabilities. This particular node is called the *sink node*. WSNs are organized often into clusters managed by cluster heads, that is, sink nodes. In WSN, nodes use multihop routing protocols for route construction.

- MANETs: MANETs are self-managing and self-configuring networks. They are formed by very dynamic nodes that have high-power constraints. In addition, MANETs have no infrastructure and are frequently used for military applications.

- WMNs: WMNs are formed by MRs, which are almost static and so have no energy constraints. However, the mesh clients can have processing, memory, and energy constraints especially when the mesh clients are sensors in a WSN or mobile phones in a cellular network. In general, WMNs rely on the same routing protocols as MANETs. These routing protocols use radio-aware metrics, which are more adapted to wireless networks. The new routing metrics consider the signal strength, the interference level, and the frame expected transmission time as criteria for route construction.

Thanks to their interesting properties, many WMNs testbeds and industrial solutions emerged. For example, the MIT-Roofnet is an experimental mesh network developed by the MIT Computer Science and Artificial Intelligence laboratory. MIT-Roofnet relies on twenty nodes to offer connectivity to users on the Cambridge campus. In addition, many standardization institutions had a big interest in WMNs. For example, the Institute of Electrical and Electronics Engineer (IEEE) have specified the following mesh amendments IEEE802.11s [1], IEEE802.16a [2], and IEEE802.15.5 [3] to its famous standards IEEE802.11 [4], IEEE802.16 [5], and IEEE802.15 [6], respectively.

As an example, we consider the IEEE802.11s standard [1] that modifies the well-known IEEE802.11 architecture in order to support mesh networking (Figure 16.2) [1,4]. First, the wired *Distribution System* is replaced by a wireless backbone formed by Mesh Points (MPs). These MPs act as MRs and provide multihop paths and peer-to-peer communications to mesh APs (MAPs). MAP has the same capability as a traditional AP combined with mesh functionality. MPs which offer connection to external networks (either 802 LANs or layer-3 networks) are called Mesh Portals (MPP). All these components (MPs, MAPs, and MPPs) form the Mesh Basic Service Set (MBSS).

16.1.2 Authentication in WMNs

The International Organization for Standardization [7] defines *authentication* as a mean for confirming the identity of a claimed entity. Authentication methods serve to prove the ownership of an identity. Authentication is realized in practice using shared secrets (such as pass-phrases), identification cards (such as digital passports), certificates (such as amazon.com certificate), or biometric data (such as fingerprints

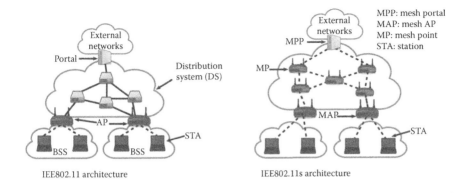

Figure 16.2   IEEE802.11 and IEEE802.11s architectures.

or DNA). That is, authentication is realized by any mean that bind an identity to its owner.

In wireless networks, a client station generally authenticates to a special entity called the Authentication Server (AS). Authentication is the first step for a station to be authorized to access the network. As such, authentication serves to filter legitimate users from malicious ones. The clients have to authenticate when they join the network for the first time or after being disconnected for a while.

As WMNs serve to interconnect mesh clients that are using different technologies, some criteria have to be respected when defining an authentication method. These criteria can be summarized in the following rules:

- *Low computation costs*: The authentication scheme has to take into consideration that STAs, such as sensor nodes or mobile phones, have limited processing power and battery constraints.
- *Low overhead*: The authentication scheme has to reduce the number of exchanged messages.
- *Low memory consumption*: The authentication scheme has to require small-storage space in the station memory.
- *Mutual authentication*: The authentication scheme serves to authenticate the two communicating entities simultaneously. As such, it avoids that a station authenticates to a rogue MR.
- *Valid for* MR*s and clients*: The authentication scheme can be started either by MRs or by mesh clients when they need to authenticate to the network.
- *Adapted to hand-off situation*: The authentication can be used by a mesh client who is moving from one MR to another MR.

In this chapter, we describe some of the protocols that were proposed for authenticating clients and MRs in WMNs, both in standards and literature. We start by describing the extensible authentication protocol (EAP) as it is a fundamental, flexible, and well-spread standard supporting various authentication methods. EAP controls the network access between a *supplicant* entity and an *authenticator*. The supplicant is the equipment which is requesting the authentication, while the authenticator is the equipment that is leading the authentication. That is, the authenticator authenticates the supplicant station relying on a specific EAP authentication method.

When the authenticator does not support the authentication method selected by the supplicant station, it can ask for the help of a back-end AS. The AS implements most of the authentication methods and is able to verify the validity of the authentication data of supplicants. As such, the authenticator serves as a pass-through server between the supplicant and the AS.

Thanks to its flexibility, EAP has been used in many network security standards to provide STAs' authentication. EAP was adapted to IEEE wired LANs in the standard IEEE802.1X. Then, the IEEE community specified in the IEEE802.11i standard how to encapsulate EAP messages in IEEE wireless environments.

The remainder of this chapter is organized as follows. First, we describe EAP in Section 16.2. Then, we give an introduction to IEEE802.11 network security (Section 16.3) and to IEEE802.11s network security (Section 16.4). Finally, we summarize some of the methods that were proposed in literature, to authenticate mesh STAs (Section 16.5).

16.2 Extensible Authentication Protocol

Extensible Authentication Protocol (EAP) was originally defined as an extension of the point-to-point protocol (PPP) [8] in order to provide a mechanism for selecting authentication methods. First, EAP has been defined as an amendment to PPP [9] to separate the PPP authentication phase from the PPP link establishment phase. Then, EAP [10] evolved to become the reference when talking about authentication in networks, thanks to its flexibility.

Flexibility is the result of EAP's organization into layers (Figure 16.3) [10]. That is, EAP dissociates authentication methods from the lower transport protocols. The

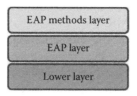

Figure 16.3 The EAP layers structure.

lower layer controls the transmission of EAP packets from the authenticator to the supplicant. It is usually controlled by protocols such as PPP, IEEE802.1X, and IEEE802.11i. It is responsible for error detection, packet fragmentation and reassembly, and packets reordering. Packet reordering is an important matter because EAP is a lock-step protocol. Meanwhile, the *EAP layer* performs duplicated EAP message detection and controls the retransmission of the EAP request message. In addition, it ensures the encapsulation of EAP messages during an authentication. Finally, the *EAP methods layer* specifies the different EAP methods that can be used during an authentication. Examples of EAP methods are presented in Section 16.2.1.

EAP defines three authentication parties: the supplicant, the authenticator, and the back-end AS. The supplicant is the entity being authenticated and the authenticator is the entity that starts and performs the authentication. Through the exchange of some *request* and *response* messages, the authenticator and the supplicant negotiate and execute the authentication method.

When the authenticator does not support the authentication method proposed by the supplicant, it acts as a pass-through server to transmit the authentication message to a back-end AS. Generally, the back-end AS implements most of the well-known authentication methods. The link between the authenticator and the AS is secured using AAA protocols such as RADIUS [11,12] and Diameter [13,14].

In practice, the supplicant can be a WLAN STA, which needs to authenticate to an AP, connected to an AS (Figure 16.4). The STA sends its EAP Over LAN [15] packets to the AP. Then, the AP encapsulates the received EAP packets into RADIUS packets and forwards them to the AS. The result of the authentication is conveyed in an authentication *success* or an authentication *failure* message. In peer-to-peer networks, a node acts as the supplicant and authenticator at the same time in order to mutually authenticate with its neighbors.

EAP exchanges are always started by the authenticator, which sends an EAP request to the supplicant. Then, the authenticator waits for the supplicant response as EAP is a lock-step protocol. When no EAP response is received, the authenticator retransmits the request packet to the supplicant. The retransmission is stopped when a retry counter runs out or when a lower layer failure indication is received. The supplicant responses must only be sent in reply to request messages received from the authenticator.

Figure 16.4   The EAP encapsulation.

16.2.1 EAP Authentication Methods

With EAP, many types of authentication methods can be negotiated between the authenticator (or the AS) and the supplicant station, but only one method is used to perform the authentication. We classify these methods into two groups based on the type of the authentication. On one hand, we have methods that rely on the use of a preshared secret between the supplicant STA and the AS. On the other, we have methods that use certificates to authenticate either the AS or the supplicant STA.

EAP–PSK [16], EAP–SAKE [17], and EAP–GPSK [18] are examples of EAP methods which rely on a Preshared Key (PSK) to support mutual authentication between clients and servers. The shared key can also be used as a Master Session Key (MSK) for deriving authentication keys, integrity check keys, and encryption keys during the authentication.

EAP–TLS [19], EAP–TTLS [20], and EAP–FAST [21] are three authentication methods relying on certificates and the transport layer security (TLS) protocol [22]. TLS supports mutual authentication between clients and servers, based on certificate exchange, and it enables the secure transmission of all the information elements needed for the computation of the preshared MSK. Figure 16.5 shows

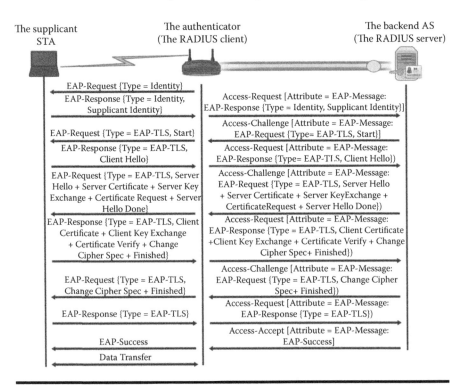

Figure 16.5   An example of EAP–TLS authentication.

an example of EAP–TLS authentication between the supplicant STA and the AS. Note that the EAP–TLS packets are encapsulated using RADIUS between the authenticator and the AS.

16.2.2 IEEE802.1X Standard

The IEEE802.1X standard [15] specifies the security guidance for wired IEEE LANs. In addition, it proposes an encapsulation protocol for EAP packets over IEEE802 protocols that are different from PPP. It defines two logical entities at every point of attachment to the network. These two entities are the controlled port and the uncontrolled port. In order to get access to the services offered by the controlled port, a station must authenticate to a Port Access Entity (PAE) through the uncontrolled port.

IEEE802.1X defines the Port Access Control Protocol (PACP) to manage the state of the ports of an authenticator and a supplicant STA. Every entity, either a supplicant or an authenticator, has a controlled port that can authorize or unauthorize the access to the entity services. Consequently, the port status can be either authorized or unauthorized.

The IEEE802.1X standard defines the EAP Over LAN encapsulation (EAPOL). It extends EAP with new messages to enhance its usage in IEEE LAN environments. The first added frame is the *EAPOL-start* frame. It is used by the supplicant to inform the authenticator that it wants to start a new EAP authentication. Second, the *EAPOL-log-off* frame was defined to express the request of stopping the authentication by the supplicant or the authenticator. Third, the *EAPOL-key* frame is used to exchange key material and the packets of the authentication method between the authenticator and the supplicant. Finally, the *EAPOL-Encapsulated-ASF-Alert* carries an alert message that is always received on the uncontrolled port.

16.2.3 Protocol for Carrying Authentication for Network Access

The Protocol for carrying Authentication for Network Access (PANA) [23] is another lower layer protocol for the encapsulation of EAP authentication methods. PANA has the property of allowing EAP frames to be carried over the IP. That is, the supplicant STA is not obliged to be located one-hop away from the authenticator, as the EAP method is no more carried over a link layer protocol but over a network layer protocol. The only requirement is that the PANA Authentication Agent (PAA) and the PANA Client (PaC) must be exactly one IP hop away from each other [24].

PANA defines the PaC as the supplicant STA which is going to carry out an EAP session with a back-end AS. The PaC can be one or more hops away from the network access server called the PANA Authentication Agent (PAA). The PAA acts as the authenticator and as a pass-through server for the EAP frames that are

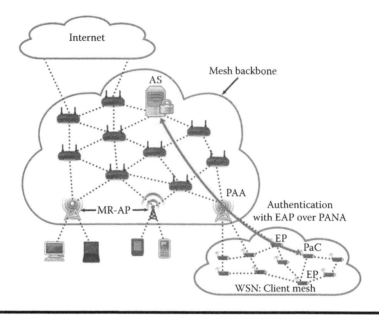

Figure 16.6   An example of PANA usage in a WMN.

exchanged between the PaC and the AS. When a PaC authentication ends success-fully, the PAA relies on an Enforcement Point (EP) to provide access control for an authenticated (i.e., authorized) PaC. The EP is preferred to be a one-hop neighbor of the PaC. The EP is in charge of traffic filtering and allows only the traffic of authorized PaCs that have already authenticated to the AS. However, unauthen-ticated nodes will be allowed to transmit only PANA, DHCP, and configuration traffic to the access network.

Thanks to the ability of authenticating a PaC that is n-hops away from the PAA, PANA seems to be well suited for nodes authentication with EAP methods in a WMN. Let us consider, for example, a WSN that is connected to the mesh back-bone via an MR as presented in Figure 16.6. The MR–AP will be used as the PAA. Meanwhile, every sensor will act as a PaC and starts an EAP authentication with the AS by passing through the PAA. When a sensor authenticates successfully to the AS, the latter informs the PAA of the authentication results. Then, the PAA pro-vides the EP with the attributes required for controlling the access of the PaC. The EP is a one-hop neighbor of the PaC, which has already authenticated to the AS.

16.3 IEEE802.11 Network Security

IEEE802.11 [4] network security has been specified in the IEEE802.11i stan-dard [25] that adapts the security requirements and methods proposed in the

IEEE802.1X standard to wireless environments. IEEE802.11i standard introduces the concept of Robust Security Network Association (RSNA). The RSNA is an association either between an STA and an AP or between two STAs. This association serves to ensure data confidentiality, integrity and authentication. The RSNA association relies on the use of encryption keys.

In IEEE802.11 networks, the keys are classified into three different categories. The *pairwise key* secures the unicast traffic between an AP and an STA. The *group key* secures the multicast or broadcast traffic between an AP and a group of STAs and the *STA to STA link key* (STSL key) secures the link between two STAs that communicate directly in an ad hoc mode, that is, not through an AP.

In practice, the RSNA capabilities of an STA or an AP are carried in a Robust Security Network (RSN) information element that is included in beacon, probe response, or association request frames. The RSN information element describes not only the encryption and the authentication algorithms supported by an STA but also the key management procedures that an STA implements. That is, the RSN information element contains information about the group cipher suite, the pairwise cipher suite, and the Authentication and Key Management algorithm (AKM). The group cipher suite specifies the encryption algorithms which secure the multicast and the broadcast traffic, while the pairwise cipher suite indicates which encryption schemes serve to secure the unicast traffic.

In practice, an RSNA is established between an AP and an STA (or between two STAs) only after their mutual authentication. First, a supplicant STA authenticates to an authenticator (an AP or another STA) using an IEEE802.1X authentication or a PSK authentication. This authentication usually happens between the supplicant STA and an AS which is trusted by the authenticator. That is, the authenticator acts as a pass-through server between the supplicant STA and the AS. If the authentication succeeds, the supplicant STA and the AS share an MSK. Then, they derive a *Pairwise Master Key* (PMK) from the MSK using a pseudorandom derivation function. Afterwards, the AS transfers the PMK to the authenticator. As such, the supplicant STA and the authenticator execute a *handshake* to derive from the PMK a set of Temporal Keys (TKs) needed for data encryption. When the TKs are finally derived, we say that an RSNA has been established between the supplicant and the authenticator.

TKs derivation and exchange happen during the *4-way handshake* between an STA and an AP (or between two STAs). TKs are used with the *temporal key integrity algorithm* (TKIP) or the *CTR with CBC–MAC protocol* (CCMP) to provide data confidentiality and integrity. The IEEE802.11i standard also defines the *Wired Equivalent Privacy* (WEP) to be used by RSNA-disabled equipment.

In the following sections, we describe the different types of RSNA that have been defined for IEEE802.11 (Section 16.3.1). Then, we present the different methods that were specified for STA authentication (Section 16.3.2). Finally, we describe the different key handshakes that serve to exchange a set of keys between an AP and STA or between two STAs (Section 16.3.3).

16.3.1 Robust Security Network Associations

RSNA is based on the concept of Security Association (SA). The SA defines a set of security policies that aim to protect the exchanged data between STAs. The SA types depend on the kind of key being used or created. IEEE802.11i defines the following types of SAs:

■ *Pairwise Master Key Security Association* (PMKSA) is the result of a successful IEEE802.1X authentication or a PSK authentication between an AS and a supplicant STA.

When an IEEE802.1X authentication is performed, the AS and the supplicant STA exchange authentication and keying information. That is, the AS and the supplicant STA collaborate to compute an MSK, which serves to derive a PMK. Then, the AS securely sends the PMK to the authenticator. As such, the authenticator and the supplicant STA execute a 4-way handshake to derive a Pairwise Transient Key (PTK) from the same PMK.

If a PSK is used for the authentication, the AS and the supplicant STA use the PSK as an MSK. That is, the PSK will serve to derive the PMK.

■ *Pairwise Transient Key Security Association* (PTKSA) is the result of the 4-way handshake between the authenticator (an AP) and the supplicant STA. It creates a PTK to protect the traffic between the authenticator and the supplicant. The PTK is derived from the PMK, which is created during the authentication of the supplicant to the AS. That is, the AS securely transmits the PMK to the authenticator. Then, the authenticator collaborates with the supplicant STA to derive the PTK from the PMK.

■ *Group Temporal Key Security Association* (GTKSA) is the result of a successful 4-way handshake or a *group key handshake*. It serves to create a group temporal key (GTK) to secure the multicast or broadcast traffic between an AP and a set of STAs.

■ *STSL Master Key Security Association* (SMKSA) is the result of the *STSL Master Key (SMK) handshake* in environments where the AP supports the *peer-key* usage. The *peer-key handshake* is executed when two STAs, which communicate through the same AP, decide to communicate directly using a shared secret key.

■ *STSL Transient Key Security Association* (STKSA) is the result of a 4-way handshake that follows an *SMK handshake*. It aims to create an STSL Temporal Key (STK) to secure the traffic between two STAs.

16.3.2 IEEE802.11 STA Authentication

The IEEE802.11 standard relies on different methods for STA authentication. These methods depend on STA's capability of supporting an RSNA creation. In the following, we make a quick presentation of the different IEEE802.11 authentication methods:

■ *Open system authentication* is an authentication method that was used before RSNA emergence. It provides the weakest level of security. It is based on the exchange of two authentication frames using IEEE802.11 management frames. In the first message, the supplicant inserts its identity. In the response message, the authenticator indicates the result of the authentication based on the verification of the supplicant identity.

■ *Shared key authentication* is used by RSNA-disabled STAs. It uses a shared secret between the authenticator and the supplicant to exchange an encrypted challenge message. It is only used when WEP is selected to provide confidentiality. The shared key authentication contains four steps. First, the supplicant asks the authenticator to start a shared key authentication. So, the authenticator creates a challenge message with the WEP algorithm and sends it as a response to the supplicant. Then, the supplicant copies the challenge in a third frame which is encrypted using the WEP algorithm and the shared key. Upon receiving the third message, the authenticator deciphers its content and compares the value of the received challenge to its own. If the two challenges are equal, the authentication ends with a success frame sent to the supplicant, otherwise it ends with a failure frame.

■ *IEEE802.1X authentication* relies on EAP authentication methods. The EAP–TLS is one of the methods that were proposed to make a mutual authentication between an STA and its AS. That is, IEEE802.1X authentication is an EAP-based authentication.

16.3.3 IEEE802.11 Handshakes

The IEEE802.11 standard relies on three different handshakes to exchange keys between a supplicant STA and its AP or between two STAs after the authentication phase.

■ *4-way handshake* includes four steps (Figure 16.7) [1]. During this handshake, an AP and a supplicant STA collaborate to create a PTK from the PMK or two STAs collaborate to derive an STK from an SMK. The AP and STA exchange random nonces in messages 1 and 2 of the handshake. Then, they use these nonces, their Media Access Control (MAC) addresses, and the PMK as inputs to a pseudorandom function that outputs the PTK. The same operations are done by two STAs to derive an STK from an SMK.

The PTK and the STK are then decomposed into three keys. The first key is the EAPOL–key Confirmation Key (KCK). The supplicant STA and the AP use their KCK with HMAC–MD5 or HMAC–SHA1 to compute a Message

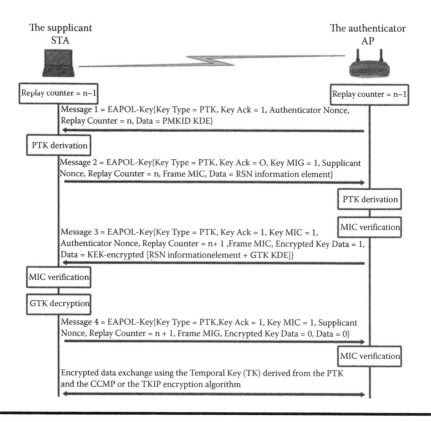

Figure 16.7 PTK generation during a 4-way handshake between an STA and an AP.

Integrity Check (MIC) sequence for their messages 2, 3, and 4. The second key is the EAPOL–key Encryption Key (KEK). The AP uses KEK to encrypt its GTK in message 3. The third key is the TK. The AP and STA use their TK to exchange encrypted data either with TKIP or CCMP at the end of the 4-way handshake [4].

The IEEE802.11 standard relies on two different functions for the computation of the keys. The first function is denoted by $L(Str, F, L)$ and, from Str, extracts L bits starting from bit number F. The bit counting starts from left to right. The second function is called *PRF-n* and generates n-bits long pseudorandom number. Table 16.1 summarizes the different keys that are derived from the PMK and the SMK during the 4-way handshake. Note that each encryption algorithm TKIP or CCMP specifies a different length for the TK.

■ *Group key handshake* is used by an AP to convey a new GTK to STA. The new GTK is encrypted using the KEK, which was derived during the 4-way handshake. The AP has a group master key (GMK) that is used for the GTK derivation (Table 16.1).

Table 16.1   IEEE802.11 Key Hierarchy

PMK and PTK key derivation:
PMK = L(MSK, 0, 256)
• PTK_{TKIP} = PRF-512(PMK, pairwise key expansion, authenticator address, supplicant address, authenticator nonce, supplicant nonce)   • or PTK_{CCMP} = PRF-384(PMK, pairwise key expansion, authenticator address, supplicant address, authenticator nonce, supplicant nonce)   • KCK = L(PTK, 0, 128)   • KEK = L(PTK, 128, 128)   • TK_{TKIP} = L(PTK, 256, 256) or TK_{CCMP} = L(PTK, 256, 128)
GTK key derivation: GTK_{TKIP} = PRF-256(GMK, group key expansion, authenticator address, group nonce)   or GTK_{CCMP} = PRF-128(GMK, group key expansion, authenticator address, group nonce)   • TK_{TKIP} = L(PTK, 0, 256) or TK_{CCMP} = L(PTK, 0, 128)
STK key derivation: STK_{TKIP} = PRF-512(SMK, peer key expansion, initiator address, peer address, initiator nonce, peer nonce)   or STK_{CCMP} = PRF-384(SMK, peer key expansion, initiator address, peer address, initiator nonce, peer nonce)   • SKCK = L(STK, 0, 128)   • SKEK = L(STK, 128, 128)   • TK_{TKIP} = L(STK, 256, 256) or TK_{CCMP} = L(STK, 256, 128)

■ The *SMK handshake* aims to create an SMK between two STAs that are associated to the same AP. The AP must have an RSNA established with the two STAs before allowing them to connect directly and establish an STKSA. The SMK is used by the two STAs to derive an STK, which is used to secure their directly exchanged traffic. The SMK handshake is the first part of the *peer-key handshake*, which is conducted by two STAs wishing to create a secure STA-to-STA link. The SMK handshake traffic transits through the AP using the PTK created by the AP and each of the STAs. The second part of the peer-key handshake consists in creating the STK from the SMK using the 4-way handshake between the initiator STA and the peer STA (Table 16.1).

16.4 IEEE802.11s Network Security

IEEE802.11s security is based on SAs as in IEEE802.11. The SAs are created after the mutual authentication and key distribution between two mesh STAs (MRs) or between a mesh STA and a client STA.

The supplicant mesh STA authenticates to the AS by passing through another mesh STA acting as a mesh authenticator (MA). At the end of the authentication, the AS delegates the key derivation to some mesh key distributors (MKDs). That is, the burden of computing PMKs is removed from the AS and is given to the MKDs. The AS is assumed to be connected via secure channels to the MKDs. Every MKD controls an MKD domain where a mesh STA may initiate its authentication. The MA is only connected to one MKD while an MKD domain may contain many MAs (Figure 16.8). The MKDs and MAs are called key holders and are MRs (mesh STAs) that implement either the MA or the MKD functions.

When a mesh STA joins the mesh backbone (MBSS) for the first time, it has to authenticate to the AS using the *Mesh Security Association (MSA) initial authentication*. The MSA initial authentication is composed of a peer link establishment phase between the supplicant and the MA, and an IEEE802.1X authentication between the supplicant and the AS.

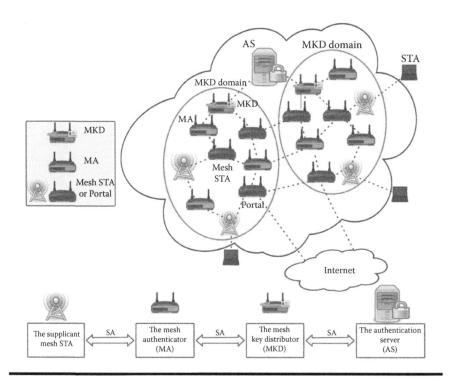

Figure 16.8   IEEE802.11s security architecture.

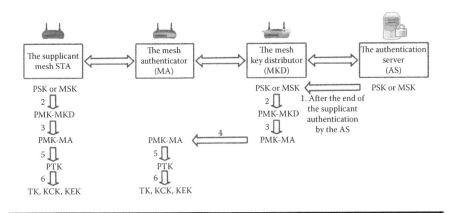

Figure 16.9 Key hierarchy resulting from an MSA initial authentication.

After a successful initial IEEE802.1X authentication, the AS sends the MSK to the MKD. As such, the MKD and the supplicant derive the PMK corresponding to the MKD (PMK–MKD). Then, they derive the PMK corresponding to the MA (PMK–MA) from the PMK–MKD (Figure 16.9). At this moment, the MKD sends the PMK–MA to the MA, which starts a 4-way-handshake with the supplicant mesh STA. The 4-way-handshake results in the creation of the PTK between the supplicant mesh STA and the MA.

After its initial authentication, any mesh STA may become an MA by executing the *key holder security handshake* with the MKD. When it becomes an MA, a mesh STA needs a set of cryptographic keys to securely exchange the PMK–MA of other mesh STAs with the MKD. These keys are derived from the PMK–MKD as presented in Figure 16.10.

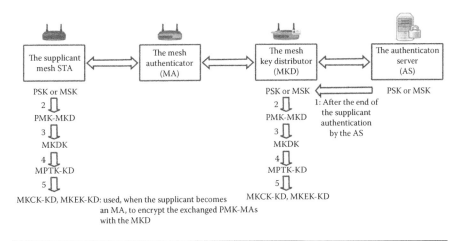

Figure 16.10 Key hierarchy resulting from a key holder security handshake.

First, the mesh STA and the MKD derive the Mesh Key Distribution Key (MKDK) from the PMK–MKD. Second, they derive the mesh PTK for key distribution (MPTK–KD) from the MKDK. The MPTK–KD is divided into two components: the Mesh Key Confirmation Key for Key Distribution (MKCK–KD) and the Mesh Key Encryption Key for Key Distribution (MKEK–KD). These keys are used when the supplicant mesh STA finishes its authentication and becomes an MA. These keys serve to encrypt the PMK–MA of other supplicant mesh STAs when they are sent from the MKD to the MA. The MKCK–KD with an HMAC algorithm provides data authenticity; the MKEK–KD is the encryption key of the PMK–MAs.

Two peer mesh STAs can make an *abbreviated handshake* to exchange a TK. The abbreviated handshake assumes the existence of a PMK–MA between the two mesh STAs. That is, one of the mesh STA acts as an MA for the other mesh STA during its MSA initial authentication.

We present in Section 16.4.1 the protocols that were specified for mesh STAs authentication in IEEE802.11s. Then, we describe the handshake processes that serve to share the different keys between mesh STAs (Section 16.4.2).

16.4.1 IEEE802.11s Authentication Protocols

IEEE802.11s mesh STA or client STA authentication process relies either on the IEEE802.1X authentication or on the Simultaneous Authentication of Equals (SAE) [26]. IEEE802.1X authentication implies the use of an authentication server and involves an MA and a supplicant STA. However, the SAE authentication is based on the use of a shared secret between two STAs. That is, it does not need the specification of a supplicant or an MA.

16.4.1.1 MSA Initial Authentication Protocol

The MSA initial authentication protocol (Figure 16.11) includes the peer link management phase, the IEEE802.1X authentication, and the 4-way handshake, which establishes a PTK between two mesh STAs and allows them to exchange their GTK.

When a mesh STA joins the MBSS for the first time, it chooses an MKD domain to belong to by analyzing the RSN information elements that it receives from its one-hop neighbors. That is, the mesh STA chooses the MA that will act as a pass-through server for its authentication messages to the AS. Recall that an MA is only connected to one MKD. As such, the supplicant mesh STA selects its MKD domain and its future MKD when it chooses its MA. At the end of the authentication, the supplicant mesh STA and its MKD collaborate to create the key hierarchy.

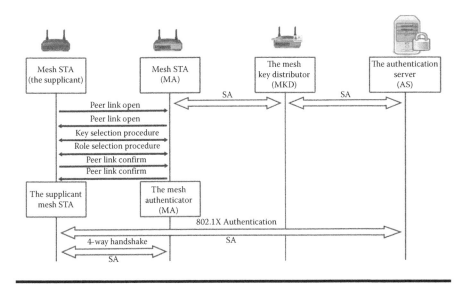

Figure 16.11   The MSA initial authentication.

The MSA initial authentication starts when the arriving mesh STA (local mesh STA) exchanges a couple of *peer link open* frames with one of its peer mesh STAs. The peer link open frames serve to negotiate the link establishment parameters and the cipher suites between the two mesh STAs. Then, the two STAs perform the *key selection procedure* to decide either an initial MSA authentication is needed or the STAs will go through an abbreviated handshake.

Two mesh STAs, which have some valid PMK–MAs from a previous MSA initial authentication, can decide to make an abbreviated handshake and avoid the initial MSA authentication. During the abbreviated handshake, the mesh STAs select one of their recent PMK–MA to be used for the authentication and key derivation.

When two peer mesh STAs belonging to the same MKD domain decide to make an IEEE802.1X authentication, they must run the *802.1X role selection proce-dure*. This procedure aims to identify which mesh STA is going to be the supplicant STA and which one will be the MA. If one of the mesh STAs is connected to the MKD, it is used as the MA. However, if the two STAs are connected to the MKD, the one who requests the authentication will be the supplicant mesh STA.

Finally, the two mesh STAs exchange two *peer link confirm* frames and start the IEEE802.1X authentication. That is, the supplicant mesh STA authenticates itself with the AS using an EAP method. The EAP frames are encapsulated using the EAPOL protocol between the supplicant mesh STA and the MA.

However, the MA uses the *mesh EAP message transport protocol* to transmit the EAP frames to the MKD which relays them to the AS.

When the message exchange results in a successful authentication, the AS securely sends the MSK or PSK corresponding to the supplicant mesh STA to the MKD. Then, the MKD and the supplicant mesh STA creates the key hierarchy of the supplicant (Figure 16.9). In addition, the MKD sends the PMK–MA to the MA, which starts the 4-way handshake with the supplicant mesh STA to derive a PTK.

16.4.1.2 Simultaneous Authentication of Equals

The Simultaneous Authentication of Equals (SAE [26]) permits two mesh STAs that share a PSK to make a mutual authentication, which results in the creation of a PMK between the two STAs. The SAE can use either an elliptic curve (EC) group or a multiplicative group. The SAE is started when a mesh STA receives an SAE frame from one of its neighbors. The SAE frames are IEEE802.11 authentication frames that have been adapted to IEEE802.11s requirements. The SAE relies on two types of messages: the *commit* message and the *confirm* message.

When a local mesh STA wants to authenticate with a peer mesh STA, it generates a password *PWD* using the shared PSK and computes a scalar *m* from the two STAs MAC addresses. *PWD* can be either a scalar or an EC point while *m* can only be a scalar point. Then, the local mesh STA computes a secret value *N* as the product of *PWD* and *m*. Then, the local mesh STA begins the authentication by creating the *commit* message. It generates a secret random value $rand_l$ and a temporary secret random value $mask_l$. The *commit* message contains two parts:

- *local-commit-scalar* = $(rand_l + mask_l)mod[r]$, where *r* is the order of the chosen cyclic group (either the EC points group or the multiplicative group)
- *local-commit-element* = $inverse(mask_l.N)$, where *inverse()* is defined as the function that associates to an element *X* its inverse in the chosen group *inverse(X)*

When the peer mesh STA receives the *commit* message from the local mesh STA, it computes:

$$K = rand_p.(\text{local-commit-scalar}.N + \text{local-commit-element})$$
$$K = rand_p.rand_l.N$$

The local and peer STAs compute the *K* value based on their random secrets $rand_l$ and $rand_p$, and the received *commit* messages. Then, they compute the key *k* as $k = F(K)$, where *F()* is a function that returns the *x* coordinate of an EC point. That is, *F()* is only used with EC points. Finally, each mesh STA generates the *confirm* message as follows:

confirm = hash(k||send-confirm||local-commit-scalar||local-commit-element||peer-commit-scalar||peer-commit-element)

Upon receiving the *confirm* message, each mesh STA computes a verifier to validate the value of the *confirm* message:

verifier = hash(k||send-confirm||peer-commit-scalar||peer-commit-element||local-commit-scalar||local-commit-element)

If the verifier is not equal to the *confirm* message received from the peer, the authentication of the peer is rejected. If the authentication succeeds, the two mesh STAs generate a PMK as follows:

PMK = hash(k||(local-commit-scalar + peer-commit-scalar)mod[r]||F(local-commit-element + peer-commit-element))

Note that the SAE is an ElGamal [27] type algorithm, which serves to exchange a Diffie–Hellman key namely *K* between the two mesh STAs.

16.4.2 IEEE802.11s Handshake Protocols

In this section, we give a quick presentation of the IEEE802.11s mechanisms that aim to create and exchange keys either between two mesh STAs or between a mesh STA and an MKD.

- *Abbreviated handshake*, also called the abbreviated MSA authentication protocol, is based on the assumption that a PMK–MA has already been established between two mesh STAs either during an initial MSA authentication or by executing the SAE. It relies on the peer link frames to establish a key hierarchy between two mesh STAs. That is, the handshake ends with the derivation of a TK by the two mesh STAs.
- *Mesh key holder security handshake* occurs at the end of an initial MSA authentication. The new authenticated mesh STA may decide to become an MA, so it contacts the MKD with which it has created a key hierarchy and starts the key holder security handshake. At the end of this handshake, the mesh STA becomes an MA.

16.5 WMNs Authentication: Related Works

In the literature, many protocols for STA and MR authentication in WMNs were proposed. In 2006, Li et al. [28] used IEEE802.1X for client mesh STA authentication. They made the AS send different PMKs of the authenticated STA to a set of neighboring mesh STAs. As such, STA has just to make a 4-way handshake to authenticate itself with its neighboring mesh STAs. They did so to avoid for STA an IEEE802.1X authentication each time STA reconnects to a new mesh STA.

In 2006, Zhang and Fang [29] presented the first hybrid authentication architecture that combines certificates with ID-based cryptography (IBC). They used a user-broker-operator model. A broker is an independent trusted entity that issues an ID-based private key and a user pass (UPASS) for the mesh client. Meanwhile, any operator creates an ID-based private key and a router pass (RPASS) for its MRs. When a client mesh STA wants to connect to an MR, it starts a mutual authentication with the chosen MR using its UPASS and private key. The authentication is based on the use of signatures. If the authentication succeeds, the MR orders its operator to generate a temporary $UPASS_{operator}$ and ID-based private key for the authenticated client. It is the MR which transmits the new ID-based private key and $UPASS_{operator}$ to the client. As such, the MR can impersonate as the legitimate client knowing its private key and pass.

In 2007, Nguyen and Rong [30] were the pioneers to propose IBC for authentication in ZigBee mesh networks. They use the ZigBee *trust center* as an ID-based private key generator (PKG). Later, in 2009, Wenju et al. [31] propose an ID-based authentication method for the EAP. They added a dedicated entity (PKG) to the backbone in order to manage the ID-based key derivation and secure transmission to its owner. The weakness of these protocols is that they never discuss the authentication between STAs and the PKG. They either suppose that the PKG has securely transmitted the private keys and the IBC domain parameters to STAs or they assume that these elements are hard coded in advance in STAs' memories. That is, they never specify how to ensure a secure communication between the PKG and the different STAs.

In 2008, Santhanam et al. [32] proposed a method relying on Merkle trees for mesh clients' authentication. First, the client mesh STA registers in advance with the AS. During the registration phase, the mesh STA generates a Merkle tree, which has a secret random as a root and some children leaves. The mesh STA securely transmits the tree root (the secret) to the AS. For its future authentication with an MR, the client mesh STA uses a one-time token computed from the leaves of its Merkle tree. That is, STA sends the token to the MR, which transmits it to the AS for verification. The STA token serves to compute the tree root that STA has already exchanged with the AS.

In 2008, Qazi et al. [33] proposed a ticket-based authentication for mesh STAs. They assumed that each STA authenticates with its certificate to the AS. Then, the AS creates a token for the STA. Each STA combines its token with a signature to secure its AODV route request and route response messages. Lee et al. [34] proposed an authentication scheme that relies on a PSK between the MAP and the client mesh STA. Their idea consists of delegating STA authentication with an unknown AP to a trusted AP. That is, the unknown AP requires prom the trusted AP the authentication of the STA with whom he shares a secret key. The trusted AP and STA use their shared secret for the authentication.

In 2008, Feng et al. [35] proposed an authentication scheme that uses blind signature and a token to make STA authentication while preserving STA privacy.

First, STA registers to a Registration and Signing Center (RSC) with its true identity. Then, STA makes the RSC blind-sign a token containing a pseudonym of STA. STA uses this token to authenticate to a Key Management Center (KMC) that acts as a certification authority (CA). KMC derives a pair of keys that correspond to the STA pseudonym.

In 2008, Lin et al. [36] presented a password-based authentication for mesh STA. The mesh STA authenticates to a set of ASs with its password, while the ASs authenticate the STA to the MA using a threshold-based signature. That is, STA securely transmits its password to the group of ASs. If the password is valid, the ASs make a group signature, not only to authenticate them to the MA but also to certify that the STA password has been verified.

In 2008, Yang et al. [37] also presented a (k,n) threshold scheme for STA authentication. They assumed the existence of a trusted third party that maintains a long-term secret between n ASs and a mesh STA. To start an authentication, the mesh STA sends k shares of a temporary secret to k different ASs. Then, the k ASs collaborate to recover the long-term secret of STA which they combine to the temporary secret. The trusted third party compares the resulting long-term secret of the ASs to the secret of STA to complete the authentication. Zhu et al. [38] also presented a threshold-based authentication for STA fast handoff situations. Their authentication method relies on a distributed PKI and on splitting the private keys between the MRs. When the MRs authenticate a mesh client, they keep track of the authentication to avoid the reauthentication of this client when it makes a handoff from one MR to another.

In 2009, Martignon et al. [39] proposed security architecture for WMNs. They relied on IEEE802.11i standard for STA authentication. That is, they proposed to use EAP authentication methods for client STA authentication. Then, they proposed to use a TLS to authenticate a client STA to a key server when the client wants to become an MR.

In 2010, Buttyan et al. [40] defined an authentication scheme for a multioperator WMN. They assumed that each operator is controlling its own CA, which is delivering certificate for its clients. Mesh STAs authentication is done in a challenge-response fashion using signatures. In the same year, He and Agrawal [41] proposed an authentication scheme for mesh clients and routers relying on IBC. They proposed an authentication scheme that makes an ID-based signature and encryption of every exchanged message.

In 2011, Li and Nguyen [42] proposed a fast authentication scheme for mesh clients and routers. The protocol relies on tickets that are inspired from Kerberos ones. Tickets are generated by trusted ticket agents. Then, they are securely transmitted to mesh clients and routers. The tickets serve to authenticate mesh client and routers during a client's initial login or handover. In the same year, Boudguiga and Laurent [42] proposed an authentication scheme that relies on IBC, solving the problem of key escrow, and proposes a simple way for ID-based parameters refreshment.

16.6 Conclusion

In this chapter, we present several authentication methods that have been proposed for WMNs. These authentication methods can be classified into three different types. First, the PSK authentication methods rely on a shared secret between two STAs or between an STA and an AS. In the first case, STA has to store a password or a PSK for each of its peer STAs, while in the second case STA has to pass through the AS each time it authenticates to another STA. The second type of authentication is the certificate-based authentication. It relies on the use of certificates and requires the heavy deployment of a public key infrastructure (PKI) that is controlled by a CA. In general, the management of certificates in wireless environments is a cumbersome task because it requires the definition of a certificate management policy to control the generation, transmission and revocation of certificates. That is, the network STAs will have to periodically download a Certificate Revocation List (CRL) in order to update their local lists of revoked certificates and expired keys. However, the CRL request and response messages are bandwidth consuming. Moreover, in the period separating two CRL updates, STAs do not know the newly revoked certificates, and consequently, attackers that successfully compromised a private key (recently revoked), can impersonate as a legitimate network user. Finally, the third type of authentication relies on IBC. IBC seems to be a well-suited solution for authentication issues related to WMN. Indeed, IBC permits the use of public key cryptography with no need for certificates and PKI deployment, as keys are directly derived from nodes identities. However, IBC algorithms, especially those relying on pairing functions, are less efficient in terms of time compared to classical encryption and signature algorithms (e.g., DES, AES, RSA, ECDSA, ElGamal) [44]. Thanks to the effort done by cryptographers such as Beauchat et al. [45] leading to defining the fastest existing pairing in less than 1 ms, efficient IBC schemes are emerging and will emerge in the next few years. These new IBC schemes will define more secure, competitive, and faster authentication protocols than the actual ones. In addition, researchers are getting more interested in lightweight cryptography, which aims at offering better security properties at a less power cost for constrained devices such as sensors or objects in the Internet. Lightweight cryptography is a challenging field that will have direct consequences on the performances of authentication protocols.

References

1. IEEE Std 802.11s-2011. IEEE Standard for Information Technology-Telecommunications and Information Exchange between Systems Local and Metropolitan Area Networks—Specific requirements-Part 11: Wireless LAN Medium Access Control (MAC) and Physical Layer (PHY) Specifications Amendment 10: Mesh Networking, October 2011.
2. IEEE Std P802.16/REVd/D5. Unapproved Draft IEEE Standard for Local and Metropolitan Area Networks Corrigendum to IEEE Standard for Local and

Metropolitan Area Networks-Part 16: Air Interface for Fixed Broadband Wireless Access Systems (Revision of IEEE Std 802.16-2001; IEEE Std 802.16c-2002, and IEEE std 802.16a-2003), 2004.

3. IEEE Std 802.15.5-2009. IEEE Recommended Practice for Information Technology-Telecommunications and Information Exchange between Systems—Local and Metropolitan Area Networks-Specific Requirements-Part 15.5: Mesh Topology Capability in Wireless Personal Area Networks (WPANs), August 2009.

4. IEEE Std 802.11-2007. IEEE Standard for Information Technology—Telecommunications and Information Exchange between Systems—Local and Metropolitan Area Networks—Specific Requirements-Part 11: Wireless LAN Medium Access Control (MAC) and Physical Layer (PHY) Specifications, December 2007.

5. IEEE Unapproved Draft Std P802.16Rev2/D9. IEEE Draft Standard for Local and Metropolitan Area Networks-Part 16: Air Interface for Broadband Wireless Access Systems, January 2009.

6. IEEE Std P802.15.1/D6. Approved Draft Standard for Information technology—Telecommunications and information exchange between systems—Local and metropolitan area networks—Specific requirements-Part 15.1REVa: Wireless Medium Access Control (MAC) and Physical Layer (PHY) Specifications for Wireless Personal Area Networks (WPANs) Replaced by IEEE 802.15.1-2005. 2004.

7. International Organization for Standardization ISO. Information processing systems-OSI reference model-Part 2: Security architecture. Standard, (7498-2), 1989.

8. W. A. Simpson. The Point-to-Point Protocol (PPP). RFC 1661 (Standard), July 1994. Updated by RFC 2153.

9. L. Blunk and J. Vollbrecht. PPP Extensible Authentication Protocol (EAP). RFC 2284 (Proposed Standard), March 1998. Obsoleted by RFC 3748, updated by RFC 2484.

10. B. Aboba, L. Blunk, J. Vollbrecht, J. Carlson, and H. Levkowetz. Extensible Authentication Protocol (EAP). RFC 3748 (Proposed Standard), June 2004. Updated by RFC 5247.

11. C. Rigney, A. Rubens, W. Simpson, and S. Willens. Remote Authentication Dial in User Service (RADIUS). RFC 2138 (Proposed Standard), April 1997. Obsoleted by RFC 2865.

12. B. Aboba and P. Calhoun. RADIUS (Remote Authentication Dial in User Service) Support for Extensible Authentication Protocol (EAP). RFC 3579 (Informational), September 2003. Updated by RFC 5080.

13. P. Calhoun, J. Loughney, E. Guttman, Glen Zorn, and Jari Arkko. Diameter Base Protocol. RFC 3588 (Proposed Standard), September 2003.

14. P. Eronen, T. Hiller, and G. Zorn. Diameter Extensible Authentication Protocol (EAP) Application. RFC 4072 (Proposed Standard), August 2005.

15. IEEE Std 802.1X-2010. IEEE Standard for Local and metropolitan area networks—Port-Based Network Access Control, May 2010.

16. F. Bersani and H. Tschofenig. The EAP-PSK Protocol: A Pre-Shared Key Extensible Authentication Protocol (EAP) Method. RFC 4764 (Experimental), January 2007.

17. M. Vanderveen and H. Soliman. Extensible Authentication Protocol Method for Shared-secret Authentication and Key Establishment (EAP-SAKE). RFC 4763 (Informational), November 2006.

18. C. Clancy and H. Tschofenig. Extensible Authentication Protocol—Generalized Pre-Shared Key (EAP-GPSK) Method. RFC 5433 (Proposed Standard), February 2009.

19. D. Simon, B. Aboba, and R. Hurst. The EAP-TLS Authentication Protocol. RFC 5216 (Proposed Standard), March 2008.

20. P. Funk and S. Blake-Wilson. Extensible Authentication Protocol Tunneled Transport Layer Security Authenticated Protocol Version 0 (EAP-TTLSv0). RFC 5281 (Informational), August 2008.

21. N. Cam-Winget, D. McGrew, J. Salowey, and H. Zhou. The Flexible Authentication via Secure Tunneling Extensible Authentication Protocol Method (EAP-FAST). RFC 4851 (Informational), May 2007.

22. T. Dierks and E. Rescorla. The Transport Layer Security (TLS) Protocol Version 1.2. RFC 5246 (Proposed Standard), August 2008.

23. P. Jayaraman, R. M. Lopez, Yoshihiro Ohba, Mohan Parthasarathy, and Alper Yegin. Protocol for Carrying Authentication for Network Access (PANA) Framework. RFC 5193 (Informational), may 2008.

24. A. Yegin, Y. Ohba, R. Penno, G. Tsirtsis, and C. Wang. Protocol for Carrying Authentication for Network Access (PANA) Requirements. RFC 4058 (Informational), may 2005.

25. IEEE Std 802.11i-2004. IEEE Standard for Information Technology—Telecommunications and Information Exchange Between Systems—Local and Metropolitan Area Networks—Specific Requirements-Part 11: Wireless LAN Medium Access Control (MAC) and Physical Layer (PHY) Specifications Amendment 6: Medium Access Control (MAC) Security Enhancements, 2004.

26. D. Harkins. Simultaneous Authentication of Equals: A secure, password-based key exchange for mesh networks. *In Proceedings of the 2008 Second International Conference on Sensor Technologies and Applications*, SENSORCOMM '08, pages 839–844, Washington, DC, USA, 2008. IEEE Computer Society.

27. T. ElGamal. A public key cryptosystem and a signature scheme based on discrete logarithms. In *Proceedings of CRYPTO 84 an Advance in Cryptology*, pages 10–18, Santa Barbara, 1985.

28. Y. Li, X. Cui, L. Hu, and Y. Shen. Efficient security transmission protocol with identity-based encryption in wireless mesh networks. In *2010 International Conference on High Performance Computing and Simulation (HPCS 2010)*, pages 679–685. IEEE, 2010.

29. Y. Zhang and Y. Fang. A secure authentication and billing architecture for wireless mesh networks. *Wireless Networks*, 13(5):663–678, 2007.

30. S. Nguyen and C. Rong. ZigBee security using identity-based cryptography. In Bin Xiao, Laurence Yang, Jianhua Ma, Christian Muller-Schloer, and Yu Hua, eds, *Autonomic and Trusted Computing, volume 4610 of Lecture Notes in Computer Science*, pages 3–12. Springer, Berlin/Heidelberg, 2007.

31. W. Liu, Y. Shang, and Z. Wang. A wireless mesh network authentication method based on identity based signature. In *5th International Conference on Wireless Communications, Networking and Mobile Computing*, 2009 (WiCom '09), pages 1–4, 2009.

32. L. Santhanam, B. Xie, and D. Agrawal. Secure and efficient authentication in Wireless Mesh Networks using merkle trees. *In 33rd IEEE Conference on Local Computer Networks 2008 LCN 2008*, pages 966–972. IEEE, 2008.

33. S. Qazi, Y. Mu, and W. Susilo. Securing wireless mesh networks with ticket-based authentication. *2nd International Conference on Signal Processing and Communication Systems (ICSPCS 2008)*, pages 1–10, 2008.

34. I. Lee, J. Lee, W. Arbaugh, and D. Kim. Dynamic distributed authentication scheme for wireless LAN-based mesh networks. In Teresa Vazão, Mário Freire, and Ilyoung

Chong, eds, *Towards Ubiquitous Networking and Services Information Networking, volume 5200 of Lecture Notes in Computer Science*, pages 649–658. Springer, Berlin/Heidelberg, 2008. 10.1007/978-3-540-89524-4_64.

35. Y. Feng, M-Y. Fan, and C-P. Liu. A new privacy-enhanced authentication scheme for wireless mesh networks. *In 2008 International Conference on Apperceiving Computing and Intelligence Analysis (ICACIA 2008)*, pages 265–269, December 2008.

36. X. Lin, R. Lu, P-H Ho, X. Shen, and Z. Cao. TUA: A novel compromise-resilient authentication architecture for wireless mesh networks. *IEEE Transactions on Wireless Communications*, 7:1389–1399, 2008.

37. Y. Yang, Y. Gu, X. Tan, and L. Ma. A new wireless mesh network authentication scheme based on threshold method. *In the 9th International Conference for Young Computer Scientists (ICYCS 2008)*, pages 2260–2265, 2008.

38. H. Zhu, X. Lin, R. Lu, P-h. Ho, and X. Shen. SLAB: A secure localized authentication and billing scheme for wireless mesh networks, 2008.

39. F. Martignon, S. Paris, and A. Capone. Design and implementation of MobiSEC: A complete security architecture for wireless mesh networks. *Computer Networks*, 53(12):2192–2207, 2009.

40. L. Buttyán, L. Dóra, F. Martinelli, and M. Petrocchi. Fast certificate-based authentication scheme in multi-operator maintained wireless mesh networks. *Computer Communications*, 33(8):907–922, 2010.

41. B. He and D. Agrawal. An identity-based authentication and key establishment scheme for multi-operator maintained Wireless Mesh Networks. In 2010 *IEEE 7th International Conference on Mobile Adhoc and Sensor Systems (MASS)*, pages 71–78, San Francisco, USA, November 2010.

42. C. Li, and U. T. Nguyen, Fast authentication for mobility support in wireless mesh networks, *Wireless Communications and Networking Conference (WCNC), IEEE* 2011, vol., no., pp.1185,1190, 28–31, doi: 10.1109/WCNC.2011.5779299, March 2011.

43. A. Boudguiga and M. Laurent. Key-escrow resistant ID-based authentication scheme for IEEE 802.11s mesh networks. In *2011 IEEE Wireless Communications and Networking Conference (WCNC 2011)*, pages 784–789, Cancun, Mexico, March 2011.

44. A. Boudguiga, and M. Laurent, An EAP ID-based authentication method for wireless networks, *Internet Technology and Secured Transactions (ICITST), 2011 International Conference for*, vol., no., pp.232,239, 11–14 Dec. 2011.

45. J.-L. Beuchat, J. E. González-Díaz, S. Mitsunari, E. Okamoto, F. Rodríguez-Henríquez, and T. Teruya, High-speed software implementation of the optimal ate pairing over barreto-naehrig curves, in *Proceedings of the 4th international conference on Pairing-based cryptography, ser.* Pairing'10. Berlin, Heidelberg: Springer Verlag, 2010, pp. 21–39.

46. Y. Zhang, J. Zheng, and H. Hu, *Security in Wireless Mesh Networks, ser. Wireless Networks and Mobile Communications*. Taylor & Francis, 2008. [Online]. Available: http://books.google.fr/books?id=VyoE7LmsHEAC

Chapter 17

Wireless-Based Application-Layer Cross-Layer Design with Simultaneous Quality of Service and Security Support

Sasan Adibi

Contents

447

17.1 Introduction

Security requirements associated with current communication systems will be increasingly more demanding because the nature of attacks are becoming more complex, which requires more versatile security measures to combat such threats. Adding QoS to the picture increases the complexity, since security measures normally add redundant information (resulting with extra overhead and added delay figures, which may impact QoS schemes in negative ways. Therefore, a system where both security and QoS can coexist may be a challenge and involving power consumption constraints to the picture will add more dimensions to the mentioned challenges. The challenges involving security, QoS, and power consumption figures will always exist, as long as wireless broadband communication systems (with limited power resources) face increasing levels of attacks and demands for higher quality.

17.1.1 Rationale for Application-Layer Provisioning

An application-layer provisioning provides flexibility and increases performance in comparison with lower layer provisioning schemes [1–3]. From the QoS point of view, after data packets have moved down the layers, following the application layer down to the MAC/physical layers, it would strategically be "too late" to apply best QoS practices. Therefore, it becomes apparent that efficient QoS practices to be initiated at the highest layer; application layer.

Topologically speaking, the application layer is the only layer where the processing can be applied to very large amount of data blocks. The processing unit sizes deployed in other layers are limited to the number of bits per unit frame. Thus, higher efficiency can be achieved at the application layer. From the security point of view, application-layer provisioning may be very beneficial when large amounts of data blocks are processed at once.

The challenge here is to design a cross-layer-based application-layer scheme, which supports simultaneous QoS and security provisioning within acceptable power consumption figures.

17.1.2 Rationale for the Cross-Layer Design Deployment

Deploying cross-layer design has a number of advantages, such as the performance increase in a number of wireless networking scenarios [4]. In particular, knowledge provisioning regarding certain layered details may increase visibility to the wider networking visibility, which may yield to better decision making. This is particularly beneficial for far layers (i.e., application/physical layers) where early consideration of the conditions in the far layers may offer early tweaking, which can help optimize the performance and data flow in the functions related to the entire layers.

17.1.2.1 QoS-Based Cross-Layer Design

In a cross-layer QoS-based system, a number of QoS-related parameters are gathered from selected layers and imported in other layer(s) for further processing. As mentioned, in this chapter we are interested in importing cross-layer data in to the applications layer. The current values of these imported cross-layer parameters may define the quality of the wireless link and may partially be used to determine the quality of QoS-dependent functions (i.e., multimedia encoders) located at the application layer. In the case of application-layer multimedia encoding, an adaptive multimedia encoder/decoder system can be designed and deployed, which adapts its coding quality to the quality of the environment and adjusts the encoder bit-rate accordingly.

17.1.2.2 Security-Based Cross-Layer Design

In the cross-layer security-based scheme, again a number of layered-based security parameters are gathered from various layers, which are accompanied with the rest of the gathered QoS-related parameters and fed into an application-layer security-based module. Security and QoS cross-layer feedback information may be used for a common or an independent process (or a mix-and-match of both), depending on the specifics of the implementations.

17.2 Quality of Service

In the best-effort scenario, all traffic flows have the same priority. QoS provisioning, on the other hand, offers priority for flows, which fall in predefined groups or criteria. QoS has become a de-facto requirement for enterprise networks as well for public and private wireless and wired networks. This includes QoS mechanisms for transmission of aggregation of packets, admission control, and other types of traffic where flows and packets may compete over the channel access.

QoS comes in two main categories: network-perspective-QoS [5] and user-perspective-QoS, which is also known as quality of experience (QoE) [6]. Examples of network-perspective-QoS parameters may include end-to-end delay, jitter, and throughput, which are further explained in the following subsections.

17.2.1 QoS versus QoE

As mentioned earlier, QoS is based on network-centric performance. Provisioning includes bandwidth, delay, and jitter. QoS mechanisms are used to quantify the quality schemes administered at each layer and the effects of the interaction between layers and the impact on the overall throughput/jitter/delay figures for the purpose of increasing the efficiency of the data flows.

QoE, on the other hand, is the quality measure from the human experience perspective. Therefore, what the user sees or hears is directly related to QoE, thus fidelity and latency (directly affected delay and jitter) are two very important factors in characterizing QoE.

There is normally a direct connection between QoS and QoE, which means that improving QoS figures may result in the user's better quality perception. This is due to the fact that both QoS and QoE share common some parameters (e.g., delay and jitter) to measure the quality, however, using different criteria. The focus of this chapter is on the QoS measurement.

QoS is actively used for measuring the quality of voice in voice over IP (VoIP) systems, based on delay and jitter figures, which are used to calculate the mean opinion score (MOS). QoE, unlike in QoS, uses nonintrusive voice quality-analyzing methods for predicting the MOS value based on passive voice clarity-evaluation techniques. This involves deploying audio recording and playback, measuring delays based on vocal methods, and simple analyses, which identify voice-quality impairments.

17.2.2 Bandwidth or Throughput

The definitions for bandwidth and throughput differ due to the fact that bandwidth is the maximum number of bits transmitted among two end-points per unit of time (second), including payload data, control, and redundant bits [7], whereas throughput (effective bandwidth), is the actual payload transmission per unit of time [8]. The value of throughput is typically lower than the value of the associated bandwidth.

When noise and other unwanted physical and logical issues start to rise, the difference between bandwidth and throughput becomes increasingly noticeable. The increase in the noise level degrades the throughput figures, while the bandwidth may remain unchanged.

Throughput at the physical layer is based on the payload transmitted at this layer, which is comprised of the actual data captured at the application layer plus all the headers/trailer added by the subsequent layers. Goodput, however, is defined as the actual useful information bits (entered from at the application layer) per second, which is synonymous to the application-layer throughput [9].

17.2.3 End-to-End Delay or Round-Trip Time

Delay is the prominent factor in directly degrading QoS and QoE, particularly for real-time applications, where data should be delivered from the source to the destination in a timely manner (minimum round-trip delay is required). Long delays may increase the possibility for error messages and network faults, including transmission control protocol (TCP) timeouts and Internet control message protocol (ICMP) error messages (e.g., destination unreachable). Such timeouts and errors increase the number of retries and, thus, reduce the efficiency of multimedia traffic transmission that results in audio/video fidelity reduction [10]. From the QoE perspective, this can cause user frustration, especially during interactive communications. While data traffic is transmitted between various segments and devices (e.g., routers, switches) between the source and the destination (in a local or global sense), each segment/device adds a specific amount of delay, including [11] *source-processing delay* (e.g., *packetizing, digitization delays*), which is a delay introduced at the source where the packets are created, depending on the number of concurrent applications running and the source hardware configuration (e.g., CPU, memory, clock speed). *Transmission delay,* the packet transmission time, is a function of the transmission speed and packet size. *Network delay* is comprised of a number of delays, including *protocol, propagation, queuing,* and *destination processing* delays. *Protocol delay* is caused by the execution of the communication protocols running in various parts of the network. This includes switches, routers, gateways, and network interface cards, depending on the complexity of the protocols, the load size, and the hardware configurations. *Propagation delay* is a function of how far the source and the destination are physically separated from one another and the number of hops between them. The *output queuing delays* are the delays caused by the intermediate and destination queuing systems. The *destination processing*, is normally introduced by the processing components at the destination system, including the packet reconstruction process. This is also dependent on the delivered load and hardware configuration of the destination system.

17.2.4 Jitter

As mentioned, latency (delay) is the prominent parameter in the network-centric quality measurements (e.g., QoS-related), which is normally presented as an average value. Connectionless protocols (e.g., multimedia) often rely on IP for data delivery and control signaling. The nature of the data delivery mechanism for IP does not provide any guarantee that every data packet will travel via the same path, unlike the circuit-switched network scenario. This is due to the fact that packet switched (e.g., IP-based) networks are comprised of a number of hops between any source–destination pairs and at each point of time, packets may be forced to travel different sets of hops, based on the hop availabilities, network conditions, routing protocol in use, and various metrics used to calculate the best possible path from source to destination. Therefore, it is quite possible for the current packets to reach the destination having traveled more or less hops than the prior and later packets, in which case, arrived packets may encounter variable delays, causing another network-centric, which is called jitter [12].

Another network issue that contributes to higher jitter figures is related to the intermediate gateways/routers becoming overloaded with incoming packets, which may cause an increase in the intermediate packet processing delays, creating additional jitter effects.

17.2.5 QoS Requirements in the Absence of Security Mechanisms

As mentioned, security provisioning normally affects QoS mechanisms in a negative way. Therefore, there are two approaches to consider QoS: with and without the intervention of security mechanisms. We have so far discussed network-centric QoS parameters without the intervention of security and in this section we focus on parameters affecting user-centric quality (QoE), such as

- *Bit error rate (BER):* BER is the ratio of the number of the bits sent in error (e.g., sending 1 and receiving 0) to the total number of transmitted information bits. Channel condition is one of the most effective factors contributing to the value of BER. The noise level (e.g., signal-to-noise ratio "SNR") also plays a crucial role in affecting BER as well. The higher the value of SNR, the lower the BER tends to get.
- *Connection drop (CD):* When the delay or jitter levels pass certain limits, the quality of the communication link becomes unbearable to the user or the

underlying application mechanism drops the connection, causing the user to hang-up or the connection termination, which are both considered the worst enemy of QoE.

Depending on the multimedia application in use (e.g., audio, voice, or video), the QoE requirements may be different. For example, the end-to-end delay for voice conversation should be less than 150 ms and if it increases this limit, the user comfort level will start to decrease dramatically.

■ *Packet loss ratio (PLR):* PLR represents the ratio of the number of lost packets to the total number of sent packets. The same factors contributing to BER may contribute to PLR as well. The higher the value of PLR, the less trustworthy the communication path is going to be.

■ *Packet drop ratio (PDR):* PDR is mainly dependent on the receiver's input buffer. When the input buffer closes in the overflow point, a mechanism may be engaged to discard (drop) the extra incoming packets until the queued packets are processed and the queue starts to get lower than the overflow point. The lower the PDR value, the more efficient is the operation of these buffers.

17.3 Security

Security is a combination of mechanisms, which offers the capability for supporting privacy, integrity, confidentiality, availability, authenticity, nonrepudiation, and access control, for the legitimate users as well as the transmission of information among them.

As mentioned before, traditionally speaking, there are normally negating interactions between QoS and security, in which the strength of one may often result in the weakness of the other [13], especially for wireless technologies, where resources are scarcer compared to the wired network scenarios. This used to be the case as wireless systems had relatively limited power and resources to perform extensive computational mechanisms to support the best QoS and security simultaneously practices. However, thanks to the recent advances in the wireless broadband technologies, networks are now more resourceful (e.g., faster processors, bigger memory chips, higher computational power capabilities, and higher allocated bandwidth), which has enabled us to design network structures with coexisting QoS and security schemes with less negating effects. From the infrastructural perspective (regarding the bandwidth allocation), the realization of most recent cellular (e.g., 4G) Wi-Fi (e.g., 802.11n) access technologies contributed to a much higher bandwidth allocation, which can be used for accommodating simultaneous security—and QoS-enabled multimedia traffic treatments.

The QoS-security coexistence and its possible implications are important issues, which need to be addressed. In this chapter, such coexistence is considered at the application layer with a cross-layer approach. The cross-layer mechanism assumes security/QoS parameters, which are created in different layers, are accessible at the application layer.

The following subsections cover the fundamentals of security. The most basic supports found in any secure system should offer capabilities to prevent variety of passive and active attacks, through the following mechanisms [14–18].

17.3.1 Confidentiality

Confidentiality or *privacy* is the capability to secure the content of the information communicated between authorized parties in the communication channel. Enabling confidentiality (e.g., using encryption) should prevent an adversary from recovering any data, which is also known as "data confidentiality." In a broader term, an adversary should not be able to figure out the parties involved in the communication (user confidentiality) and the fact that a communication session has even been established. These objectives can be achieved by using source/destination address encryption offered by virtual private network (VPN) tunneling mechanism, such as that offered in IP security (IPSec) [19].

17.3.2 Data Integrity

Integrity offers reliability, accuracy, and completeness of information. Data integrity, in its simplest form, is expected to prevent unnoticed modification of the transmitted data. In a broader term, integrity ensures the data is unaltered and current (avoiding an unduly delay), therefore, it safeguards against unplanned alteration of data (e.g., deletion, addition, and unduly delays) [20].

17.3.3 Authentication

Authentication is one of the most important security mechanisms, which offers the capability of verifying the identity of the involved parties taking part in a communication.

There are three approaches to authentication, *end-user authentication*, in which the identity of a user or a device is provisioned to ensure legitimacy before the start of a communication. *Geo-authentication* is an approach in which the location of a device or user is considered and checked before the actual communication takes place. *Attribute authentication* is another approach for establishing confidence based on an action, attribute, or a specific property associated with a user. *Data authentication* is the ability of the authorized users to ascertain the authenticity of the received data.

In a broader sense, having a mutual authentication mechanism will provide two-sided authentication coverage, where both communicating users are required to be authenticated to one another. This is required to reduce the possibility of a man-in-the-middle-attack (MitMA) [21]. In a MitMA scenario, an adversary gains control of a newly established link between two legitimate users by masquerading as a legitimate user and stealing the credentials of other communicating users, either by taking over the already established links or establish new links.

17.3.4 Authorization

Authorization is a mechanism, which checks the rights for access and control of a device, process, or user before the start of a communication. In a client–server scenario, authorization provides the server access rights to request credentials from the client, however authentication is used to prove the identity of the client to the server. It may be possible to have authentication without authorization (authenticated users without any other rights), however, having authorization without authentication does not make sense [14,22].

17.3.5 Nonrepudiation (Accountability)

A nonrepudiation (accountability) scheme provides the ability to prevent an authorized user from denying its involvement in taking part in a previous activity or communication. Various flavors of accountability include *sender's accountability*, which prevents the sender from denying its action on the transmitted information. *Forwarder's accountability* that prevents the forwarding entities on the communication path from denying their involvements (receiving the data and forwarding it). *Receiver's accountability*, which prevents the receiver from denying it has received the final information [14]. Nonrepudiation mechanisms may utilize hash and digital signature schemes, which are defined in the following subsections.

17.3.5.1 Hashing

A hash (message digest [MD]) is a mathematical function that takes a random number message with a limited size (bounded by the minimum and maximum input bits) and creates a fixed-length message-digest output string, which satisfies with the following general specifications:

1. The output string is independent of the input string length and has a fixed length.
2. A single bit difference between two data input strings should result in close to 50% change in the hash's output strings.
3. Given the hash value, it must be computationally impossible (within the time frame that the message content is useable) to find the input string.
4. A collision is a rare mathematical situation where two different input strings produce identical hash output strings. The probability of such a collision occurring depends on the number of output hash length and the type of hashing function and, it is generally very low [23].

A hash value derived from the input string m is denoted by $H(m)$.

17.3.5.1.1 Keyed versus Unkeyed Hashes

The two different types of hash functions are *keyed* and *unkeyed* [24]. A keyed hash function (such as the message authentication code "MAC") accepts an input secret key as a secondary input, such as hash-based MAC (HMAC) or block cipher-based MAC (DES–CBC–MAC) schemes.

An unkeyed hash function (such as the manipulation detection code "MDC") requires no secret key input string, such as collision-free hash function (CFHF) or one-way hash function (OWHF).

A few hashing functions that are used in the structure of both keyed and unkeyed hash schemes are MD family (e.g., MD2, MD4, MD5) and secure hash algorithm (SHA) family (SHA-1, SHA-256, SHA-384, and SHA-512). Almost all MD family members have been broken [25–27]. They are all prone to internal collisions, which is a situation where the cryptographic algorithm computes two different input messages, however returns the same output strings.

All hash functions are subject to an attack, which is called the birthday attack. In this type of attack, given a function $h(x)$, the aim of the attack is to run into the possibility that two different inputs; x_1, x_2 would yield $h(x_1) = h(x_2)$. Then the pair x_1 and x_2 is called a collision. In the cryptographic term, the hashed value is usually known and for different scenarios, the message m, may or may not be known. In the worst case scenario, the message m is also given. The strength of the hashing algorithm is normally determined against a worst-case scenario.

Assuming function $h(x)$ has an output space of S and S is large enough (larger than 10^9), for an unknown x_1 and known $h(x_1)$, it is expected to find x_2, which satisfies the equation $h(x_1) = h(x_2)$, after computing $h(x_i)$ for about $\sqrt{(\pi/2)S} \approx 1.25 \times \sqrt{S}$ different arguments on average.

Brute-force attack is another type of cryptographic attack, which features exhausting the search of all possible keys space until the correct key is found. This is usually used to break the encrypted information by searching every key to decipher the ciphertext. On average, half of the key space should be tried before a matched key is found, therefore the complexity of a brute-force attach increases exponentially when the number of the key size increases linearly. The strength associated to the brute-force attack lies in the simplicity of the attack algorithm. The weakness related to the brute-force attack comes from its inefficient algorithm, which is relatively very slow and requires a long period of time before a result is yielded [28].

17.3.5.1.2 Secure Hash Algorithm (SHA)

SHA (SHA-0) was originally introduced in 1993 (based on FIPS PUB 180), which was initiated by the National Institute of Standards and Technology (NIST) and promulgated by the National Security Agency (NSA). SHA-1 was introduced later on, which differs in a single bitwise rotation in its compression function of the message schedule compared with SHA-0. These two schemes take any arbitrary

Table 17.1   Security Strengths of SHA-0 and SHA-1

	SHA-0	*SHA-1*
Secure or vulnerable	Broken in 2004	Broken in 2010
Number of operations	2^{39}	2^{63}
Collision probability	2^{-43}	2^{-69}

Source: S. Adibi, An application layer non-repudiation. Wireless system: A cross-layer. Approach, PhD Thesis at the Electrical and Computer Engineering Department, University of Waterloo, September 27, 2010.

message input length (of less than $2^{64}-1$ bits or less than 2 billion GB) and produce 160 bit (fixed length) hash codes [29].

The newer family of SHA (SHA-2) has been developed since 2001, such as SHA-224, SHA-256, SHA-384, and SHA-512. The development of SHA-3 family has not yet been finalized by NIST as of March 2013 [29].

SHA-0 is considered to be the least computationally intensive, thus the weakest of all SHA family and it was eventually proven to be broken in 2004. SHA-1 was considered secure till 2010 [29]. Table 17.1 shows the security strengths of SHA-0 and SHA-1.

17.3.5.2 Digital Signature

A digital signature is a mathematical function, which offers integrity of data and authenticity of the person generating it. The originator of the message generates the digital signature and attaches it to the ongoing blocks of transmitting data. This enables the receiver of the message to correctly identify the true originator of the message and to ensure that the received message has not been altered by an intermediate adversary. In some security mechanisms, encryption is needed to ensure that the transmitted data was not revealed to unauthorized users (privacy).

Before continuing with the introduction to other digital signature schemes, it is important to discuss public and private key systems. Key exchange mechanisms are used when two users (i.e., *Alice* and *Bob*) require exchanging keys privately and exclusively in an insecure channel. There are two key-exchange cryptographic schemes, *symmetric and asymmetric* key exchange systems. A symmetric key system (e.g., data encryption standard "DES") is an encryption-based system, which requires the sender and receiver to use a single key for encrypting and decrypting messages.

Key exchange is an important part of symmetric key systems. One method of performing key exchange is through the Diffie–Hellman (D–H) key exchange algorithm that goes back to 1976 [30]. A D–H key exchange algorithm, between *Alice* and *Bob,* has been shown in Figure 17.1, where a is Alice's secret key and m, p are public key parameters, where p is a very large prime number and m is a primitive root mod p. *Alice* calculates the value F based on m, a, and p and transmits F

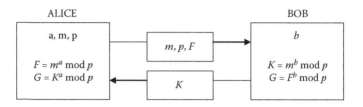

Figure 17.1 Diffie–Hellman key exchange algorithm. (Adapted from S. Adibi, An application layer non-repudiation. Wireless system: A cross-layer. Approach, PhD Thesis at the Electrical and Computer Engineering Department, University of Waterloo, September 27, 2010.)

to Bob. Bob takes F along with the public key parameters (m and p) and its own private key, b, and calculates K and transmits E to *Alice*. Now both *Alice* and *Bob* can calculate G based on the numbers received from one another. The values of a and b are chosen in such a way that even if a third party intercepts both F and K, it would be computationally infeasible to calculate a, b, or G in a limited time with limited resources.

An *asymmetric key* system, on the other hand uses two different sets of keys for encryption/decryption. These keys are mathematically related to one another, forming the public/private key pair. The private key should be kept private, while the public key can publicly be available. An example of an asymmetric key system is the public key cryptosystem.

A public key is typically used to encrypt and a private key is used to decrypt messages.

Public key infrastructure (PKI) is an example of public-key-based (asymmetric) system, which associates public keys with user identities via a certificate authority (CA).

The symmetric key algorithms are often simpler compared to those in asymmetric key algorithms, since they require only one key, however they require the key exchange to take place in a secure manner in an insecure channel. On the other hand, a public key cryptosystem do not require key exchange, however their algorithms are more complex compared to those in symmetric key systems. The asymmetric key encryption mechanisms are relatively slower than those in the symmetric key encryption mechanisms and therefore they are only used for digital signature and key exchange algorithms.

The algorithm of a digital signature normally consists of three steps, *key generation* (public/private key pair), *signing*, and *verifying* operations.

There are a number of well-known digital signature mechanisms, such as digital signature algorithm (DSA) [31], RSA [32], and elliptic curve digital signature algorithm (ECDSA, based on DSA algorithm) [33,34].

DSA: The DSA algorithm (based on the digital signature standard "DSS") was initially published in 1994 by NIST and supported by Federal Information

Processing Standard (FIPS) 186-3. DSA is based on the discrete logarithm problem that tries to find the value of E in the equation, $B^E = P \,(mod\, M)$, where B is the base value, P is the power value, and M is the modulus value. The key generation, signing, and verification algorithms in DSA are out of the scope of this chapter, which are given in [33].

RSA: RSA (taken from the names of the inventors; Rivest, Shamir, and Adleman) cryptosystem operates based on the assumption that when a very large number is the result of the product of two very large prime numbers, it is a very hard task to find (factorizing) those two prime numbers. The details of the three stages of RSA (key distribution, signing, and verifying) are out of the scope of this chapter, which can be found in [34].

DSA versus RSA: RSA cryptosystem can be used for both encryption and digital signature, whereas DSA cryptosystem is only used for digital signature. The DSA signing algorithm is faster than its verifying algorithm, whereas in RSA, verification is much faster than signing. The exact signing and verifying delays for both algorithms depend on their implementation codes, the operating system running the algorithms, and the hardware configurations on which the testbeds are tested.

ECDSA: ECDSA is based on the elliptic curve version of the DSA scheme. NIST and IEEE have both accepted ECDSA in 2000 [35,36]. The elliptic curve cryptography (ECC) uses arithmetic operations and calculations of points, which are the coordinates of an elliptic curve equation defined over a finite field. The details of the three stages of ECDSA are out of the scope of this chapter, which are provided in [36].

The bottom line is that for the same cryptographic strength, ECC involves more efficient and shorter signature messages compared to RSA and DSA. For instance, ECDSA 160 and RSA 1024 bits have the same cryptographic strength, therefore the key size required in ECDSA is almost one seventh of the key size required in RSA [37].

17.3.6 Availability

Availability is a probabilistic measure of the availability of the resources for the needed communication. The higher the probability of the resources being available, the lower the possibility is that resources become unavailable. The strength of the availability of the system can be tested through the denial of service (DoS) attack (an attack where it causes the users to have limited or not access to the sources). Therefore, availability and DoS oppose one another's effects [14].

17.3.7 Access Control

Access control is used for enabling authorized users to have access to the resources. Access control may utilize one or more of the security mechanisms for granting the access to resources, communication channel, application, or database. The following processes are part of the access control mechanism, *user identification*, which is used for granting access to authorized users, *emergency access*, where high-priority emergency access procedures take precedence (e.g., disaster relief), *data encryption–decryption*, used for privacy and data integrity procedures, *automatic logoff/logon,* such as isolating parts of the network due to security breaches, dynamically granting special access permission to specific parts of the network to specific group of users [14].

17.3.8 Impacts of QoS on Security

In Section 17.2.5, QoS parameters were discussed without the security mechanism consideration. In this subsection, we will briefly touch on the impacts of security on QoS mechanisms.

Each of the abovementioned security mechanisms (*confidentiality, integrity, authentication, authorization, nonrepudiation, access control,* and *availability*) will cause extra overhead on to the traffic, which may affect network-centric parameters in a negative way, such as an increase in latency, a decrease in the available bandwidth, and reduction of the processing power allocation.

17.4 Multilayer Wireless Simultaneous QoS/Security Provisioning

QoS and security provisioning and their coexistence issue are important aspects in the design of any protocol stacks. Such provisioning becomes more critical for wireless protocols because, traditionally, the resources were more limited for wireless systems compared to wired systems.

QoS and security protocols provisioning have traditionally been considered at lower OSI layers (e.g., PHY, MAC). However, recently there are increasing interests in shifting the provisioning from lower layers to higher OSI layers; ultimately to the application layer. The assumption is that such a transition will increase efficiency and user-awareness of the network with added adaptability of the current network demands.

This section includes a discussion on QoS and security requirements at the application layer in a multilayer, cross-layer perspective.

17.4.1 QoS/Security Parameters at the Application Layer

The Internet is based on TCP/IP protocol suite, where its application layer covers all top three of the OSI layers (application, presentation, and session layers). An application is the highest level of user–interface interactions. Traditionally, in the communication systems, less power was allocated to this layer in terms handling underlying layers processes. However, in more recent approaches, it is becoming evident that more control and intelligence have been given to the application layer for better performance especially for multimedia-rich wireless applications.

17.4.1.1 QoS at the Application Layer

QoS's objectives at the application layer are to enable QoS-ready processing and information handling at this layer, where it is possible to have application negotiations on the run and transmit QoS-enabled parameters between various layers for supporting soft-state QoS provision. In such a QoS provision, QoS parameters can be renegotiated during the application run, creating a dynamic QoS mechanism. Soft-state QoS is especially important during wireless handoff and mobility for performance adaptation.

Application-layer QoS schemes require QoS-based per-user-based traffic support [38] to enable various levels of quality assignments for different users. Table 17.2 shows the application QoS network requirements. The classes of service mentioned in this table are to enable various levels of services to various types of applications and services.

The application-layer QoS supports provisioning of QoS triggered by requests issued from the QoS policy manager in the network and those which are signaled to the QoS policy manager at the service layer [39]. Such a signaling scheme can be originated either from the subscriber station (SS), from outside of the wireless network, from an intermediate server, or from a host connected to the SS. Another scenario is the provisioning of QoS via the QoS policy manager triggered from on-path signaling, which is processed by policy enforcement points in the network, which consequently generates requests to the policy manager. Such a signaling scheme can be originated either from inside or outside of the wireless network, including the SS, or from a host connected to the SS. For backward compatibility, the application-layer QoS should conform to the best effort or differentiated QoS across active service flows and should be consistent with the service level agreements (SLA) across base stations (BSs), also during handovers.

Current application-layer QoS mechanisms include streaming audio/video and telephony applications. For video streaming, the application-layer QoS controller module resides in the streaming server and controls the transport protocols and is fed by the compressed audio/video feeds. Such an application-layer QoS controller reduces the possibility of congestion and maximizes the video quality in the presence of packet loss. This is done by error-resilient encoding mechanisms and delay-constrained retransmission techniques [40].

Application-layer QoS is used in wireless technologies for variety of applications, including session initiation protocol (SIP) [41]. The application-layer QoS is responsible for granting QoS resources and ensuring that both session border control (SBC) downstream traffic payloads and application service provider (ASP) comply with the agreed QoS level.

Table 17.2 Application QoS Requirements

Service Class #	Service Class Application	Application-Layer Throughput	End-to-End Transport Layer One-Way Delay	End-to-End Transport Layer One-Way Delay Variation	Transport Layer Information Loss Rate
1	Real-time gaming	50–85 kbps	<60 ms	<30 ms	<3%
2	Conversational (VoIP, video, phone)	4–384 kbps	<60 ms for best quality <150 ms limit	<20 ms	<1%
3	Real-time streaming (IPTV, video clips)	>384 kbps	<60 ms	<20 ms	<0.5%
4	Interactive Applications (Web browsing, email access, IM)	>384 kbps	<90 ms	N/A	Zero
5	Non-real-time download (P2P, movies)	>384 kbps	<90 ms	N/A	Zero

Source: Fundamentals of WiMAX—A Technology Primer, Telesystem Innovation Ins.,
http://www.tsiwireless.com/docs/whitepapers/Fundamentals%20of%20
WiMAX%20-%20A%20Technology%20Primer.pdf.

Dynamic rate adaptation (DRA) is an application-layer QoS scheme in which transmission rates are changed dynamically based on the application criteria. In particular, audio/video applications running on wireless links may be using DRA for higher efficient bandwidth allocations. In a multimedia application, which offers DRA, the algorithm adjusts the encoding parameters to achieve dynamic target throughputs.

Application-layer-efficient DRA provisioning could lead to congestion control. Congestion control schemes are rate- or window-based. For the rate-based congestion control scheme, the sender is expected to incorporate rate shaping and rate adaptation compression schemes. At the receiver, delay concealment, error, and jitter schemes are required for an end-to-end application-layer-based congestion control system.

Application-layer parameters include, QoS-aware applications, application QoS handler, and session-based priorities.

Current application-layer QoS-control techniques control the transmission delays and packet loss figures caused by the network congestion. This is normally done without any support from the network infrastructure. We are interested in designing a system based on a cross-layer approach that includes various feedbacks from other layers for an intelligent application-layer QoS provisioning.

Application-layer-QoS includes error control schemes and congestion control mechanisms. Congestion control mechanisms are usually further classified into rate control and rate shaping schemes. Error control mechanisms are comprised of forward error correction (FEC) coding, error-resilient coding, retransmissions, and error concealment. FEC coding schemes include *convolutional forward error correction (CFEC), Golay forward error correction (GFEC Hamming distance,),* and *Reed–Solomon forward error correction with interleaving (RSFECI)* [43]. Cross-interleaved Reed–Solomon codes (CIRCs) are used for both error detection and correction mechanisms, specifically used for countering mixture of burst and random errors.

Reed–Solomon FEC codes are based on fixed (input and output) block codes structures. Most commonly used R–S code is the (255, 223) coding structure, which is based on 223 input block of 8 bits long symbols, which creates a 255 encoded output symbols based on a systematic conversion mechanism. In a systematic conversion, some parts of the output symbol strings contain the original strings of the input symbols. The R–S (255, 223) code system can correct up to 16 R–S errors for each codewords, thus the code can correct up to 16 short errors. In general, the coding format is RS $(n, k, n - k + 1)$, where n is the length over the finite field of F and k is the dimension with the minimum distance of $n - k + 1$.

Rate control can be achieved at the source, destination, or at both. Source-based rate control techniques are either probe-based or model-based. In a probe-based approach at the source, the scheme is experimental in nature and relies on the receiver feedbacks to adapt to the sending rate for the network bandwidth, whereas in a model-based approach, it is based on the throughput model. In a receiver-based rate control scheme, the source is required to transmit data in separate channels with various qualities. When the receiver detects no congestion in the channel, it will add a channel to improve the video quality. However, if congestion is detected, the receiver drops a channel, which causes a degradation of the video's quality. Aside from these individual approaches, a hybrid technique also exists where both source and receiver cooperate together to achieve rate control. Another technique is the rate shaping, which is used to offer congestion control. The idea behind rate shaping is to perform transcoding by using rate-adapting filters for transmission among links with various bandwidth requirements [44].

17.4.1.1.1 Jitter and Delay Concealment

Multimedia transmission via wireless channel, in particular for audio applications, is especially prone to high jitter and delay values, causing quality degradations. Therefore, delay and jitter-concealment mechanisms deployed at the application layer may reduce such degradation effects. Such mechanisms utilize adaptive packet-based time-scale modifications at the application layer based on adaptive playout algorithms, which minimize variable arrival delays (jitter), packet drops, and late packet arrivals at the receiver [45,46]. These mechanisms maintain constraints on the end-to-end delay using silence interval integration and voice segment length-stretch mechanisms.

17.4.1.1.2 Program Clock Synchronization

For decoding high-quality audio and video transmissions in the application process, a recovering high-quality system clock is crucial. For instance, the MPEG2 decoder contains audio and video clock information. Program clock synchronization is used to maintain a high-quality application-layer audio and video decoding schemes.

A method used in hardware-based fast application level switching is the utilization of the network-based application recognition (NBAR) concept [47,48]. NBAR is able to recognize packets with complex fields and attributes combinations. NBAR is capable of identifying if certain packet belongs to specific traffic stream by performing a deep-packet inspection. By using special policy schemes, specific packets are dealt with accordingly. A module that works with NBAR is the protocol description language module (PDLM), which is an application-based signature scheme. Another module that NBAR works with is the Cisco express forwarding (CEF). Together with NBAR, they offer deep-packet inspection/classification.

The following mechanisms have recently been defined as application-layer QoS frameworks: (1) application-layer dynamic services (ALDS) [49], (2) MS-triggered and MS-initiated service flow creation [46], (3) QoS API over socket framework (QAoS) [46], and (5) QoS push–pull models (QPPM) [50]. In these mechanisms, application-layer agents operate closely with the agents operating in other layers (in particular to the MAC layer) and they transfer information between each other. A few other mechanisms include hybrid automatic repeat request (HARQ) and error concealment jitter concealment (ECJC, program clock sync). Figure 17.2 depicts the approach from service- to application- and to transport-levels.

17.4.1.2 Security at the Application Layer

There are many applications offering security capabilities, which operate at the user interface level, which may or may not interact with the underlying layers. The following applications offer security schemes [52]: (1) identity-based security mechanisms, such as authentication, authorization, and shared secret across security

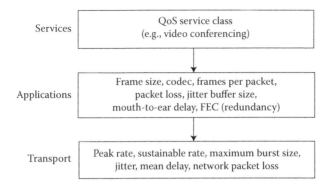

Figure 17.2 Server-to-application-to-transport level flows. (Adapted from S. Adibi, An application layer non-repudiation. Wireless system: A cross-layer. Approach, PhD Thesis at the Electrical and Computer Engineering Department, University of Waterloo, September 27, 2010.; A. Kassler, T. Guenkova-Luy, D. Mandato, and T. Robles, E2ENP: An end-to-end QoS negotiation protocol, *International Workshop on Mobile-IP-based Network Development*, London, UK, October 2002, pp 1–7.)

domains, (2) accountability and nonrepudiation mechanisms, such as archived audit trails, integrity, and message-level security, (3) content-based security mechanisms, including buffer-overflow protection scheme and application-specific security option, (4) pretty good privacy (PGP) [53], and (5) proxy firewall filters. PGP provides encryption and authentication for electronic mail services as well.

Simple object access protocol (SOAP) is another application-layer security scheme, which is used in web services. SOAP relies on extensible markup language (XML), which is used for specification exchange structure. SOAP is capable of encrypting messages at the application layer with the expense of reducing the flexibility and interoperability [54,55].

17.4.1.2.1 DoS at the Application Layer

An unauthorized user may attack the network resources with application-layer queries, causing the network to overwhelm with excessive application-based messages. This type of attack, which wastes network resources (e.g., bandwidth, battery) is called a DoS attack.

One type of application-layer DoS attack is the session hijack, where an attacker locates all ongoing sessions and the participating users, which enables him to hijack any of the ongoing sessions and to keep the sessions ongoing as long as needed. This way all resources can be wasted without any useful communication taking place.

To solve the session hijack, one should use application-layer encryption and authentication schemes. Another remedial method is to run a watchdog application to help find and cease open-ended processes aimed for resource drainage.

17.4.1.3 FEC at the Application Layer

Error control schemes are classified into four categories: (1) error-resilient encoder coding, (2) transport-layer (FEC- and delay-constrained retransmission), (3) interactive encoder–decoder error control, and (4) decoder error concealment.

Error-resilient compression schemes prevent error propagation by limiting the error damage in algorithms, which are prone to error propagation.

Application-layer FEC mechanisms incorporate content delivery protocols (CDPs) for reliable content delivery [56]. In a CDP, source blocks (SBs) are generated using transport payload multiplexing from different flows.

FEC integrates the following redundant information into the original data for packet loss compensation: (1) source-code FEC, (2) channel code, and (3) joint source/channel coding.

FEC and CDP together with service discovery data (SDD) form a complete FEC protocol suite, which cover MPEG2 (covering both multicast and unicast for RTP transport stream encapsulation). Another FEC mechanism uses raptor codes [57]. Raptor codes repair damaged data using a complex XOR operation sequencing.

Error control mechanisms use FECs with added redundant information to the data-stream. This is used to facilitate the damaged data packet reconstruction. Retransmission schemes are applicable only in scenarios where obtaining lost packet through retransmission is possible without violating its presentation deadline. Error-resilient schemes use multiple encoding methods for packet loss compensation.

17.4.1.3.1 Error Concealment

Error concealment mechanisms are often applied at the receiver side where packet loss is detected and measured, which is done using temporal and spatial interpolation algorithms. These mechanisms are able to estimate the amount of data loss to help conceal the fact that any errors happened in the first place and remain transparent from network and user perspectives.

For instance, in video applications error concealment mechanisms utilize spatial and temporal interpolation schemes to reconstruct the lost data bits within or between frames. Temporal interpolation schemes copy the pixel information at the same spatial location as in the previous frame. In spatial interpolation, the algorithm functions by estimating the missing pixels using the same frame data.

17.4.1.3.2 Processing Cost Associated to FECs

Forward error correction mechanisms, as mentioned, function with integrating a number of redundant codeword bits into the transmitting data streams. The inclusion of extraredundant bits adds to the overhead, which impacts the total end-to-end delay figures, throughput, and batter-power consumption. The strength of the

FEC code lies within the type of code used in the FEC scheme and the number of bits. Another important factor in the performance of an FEC mechanism is the types of encountering errors in the data transmission. The FEC codes suitable for bursty data streams are different than those suitable for checking random single-bit errors. Reference [58] presents a Reed–Solomon (RS) FEC coding system used for bursty errors, which features an IEEE 802.11a wireless link. FEC techniques are used to provide methods to reduce bit error rate (BER) figures on the wireless link (i.e., IEEE 802.11a) using quadruple phase shift keying (QPSK) modulation scheme, which offers a good tolerance to noise and interference. The reference [58] proves that for a bursty data of 15 symbols, the FEC scheme requires 73 machine cycles to decode. The number of machine cycles used for decoding FEC codes start to increase exponentially when the number of symbols in the burst length increases. When the number of symbols in the bursty data increases to 30, the number of machine cycles used for decoding FEC increases to 450 (exponentially increasing).

Reference [59] presents an application-layer hybrid error correction mechanism based-on Reed–Solomon mechanism, which can be used for a digital video broadcasting (DVB) service. Some applications tolerate higher degrees of packet loss compared to other applications. The number of redundant bits used in the HEC scheme (redundant information "RI") is inversed exponentially proportional to the PLR.

Though the required added code bits in FEC schemes may initially increase the end-to-end delay figures, with FEC and in the presence of noise and errors, the total amount of delay could become far greater due to the increased packet loss and retransmissions. This is due to the fact that the FEC schemes could cause a decrease in the PLR and/or retransmissions figures. However, in a good quality channel with relatively low amounts of interference and noise, the added overhead for FEC schemes will not be beneficial.

The effects of FEC on power consumption figures are also similar to those effects of FEC on the delay figures. The inclusion of redundant codeword bits, increases the power consumption, however this increase in power consumption can be normalized and even reduced when FEC is used in erroneous/noisy environments, which help correct errors and lessen the requirement of having too many retransmissions. This is because in noisy environment with relatively high amounts of BER values, the deployment of FEC schemes will add a fixed number of redundant bits, which are then used to correct error bits without the need for retransmissions. On the other hand, with this deployment an unpredictable number of data blocks may be needed to be retransmitted, depending on the channel condition, which may also increase the overhead considerably. Reference [60] presents a 4×4 64-quadrature amplitude modulation (QAM) systolic soft detector mechanism, which features a single detector that is used in a multiple input multiple output (MIMO)-based system. The performance of the mentioned system is compared against two other mechanisms without FEC mechanisms. The performance measures include power consumption based on energy per bit, which shows an extensive amount of power reduction.

Reference [61] presents an adaptive FEC mechanism used for an interactive IP streaming application. This mechanism takes aim at the loss burst sizes and packet loss rates by constructing a predictive adaptive mechanism.

17.4.1.3.3 Encryption–Decryption Interactions with FECs

A number of references [62–64] suggest that the integration of encryption–decryption techniques with FEC mechanisms optimizes both functions simultaneously. In the case of block cipher mechanisms, where a single error could propagate to the consecutive blocks, such as in advanced encryption standard (AES) cipher algorithm in cipher block chaining (CBC) mode (AES–CBC) or AES-counter with CBC–MAC (AES–CCM) [65], such integration is able to prevent the error propagation, therefore to reduce large amount of retransmissions. Reference [66] presents the idea of integrating FEC (Reed–Solomon-based) and encryption (AES-based) at the MAC layer for telemetry and communications applications.

17.5 Cross-Layer Wireless QoS/Security Provisioning

In this section, the cross-layer design is discussed as well as the interactions between various layers with the application layer.

17.5.1 Cross-Layer Design

Currently, wireless networks have maximal performance delivery, higher capacity, better QoS measures, and stronger security, all of which have to be maintained at high levels, while the complexity and energy consumption need to be kept at lowest levels. This is an extremely challenging task to do and one way is through the deployment of cross-layer design. The idea behind cross layer is similar to the idea of behind team-work, where each layer takes the current conditions of other layers into account before performing any processes.

The main objectives of cross-layer design fall into the following categories:

Bandwidth—The amount of data bits passing through each layer per unit of time is the measure of bandwidth, which is one of the performance measures. Bandwidth tends to increase as information moves from the application layer down to the physical layer. Pipelining and parallel-processing schemes at the application layer and on the sender's processing units can increase the efficiency; however, bottlenecks are centered around the PHY and MAC layers where access technologies deal with multiple access challenges and resources that are limited. Therefore, by using a cross-layer approach a more unified method can be deployed where all layers operate to enhance the lower layers efficiencies.

Bounded End-to-End Delay—To decrease the overall time delay from the application layer to the physical layer, a method can be developed to pass around delay-specific parameters among layers to speed up the process and reduce the effects of bottlenecks in the layers. This is particularly efficient for the lower layers (MAC and PHY).

Bounded packet loss—Inefficient decisions made at certain layers may contribute to relatively higher packet loss numbers while information moves down the layers. To find the most optimized loss/delay figures for the layers, cross-layer design may be helpful.

Energy resource—Energy is a scarce resource for wireless systems; therefore, any mechanisms that preserve more battery power for the same performance would be highly regarded. All layers that are prone to a DoS attack, one way or another, need to deploy mechanisms to combat DoS to avoid energy drainage. A cross-layer design helps devise a sound energy-consumption mechanism between layers.

Mobility—Mobility requires a seamless connectivity while the user is changing locations physically. Mobility could involve various layers for its operation [67].

17.5.2 Challenges in Cross-Layer Designs

One of the main ideas behind layered architectures is to assign specific tasks to a specific layer. In the cross-layer design, import and export of parameters to and from layers takes place and one of the challenges is the introduction of the additional delay because of the parameter transfer between layers. This additional delay is normally smaller than the end-to-end delay reduction for deploying the cross-layer. The other challenge is the extra overhead caused by transfer of cross-layer parameters between layers. This is also a fraction of the overhead reduction from deploying cross-layer design. The next challenge is due to the suboptimal performance of the cross-layer design. This happens because the functions performed at each layer are optimized only in a traditional sense; therefore, when those functions are affected by the cross-layer design, the result of the interactions normally increase performance in some areas, but may decrease the performance in other areas, causing suboptimal performance issue. Thus, finding a balance between the number of gathered parameters, performance, and the impact on network-centric parameters (e.g., throughput, delay), is a challenging task.

17.5.3 Cross-Layer Interactions between Various Layers and the Application Layer

In this section, the cross-layer interactions between different layers and the application layer will be discussed briefly.

The cross-layer design, centered on the application layer, is particularly important in this chapter because this layer holds the key to the new paradigm shift. In this paradigm shift, a number of important parameters, that belong to various layers can be imported efficiently to the application layer via cross-layer interactions, which is particularly known to increase efficiency for wireless multimedia applications.

There are two main application-layer cross-layer interaction flows: importing parameters from various layers to the application layer, and exporting processed parameters back to various layers.

The application layer is an important layer for multimedia processing. Encoding, decoding, application-based encryption, user-selective QoS settings, security, and many other schemes and mechanisms may take place at this layer. These schemes and mechanisms can increase the efficiency of the entire communication system when they interact with the underlying layers.

The challenge that the cross-layer design faces at the application-layer is based on the fact that user-driven applications may be impacted by the import of parameters from lower layers, which may not increase the QoE for the user. The remedy is to have a balance between the QoE-level and the feedback from lower layers.

As mentioned before, wireless multimedia applications require content-aware bandwidth usage and application-intensive processing for higher utilization, as well as processes done at the application encoder–decoder. Thus, many parameters are to be imported from various layers.

17.5.3.1 PHY-to-Application Layers

Information about the signal strength and transmission rates can inform an application-layer dynamic system to adjust the encoder quality based on to the link quality. From the security perspective, actual bit-level conversations (e.g., encryption, decryption, block-cipher) are done at PHY layer.

17.5.3.2 MAC-to-Application Layers

The MAC layer features numerous performance parameters and metrics. From the QoS perspective, the MAC layer offers, call admission control (CAC), MAC layer queuing (e.g., AC_VO, AC_VI for IEEE 802.11 systems). From the security perspective, the MAC layer offers frame structures covering security protocols, such as IEEE 802.11i and WPA (and WPA2) and working hand-in-hand with the PHY layer, where the bit-level mechanisms (e.g., encryption) are realized. A cross-layer mechanism that transfers parameters from MAC and PHY layers to the application layer is able to optimize and extend security capabilities of the lower layer at the application layer.

17.5.3.3 Network-to-Application Layers

The availability of network layer QoS (e.g., DSCP) and high-quality routes are valuable indicators that can be used at the application layer, which can optimize the application encoder's usage. From the security perspective, Internet protocol security (IPSec) and virtual private networks (VPNs)) operate at the network layer. The availability of these security mechanisms can be transported and used at the application layer.

17.5.3.4 Transport-to-Application Layers

A high-sliding window size (which may mean the ability of transmitting a relatively large number of segments at once) is an indication for a high-quality link with a relatively optimal end-to-end delay. This may trigger the source to transmit a large number of segments before the source requires an acknowledgment from the destination of the correct reception of the transmitted segments. The sliding window size may be transported to the application layer to inform the application multimedia loader to adapt itself to better encoding quality.

17.6 Future Directions

The application-layer security/QoS treatment system considered in this chapter was based on Wi-Fi wireless system (IEEE 802.11). For future work, the same analysis could be developed on the fourth-generation (4G) cellular systems using similar approach for multilayer and cross-layer design consideration at the application layer.

Another future direction is the design consideration based on the application-layer multimedia-based for wireless handhelds. Again, the same analytical and experimental approaches defined in this chapter could be carried out involving wireless handheld devices. The expected results could be used as general guidelines for the deployments of handheld devices.

17.7 Conclusion

In this chapter, the cross-layer approach centered on the application-layer was presented and studied with brief discussions on related QoS and security requirements.

The application-layer-based provisioning has been shown to offer flexibility and increased performance compared to the traditional lower layer provisioning schemes. This was discussed from both the QoS and security perspectives. From the QoS point of view, when information has already moved down the layers, starting from the application layer and reaching the PHY layer, it would topologically be "too late" to apply best QoS practices. Therefore, it is believed that efficient QoS practices should be deployed at the application layer. The application layer is the only layer at which processing can be applied to very large data blocks. The deployed processing units in other layers are bound to limited size, and much lower than that of the application layer. Therefore, higher efficiency can be achieved at the application layer. From a security point of view, this may be very beneficial since large data blocks can be processed at once.

Therefore, the challenge is to maintain a cross-layer application layer, offering an end-to-end adaptive QoS mechanism and simultaneous security support within acceptable power consumption figures, where both multimedia-based (e.g., audio, video, text) traffic payloads and security-related data are transmitted simultaneously.

One of the reasons for deploying cross-layer design is the support for QoS and security provisioning at the application layer. If no cross-layer feedback parameters were provided from lower layers, the decisions made at the application layer may not be optimized due to the lack of real-time network-based information. For instance, when the transmitter's signal strength (a physical layer parameter) drops, the system's throughput may decrease dramatically, therefore "knowing" about this drop at the application layer may be very useful to adapt the multimedia encoder quality to the channel/signal conditions. This is to avoid using high-data-rate encoding schemes when the signal strength is relatively low. Therefore, real-time cross-layer parameter delivery would provide the best decision making at the application layer.

Cross-layer design was shown to have positive impacts on the quality and performance in a wireless communication network, in particular, on the following performance metrics:

1. *Delay:* According to the literature review, VoIP applications can tolerate up to 150 ms end-to-end delay before the quality deteriorates beyond acceptable limits. Cross-layer design shows optimization of the end-to-end delay. Delay analysis in a cross-layer design requires extensive analytical, as well as experimental investigations for calculating the overall delay figures and how cross-layer design help reduce delay figures.

2. *Throughput:* Guaranteed minimum throughput is a vital QoS parameter. Delay and throughput figures are usually conversely proportional to one another. Therefore, a bounded overhead and a maximum bounded end-to-end delay are often vital requirements to guarantee a minimum throughout. Cross-layer design has been shown to increase bandwidth performance for wireless networks.

3. *Overhead:* Delay and overhead are usually directly proportional to one another and throughput and overhead are normally inversely proportional. Evaluation of the overhead for wireless networks is an important task for optimizing the performance. It would require detailed analytical, as well as experimental investigations to show how cross-layer design can reduce the overall overhead figures under various environmental and networking conditions.

The future directions include the consideration of application-layer-based QoS/security mechanisms with cross-layer/multilayer approach on top of cellular (4G) access system and for wireless handheld devices.

References

1. S. Adibi, An application layer non-repudiation. Wireless system: A cross-layer. Approach, PhD Thesis at the Electrical and Computer Engineering Department, University of Waterloo, September 27, 2010.

2. F. Guo, Traffic Analysis from Stateful Firewall to Network Intrusion Detection System, RPE Report, Jan 2004.
3. T. Szigeti, and C. Hattingh, *End-to-End QoS Network Design: Quality of Service in LAN's WANs, and VPNs*, Cisco System Press, November 2004.
4. P. Hurni, Cross-layer design in wireless networks, *Seminar Rechnernetze und Verteilte System (RVS)*, University of Bern, Switzerland, 2 April 2008.
5. E. D. Puschita, and T. P. Palade, QoS perspective for wireless scenarios, *Broadband Europe 2005*, Whitepapers, Bordeaux, 12–14 December 2005, 5 p.
6. K. Kilkki, Next generation internet and quality of experience, *EuroFGI IA.7.6 Workshop on Socio-Economic Issues of NGI*, Santander, Spain, June, 2007.
7. J. Tang, and A. Zhang, Cross-layer design of dynamic resource allocation with diverse QoS guarantees for MIMO-OFDM wireless networks, *IEEE International Symposium on World of Wireless Mobile and Multimedia Networks*, Taormina–Giardini Naxos, 13–16 June 2005, pp. 205–212.
8. S. Pasupuleti, and D. Das, Throughput and delay evaluation of a proposed-DCF MAC protocol for WLAN, *Proceedings of IEEE INDICON 2004*, December 2004, 4 p, http://dit.unitn.it/~srinivas/INDICON_IIITB.pdf.
9. W. Grote, A. Grote, and I. Delgado, IEEE 802.11 Goodput analysis for mixed real-time and data traffic for home networks, *Annals of Telecommunications*, 63(9/10), 463–471, 2008.
10. P-H. Hsiao, H. T. Kung, and K-S. Tan, Streaming video over TCP with receiver-based delay control, *IEICE Transactions on Communications*, E86-B(2), 572–584, 2003.
11. A. Ganz, Z. Ganz, and K. Wongthavarawat, *Multimedia Wireless Networks: Technologies, Standards, and QoS*, Prentice Hall Publisher, USA, ISBN: 978-0130460998, September 18, 2003.
12. O.I. Hillestad, B. Libak, and A. Perkis, Performance evaluation of multimedia services over IP networks, *Proceedings of the IEEE International Conference on Multimedia and Expo (ICME)*, Amsterdam, The Netherlands, 06–08 July 2005, pp. 1464–1467.
13. H. O. R. Thomschutz, Security in packet-switched land mobile radio backbone networks, Master of Science Thesis at Virginia Polytechnic Institute and State University, May 2005.
14. S. Adibi, and G. B. Agnew, Security measures for mobile ad-hoc network (MANETs), *Handbook of Research on Wireless Security*, IGI Global Inc., ISBN 978-1-59904-899-4, March 2008.
15. D. Zhu, Security control in inter-bank fund transfer, *Journal of Electronic Commerce Research*, 3(1), 15–22, 2002.
16. M. Sher, and T. Magedanz, Developing network domain security (NDS) model for IP multimedia subsystem (IMS), *Journal of Networks*, 1(6), 10–17, 2006.
17. L. Zhou, and Z. J. Haas, Securing ad hoc networks, *IEEE Networks Special Issue on Network Security*, 13(6), 24–30, 1999.
18. P. Kanerva, Anonymous authorization in networked systems: An implementation of physical access control system, Master's thesis, Helsinki University of Technology, Department of Computer Science and Engineering, Espoo 2001.
19. R. Ren, D-G. Feng, and K. Ma, A detailed implement and analysis of MPLS VPN based on IPSec, *Proceedings of 2004 International Conference on Machine Learning and Cybernetics*, 5(26–29), 2779–2783, 2004.
20. I. Yiakoumis, M. Papadonikolakis, and H. Michail, Efficient small sized implementation of the keyed-hash message authentication code. *Proceedings of the IEEE Eurocon Conference 2005 Computer as a Tool*, Volume 2, pp. 1875–1878, Belgrade, November 2005.

21. S-Y. Kang and I-Y. Lee, *A Study on Low-Cost RFID System Management with Mutual Authentication Scheme in Ubiquitous*, Managing Next Generation Networks and Services, Springer Link Publication, ISBN: 978-3-540-75475-6, September 18, 2007.

22. G.B. Agnew, ECE628 Computer Network Security Course Notes, Winter 2006.

23. R. Koide, T. Nagase, and T. Araki, QHF: A quaternion-based a multidimensional hash function, *Proceedings of 2007 International Symposium on Intelligent Signal Processing and Communication Systems*, Xiamen, China, 28 November–1 December 2007, pp. 830–833.

24. S. Hirose, Weak security notions of cryptographic unkeyed hash functions and their amplifiability, *Special Section on Cryptography and Information Security, IEICE Transaction of Fundamentals*, E88-A(1), 33–38, 2005.

25. X. Wang, D. Feng, X. Lai, and H. Yu, Collisions for Hash Functions MD4, MD5, HAVAL-128 and RIPEMD, Rump Session, CRYPTO 2004, Cryptology ePrint Archive, Report 2004/199, pp. 1–4, IACR 2004 Cryptography Conference, Santa Barbara, USA.

26. G. Leurent, Message freedom in MD4 and MD5 collisions: Application to APOP, *Proceedings of FSE*, 2007, LNCS, Vol. 4593, pp. 309–328. Springer, Heidelberg, Germany.

27. X. Wang, and H. Yu, How to Break MD5 and Other Hash Functions, EUROCRYPT 2005, http://www.infosec.sdu.edu.cn/uploadfile/papers/How%20to%20Break%20 MD5%20and%20Other%20Hash%20Functions.pdf.

28. A. Levitin, Brute force, Chapter 3, *Introduction to the Design & Analysis of Algorithms*, 2nd ed., Pearson Addison-Wesley, ISBN: 978-0321358288, 2007, http://www.cs.ucr. edu/~jiang/cs141/ch03n.ppt.

29. SHA hash functions, Wikipedia, http://en.wikipedia.org/wiki/SHA1, Retrieved on May 1st, 2009.

30. W. Diffie, and M. E. Hellman, New directions in cryptography, *IEEE Transactions on Information Theory*, IT-22, 644–654, 1976.

31. C. M. Gutierrez, and P. Gallagher, Digital Signature Standard (DSS), DRAFT FIPS PUB 186-3, FEDERAL INFORMATION PROCESSING STANDARDS PUBLICATION, Information Technology Laboratory, National Institute of Standards and Technology, U.S. Department of Commerce, November, 2008, http://csrc.nist. gov/publications/drafts/fips_186-3/Draft_FIPS-186-3%20_November2008.pdf.

32. D. Johnson, A. Menezes, and S. Vanstone, *The Elliptic Curve Digital Signature Algorithm (ECDSA)*, Certicom Corp., http://www.comms.scitech.susx.ac.uk/fft/crypto/ecdsa.pdf.

33. M. Medwed, and E. Oswald, Template attacks on ECDSA, WISA 2008, http://www. comms.scitech.susx.ac.uk/fft/crypto/ecdsa.pdf.

34. A. Menezes, *Evaluation of Security Level of Cryptography: RSA-OAEP, RSA-PSS, RSA Signature*, pp. 1–29, University of Waterloo, December 2001, http://www.ipa.go.jp/ security/enc/CRYPTREC/fy15/doc/1011_rsa.pdf.

35. G. B. Agnew, R. C. Mullin, and S. A. Vanstone, An implementation of elliptic curve cryptosystems over F_2155, *IEEE Journal on Selected Areas in Communications*, 11(5), 804–813, 1993.

36. M. Aydos, B. Sunar, and C. K. Koc, An elliptic curve cryptography based authentication and key agreement protocol for wireless communication, *The 2nd International Workshop on Discrete Algorithms and Methods for Mobility (DIALM 98)*, Dallas, Texas, 30 October 1998, pp. 1–12.

37. S. Seys, Lightweight cryptography enabling secure wireless networks, *Workshop on Security Issues in Mobile and Wireless Heterogeneous Networks*, Brussels, December 6, 2004.
38. T. Ozcelebi, M. O. Sunay, M. R. Civanlar, and A. M. Tekalp, Application-layer QoS fairness in wireless video scheduling, *Proceedings International Conference on Image Processing (ICIP'06)*, 1–7, 1673–1676, 2006.
39. N. Cranley, and L. Murphy, *Handbook of Research on Wireless Multimedia: Quality of Service and Solutions*, Information Science Reference, IGI Global, ISBN: 978-1599048208, July 28, 2008.
40. D. Wu, Y. T. Hou, W. Zhu, Y-Q. Zhang, and J. M. Peha, Streaming video over the Internet: Approaches and directions, *IEEE Transactions on Circuit and Systems for Video Technology*, 11(3), 282–300, 2001.
41. J. Rosenberg, H. Schulzrinne, G. Camarillo, A. Johnston, J. Peterson, R. Sparks, M. Handley, and E. Schooler, SIP: Session initiation protocol, RFC 3261, June 2002.
42. Fundamentals of WiMAX—A Technology Primer, Telesystem Innovation Ins., http://www.tsiwireless.com/docs/whitepapers/Fundamentals%20of%20WiMAX%20-%20A%20Technology%20Primer.pdf.
43. E. Alhoniemi, *Error Detection and Control in Data Transfer*, Helsinki University of Technology, 22 November 1998, http://www.tml.tkk.fi/Studies/Tik-110.300/1998/Essays/error_detection.html.
44. Cross-Layer Design of Ad-hoc Wireless Networks for Real-Time Media, http://www.stanford.edu/~zhuxq/adhoc_project/adhoc_project.html.
45. F. Liu, J. Kim, and C.-C.J. Kuo, Adaptive delay concealment for Internet voice applications with packet-based time-scale modification, *Proceedings of IEEE International Conference on Communications, Proceedings of the Acoustics, Speech, and Signal Processing (ICASSP'01)*, 3, 1461–1464, 2001.
46. B. H. Kim, and T. Kavanaugh, *QoS API for the Signalling between Upper Layer and MAC in MS*, WiMAX Forum Press, Mountain View, California, 2007.
47. F. Guo, Traffic Analysis: from Stateful Firewall to Network Intrusion Detection System, RPE Report, January 2004.
48. T. Szigeti, and C. Hattingh, *End-to-End QoS Network Design: Quality of Service in LAN's WANs, and VPNs*, Cisco System Press, San Francisco, California, November 2004.
49. Packetcable Dynamic Quality-of-Service Specification, PKT-SP-DQOS-I12-050812, http://www.packetcable.com/downloads/specs/PKT-SP-DQOS-I12-I05-050812.pdf.
50. M. Alam, R. Prasad, and J. R. Farserotu, Quality of service among IP-based heterogeneous networks, *IEEE Personal Communications*, 8(6), 18–24, 2001.
51. A. Kassler, T. Guenkova-Luy, D. Mandato, and T. Robles, E2ENP: An end-to-end QoS negotiation protocol, *International Workshop on Mobile-IP-based Network Development*, London, UK, October 2002, pp 1–7.
52. S. Tarkoma, Network Application Frameworks and XML Summary and Conclusions, T-110.5140, April 2008, http://www.tml.tkk.fi/Opinnot/T-110.5140/2008/Lectures/naf_lecture_220408.pdf.
53. Y. Xue, Email security, *CS 291 Network Security*, Vanderbilt University, 2006, http://vanets.vuse.vanderbilt.edu/~xue/cs291fall06/email.pdf.
54. T. Wennerstrom, J. Jespersen, and M Lundquist, *Simple Object Access Protocol—A Basic Overview*, University of Uppsala, September 2002, http://user.it.uu.se/~hsander/Courses/DistributedSystems/Reports/soap_report_2.pdf.

55. J. J Vargas, *SOAP (Simple Object Access Protocol), An Introduction*, University of Central Florida, CDA 5937 Fall 2002, http://www.cs.ucf.edu/~dcm/Teaching/ProcessCoordination/Fall02Class/ResearchPresentations/JuanVargas.ppt.

56. U. Kozat, and A. Begen, Pseudo Content Delivery Protocol (CDP) for Protecting Multiple Source Flows in FEC Framework, Internet Draft, July 7, 2008, ftp://ftp.mimuw.edu.pl/mirror/ftp.rfc-editor.org/internet-drafts/draft-kozat-fecframe-pseudo-cdp-00.txt.

57. M. Luby, A. Shokrollahi, M. Watson, and T. Stockhammer, Raptor Forward Error Correction Scheme for Object Delivery, RFC 5053, October 2007.

58. B. Fong, P. B. Rapajic, G. Y. Hong, and A. C. M. Fong, Forward error correction with Reed-Solomon codes for wearable computers, *IEEE Transactions on Consumer Electronics*, 49(4), 917–921, 2003.

59. G. Tan, and T. Herfet, Application layer hybrid error correction with Reed–Solomon code for DVB services over wireless LANs, *3rd International Conference on Wireless Communications, Networking and Mobile Computing (WiCOM 2007)*, Shanghai, China, 21–23 September 2007, pp. 2952–2955.

60. P. Bhagawat, R. Dash, and G. Choi, Systolic like soft-detection architecture for 4x4 64-QAM MIMO system, *IEEE Design Automation and Test in Europe (DATE'09)*, Nice, France, 20–24 April 2009, pp. 870–873.

61. F. S. Filho, E. H. Watanabe, and E. de Souza e Silva, Adaptive forward error correction for interactive streaming over the Internet, *Proceedings of the IEEE GLOBECOM*, San Francisco, California, November 2006, pp. 1–6.

62. J. Fernandez-González, G. B. Agnew, and A. Ribagorda, Encryption and error correction codes for reliable file storage, *Computers and Security*, 12(5), 501–510, 1993.

63. C. N. Mathur, K. Narayan, and K. P. Subbalakshmi, High diffusion cipher: Encryption and error correction in a single cryptographic primitive, *Proceedings of the 4th International Conference on Applied Cryptography and Network Security (American Conference on Neutron Scattering)*, Vol. 3989, Singapore, June 2006, pp. 309–324.

64. C. N. Mathur, A mathematical framework for combining error correction and encryption, Winner of the Best Dissertation Award, Stevens Institute of Technology, May 2007.

65. R. Housley, Using Advanced Encryption Standard (AES) CCM mode with IPsec Encapsulating Security Payload (ESP), *RFC 4309*, December 2005, http://www.rfceditor.org/rfc/rfc4309.txt.

66. S. Spinsante, F. Chiaraluce, and E. Gambi, Evaluation of AES-based authentication and encryption schemes for Telecommand and Telemetry in satellite applications, *Proceedings Space Ops 2006 Conference*, Paper AIAA 2006-5558, Rome, Italy, 19–23 June 2006.

67. Technical Document on Overview—Wireless, Mobile and Sensor Networks, GDD-06-14—Ver. 2.0, GENI: Global Environment for Network Innovations, Sep. 2006.

Chapter 18

Security, Trust, and Privacy in Opportunistic Multihop Wireless Networks

M. Bala Krishna

Contents

18.1 Introduction

Wireless networks have emerged as the principal contenders empowering seamless communication across homogeneous and heterogeneous networks. Opportunistic networks (OppNets) [1] are the emerging technology in mobile and wireless communications that do not necessitate a fully connected network. Multihop wireless network (MHWN) is a self-organized network with minimum infrastructure that operates based on the predefined network topology and allocated frequency spectrum. Intermediate mobile nodes reduce the fixed infrastructure and route the data toward the destination node. An OppNet is a self-organized infrastructure ad hoc network [2,3] comprising of distributed controller nodes, heterogeneous intermediate nodes, and mobile agents. Since, the routing paths are not known in advance, the forwarding nodes establish the reliable route paths based on opportunistic resources in the network. Network connectivity [4] established by the intermediate active nodes and neighboring nodes [5] form the reliable route paths in the network. Security and trust management protocols authenticate the message exchange between the nodes in the network. The features of MHWN and opportunistic multihop wireless network (OppMHWN) are explained as follows.

MHWNs are based on the design of predetermined network topology and allocated frequency spectrum. MHWN operates without the support of infrastructure and backbone network. Since the next hop is known in advance, a faulty path can be resolved using alternate route paths in the network. The size of a multihop is a function of traffic conditions, node density, and node transition state. Multihop communication consumes less power and overcomes the effect of long-distance signal propagation in wireless medium. MHWNs are classified as follows: wireless mesh networks, wireless sensor networks, and wireless ad hoc networks.

OppMHWNs operate on the stateless behavior of the network and apply broadcasting techniques to establish the route paths in the network. Since, the next hop is not predetermined, multiple intermediate nodes act as relay nodes to establish the reliable route paths in the network. The active node stores the data and waits for an opportunity (such as successful transmission nodes within the communication range, intermediate helper nodes, congestion free route paths, and so on) to forward the data toward the destination. Opportunism in multihop network is increased by using mobile-assisted intermediate nodes. The primary attributes of OppNets are network traffic conditions, number of active intermediate nodes, delay tolerant factor, node density, and the session duration. The forwarding schemes used in OppNets are classified as follows: queue-based forwarding [6], epidemic forwarding, deterministic forwarding, probabilistic-based forwarding, mobile agent-based forwarding, and social-based forwarding. The limitations of OppNets are intermittent connectivity and delay factor in the network.

18.1.1 Characteristic Features of OppNets

Energy limitations—The energy limitations in multihop OppNets are due to the following reasons:

1. More number of multihop nodes between the source node and the destination node.
2. Additional overhead to search the intermediate active nodes to forward the data.
3. Frequent changes in the selection of intermediate helper nodes to forward the data.

Performance—Performance of multihop routing in OppNets is based on the successful data delivery from end-to-end [7] nodes in the network. The node mobility can be improved by considering the parameters such as size of neighboring nodes [8], state of the network, and time stamps along the multihop route paths.

Security—Security protocols are based on routing, data aggregation, and key management techniques. Network intrusions thwart the *cooperation* and *coordination* across the OppNet nodes, compromise the route paths, and redirect the data packets to faulty destinations [5]. Cryptographic methods using public key encryption and private key encryption establish the secure channel. Intrusion detection methods based on periodic verification identify the malicious nodes in the network.

Applications—OppNets are applied in traffic monitoring, surveillance monitoring, target tracking, military applications, remote areas, interior parts of heavy machinery, retail markets, and so on. The intermediate nodes prevent random packet transmission and forward the data toward the destination [9] node. For real-time applications such as e-mail, e-commerce, and so on, the OppNets converge toward the social networks, which impose moderate mobility models and collaborate with multiple users in the network.

18.1.2 Threats in OppNets

Due to intermittent connectivity, the OppNets [10] are vulnerable to security attacks. A special node known as the *judge node* traces the unreliable behavior of noncooperative nodes [11] and eliminates them from the network. The security threats [11] are due to the following reasons:

1. The nodes only receive and do not forward the data.
2. The nodes forward the data to faulty nodes.
3. The malicious nodes generate spam messages and flood the route paths in the network.

This chapter is organized as follows. Section 18.2 explains the issues and challenges of OppMHWNs. This section highlights the security, trust, and privacy issues in OppNets and enlists the challenges based on synchronization, data storage, and key management. Section 18.3 explains the metrics of security, trust, and privacy used in OppMHWNs. Section 18.4 explains the design architecture of OppMHWN. Section 18.5 explains the security, trust, and privacy protocols used in OppMHWNs. Security protocols based on the optimistic fair-exchange, opportunistic encryption, and local key management methods are briefly explained in this section. Trust management protocols based on the composite trust, social trust, reputation-assisted data forwarding, and trust-based minimum cost-routing methods are briefly elucidated in this section. Privacy protocols based on the message content, location, group size, and interest-casting methods are expanded in this section. Section 18.6 concludes the chapter and highlights the features of security, trust, and privacy used in OppNets. Section 18.7 gives the guidelines for future research to enhance the features and applications of OppNets.

18.2 Issues and Challenges

The issues related to security and trust in the OppNets is as follows:

1. Interception of data packets from the neighboring nodes.
2. Alteration of the route paths in the network.
3. Decrease in the packet-delivery ratio and throughput rate due to compromised nodes.

Data packets received from the source node are forwarded to the intermediate helper nodes (static or mobile) based on the degree of trust. The process of electing the node with a maximum degree of trust is executed iteratively until the data packet reaches the destination node. Privacy in OppNet is based on (i) node attributes, and (ii) access rights of authorized private members.

The challenges of security and trust in OppNets are as follows:

1. Providing reliable connections between the OppNet nodes.
2. Sustaining the network connectivity for long durations across the OppNet nodes.
3. Time synchronization between the OppNet nodes.
4. Increase the storage and processing capabilities of the intermediate mobile nodes.

18.3 Metrics in OppNets

In multihop opportunistic data forwarding, the link establishment across the data forwarding nodes is based on the traffic conditions and the state of intermediate nodes. The metrics in OppNets are classified as follows: (i) types of node interaction, (ii) security metrics, (iii) trust metrics, and (iv) privacy metrics. Let N be the total number of nodes in the network and p be the probability of interaction ($p \geq 0.5$) between the nodes N_i and N_j. The node interaction is defined as follows:

1. *Node-to-node direct interaction* (*NDI*): NDI between the nodes N_i and N_j with probability P_{NDI} for successful message exchange is given as follows:

$$msg(N_i, N_j, P_{NDI}, \text{True}) \quad \text{where } 0.75 \leq P_{NDI} \leq 1 \ i, j \in N$$

2. *Node-to-node indirect interaction* (*NII*): NII between the nodes N_i and N_j with probability P_{NII} for successful message exchange is given as follows:

$$msg(N_i, N_j, P_{NII}, \text{True}) \quad \text{where } 0.65 \leq P_{NII} \leq 0.75 \ i, j \in N$$

3. *Node-to-node selfish interaction* (*NSI*): NSI between the nodes N_i and N_j with probability P_{NSI} for unsuccessful message exchange due to malicious node N_{MN} is given as follows:

$$msg(N_i, N_j, P_{NSI}, \text{False}) \quad \text{where } 0.65 \leq P_{NSI} \leq 1 \ i, j \in N$$

18.3.1 Security Metrics

Security metrics based on the cryptographic algorithms use symmetric key or asymmetric key encryptions. Encryption and digital signatures establish the secure communication channel. Various types of security metrics are defined as follows:

1. *SymEK (A, msg):* Node *A* performs symmetric key encryption with key *K* for the message *msg.*
2. *AsymEK (A, msg, Kp):* Node *A* performs asymmetric key encryption with public key *K* and private key *K$_p$* for the message *msg.*
3. *DSignatureKC (A,msg):* Digital signature by the node *A* using key certificate *KC.*
4. *EK1 (EK2 (A, msg)):* Hierarchical networks use multilayer encryptions. Node *A* initially (level 1) encrypts the message *msg* using the key *K1.* The message is further encrypted in the second hierarchical level using the key *K2.*

18.3.2 Trust Metrics

The degree of interaction between two nodes is a function of the degree of trust awarded to each node and their respective network attributes. Time-based trust degrees assigned to the authorized member nodes enhance the level of trust across the nodes in the network. The trust protocols maintain the list of trusted and non-trusted member nodes in the network and regulate the data flow. The degree of trust between the nodes *i* and *j* at time *t* based on node belief factor, uncertain factor, and the coefficient σ is given as follows:

$$\text{Trust}^t_{i,j}(msg) = \text{Belief}^t_{i,j} + \sigma * \text{Uncertain}^t_{i,j}$$

The social trust factor is based on the node familiarity index between the neighbor nodes in the network. The social trust factor between the nodes *p* and *q* are defined as follows:

$$\text{Social Trust Factor} = f_{SocialTrust} \text{ (familiarity index, degree of trust, } msg, p, q)$$

18.3.3 Privacy Metrics

The privacy models aim to protect the node identity and location from nontrusted and unauthorized access in the network. Privacy models monitor the number of intermediate nodes, hop count, and session duration. The private group *G* between the nodes *N$_i$* and *N$_j$* using the key *K* in encryption *E* is given as follows:

$$\text{PrivateGroup}_G = E_K (Ni, Nj, msg)$$

The digital signature by the private group *G* using the key from trusted third party (TTP) is given as follows:

$$DSigGroupG_{KTTP} (G, msg)$$

18.4 Design Architecture of OppMHWNs

Data forwarding in OppNets is independent of predefined network topology and end-to-end connectivity. The architecture of OppMHWN is illustrated in Figure 18.1. The source node estimates the number of neighboring nodes (based on the signal strength and node distance) and broadcasts the message. If the forwarding node is far away, then the source node waits for an opportunistic node to forward the data toward the destination. Table 18.1 indicates the node types used in OppMHWNs.

18.5 Security, Trust, and Privacy Protocols in OppMHWNs

In opportunistic data forwarding, the intermediate mobile nodes store and forward the messages toward the destination node. Since, the packets are forwarded along the intermittent route paths, the malicious node compromise the data-forwarding nodes [12] in the network. The secure multilayer credit-based incentive scheme for delay-tolerant networks (SMART) [12] address the issue of noncooperative malicious nodes in the network. This protocol awards credits for the intermediate

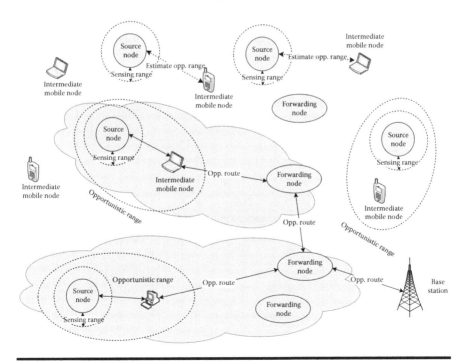

Figure 18.1 Architecture of OppMHWN.

Table 18.1   Node Types Used in Opportunistic MHWNs

Node Type	Feature
Source node	The starting node where the data is being initiated
Destination node	The end node where the data need to be sent
Adjacent node	The nodes surrounding the given node within the sensing zone
Forwarding node	The node that sends the data toward the destination node
Opportunistic node	The intermediate node (mobile/stationary) that establishes the link toward the forwarding node. The opportunistic node can be a mobile phone, laptop, or a portable computing system

forwarding nodes that do not reveal the identity of the source node. In the initial phase, the source node broadcasts the attributes such as cost of packet transfer and service requirements of the packet-forwarding node. The packet-forwarding nodes broadcast these attributes to the next layer of the protocol stack until the data reaches the destination node. This protocol saves the history of delivered packets. The credits and rewards based on the performance of forwarding node are updated periodically. Figure 18.2 illustrates the classification of security, trust, and privacy protocols in OppMHWNs.

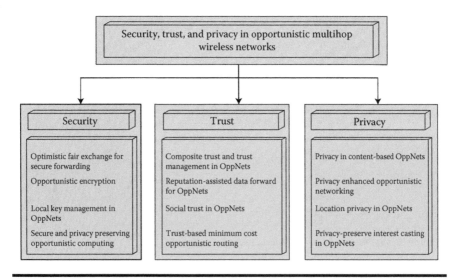

Figure 18.2   Security, trust, and privacy protocols in OppMHWN.

Opportunistic resource utilization [13] considers the resource components such as communication, computation, storage, sensing, and virtualization to enhance the performance of OppNet. The security protocols are primarily based on the reputation-based valid key exchange between the nodes. The trust protocols are primarily based on the degree of trust using the credit and reward estimates of the forwarding node. The privacy protocols are primarily based on the private attributes confined to the private node set (friends) in the network.

18.5.1 Security in OppNets

Security and trust-management techniques trace the intruders and resolve packet interceptions, latency, and route path delays in the network. The primary factors considered in the design of security protocols are intermittent connectivity of OppNets, message integrity, and delays in the forwarding route path [14]. Opportunistic broadCast protocol (OppCast) [15] aims to achieve an improved packet reception ratio (PRR) by broadcasting the warning messages between the successful packet transmissions in the network. This technique uses a two-step approach enhance the reliability of the communication link: (i) improve message warnings and (ii) reduce hop-delays for the successive broadcast acknowledgments. The warning messages are sent for long hops to achieve the required PRR. OppCast [15] protocol works in two phases: fast-forward dissemination (FFD) and makeup-for-reliability (MFR). In the FFD phase, the warning messages are forwarded with increasing number of intermediate regions toward the destination. In the MFR phase, the packet forwarding is enhanced based on multiple retransmissions along the route paths. The explicit exchange of broadcast acknowledgments for each successful packet transmission reduces the number of redundant broadcast messages and delays in the network. Table 18.2 enlists the features and adversities of security, trust, and privacy in OppNets.

18.5.1.1 Optimistic Fair Exchange for Secure Forwarding

Optimistic fair exchange for secure forwarding [16] rewards the data-forwarding nodes based on the degree of cooperation and path exchange between the adjacent nodes. The intermediate nodes receive the rewards and transmit the data packets toward the destination node. Fairness and protection of the data packet is monitored such that the malicious nodes are restrained from unauthorized access in the network. TTP nodes with valid public keys contribute as the data-forwarding nodes. The rewarding scheme is based on the *anonymity* (valid packet transmission between the authorized nodes) and *singularity* (used only for single-data transmission) of the message exchange. The sender node requests the reward that is generated by the recipient node and verified by the TTP node. After the reception of a valid signature by the recipient node, the sender node encrypts the message with the symmetric key that is further encrypted with the public key of TTP node. The

Table 18.2 Features and Adversities of Security, Trust, and Privacy in OppNets

Node/Network Features	Features of OppNets	Adversities of Security, Trust, and Privacy
Communicating nodes	It is not certain that two communicating nodes exist on the same platform or network (such as in Internet applications) [4]	The intruders or malicious nodes may be hidden in the platforms and redirect the information to the false destinations (such as Internet hackers)
Route paths	Route paths are updated to forward the data packets to the resource efficient neighbor node at each hop level	Some of these nearest neighbor nodes may be compromised in the network or the malicious nodes may act as the authentic neighbor nodes
Messages	For the off-line nodes toward the destination, the messages are queued in wait state.	Intruders attack the nodes and keep the messages in waiting mode
Message forwarding	Message forwarding is carried out by the mobile intermediate nodes	The untrusted members or intruders can increase the number of hops and delay the messages from reaching the base station

symmetric key encryption and asymmetric key encryption between the nodes A and B, based on message hashing and successful rewards for optimistic secure forwarding is given as follows:

$$SymE_K \, (A,B, \text{hash}(msg), \text{reward})$$

$$AsymE_{TTP} \, (A, B, msg, Kp, \text{hash}(msg), \text{reward})$$

While addressing the issues of key complexity, this protocol predicts the block length and quantifies the frame length (integral multiples of block length). The attacks are estimated as the function of frame length and the probability of cracking the cipher blocks, which are arranged in an ascending order of the encrypted blocks. The optimization of a cipher block is achieved by including the block length and the payload in the frame. The main limitation of this protocol is that it does not include the cost of the rewards that are assigned to the intermediate nodes.

18.5.1.2 Opportunistic Encryption

The application-specific key-distribution techniques consider the effects of noise and errors while encrypting and decrypting the message. Opportunistic encryption protocol [17] is based on addressing the trade-off between security and throughput during the encryption/decryption of data packets. The "opportunistic encryption" technique adapted by this protocol best utilizes the communication channel for the duration of relatively high signal-to-noise ratio (SNR) levels. This protocol assumes that an average of half the original message is affected by the malicious nodes. This necessitates two cipher texts that reduce the complexity of key management. Within a given time interval, if the SNR is maximum, then the opportunistic mobile node encrypts the message with reduced cipher block length and forwards the secure data. The modes of cipher blocks [17] used in the encryption are: cipher feedback (CFB) mode, output feedback (OFB) mode, and the CounTeR (CTR) mode. The security measure is defined as the logarithmic value of encrypted block length, and is normalized with a maximum block length within the given time. The significance of applying a unique key for each encryption block per session prevents the message leakage due to key revelation. The optimization of an encrypted block length depends on fixed channel conditions. In the "security quantification" phase, the security levels for each individual frame are calculated and integrated. Finally, the trade-off between the throughput and the optimized security is established in terms of transmission rate and bit error probability. The normalized throughput maximizes the transmission rate in the network. This protocols analyzes the SNR and forward error correction (FEC) codes and their effect on secure communication channel.

18.5.1.3 Local Key Management in OppNets

Privacy in content-based OppNets [18] uses intermediate nodes to exchange the encrypted advertisements and message contents between the receiver nodes and the publisher nodes. The node-transition state varies from the receiver mode to publisher mode and vice versa. The intermediate nodes use encrypted advertisements to update the forwarding routing tables. The enforced privacy models trace the malicious nodes. The privacy models are defined as follows:

1. The publisher and receiver nodes trust the intermediate nodes. The collision-free hash functions are used to exchange the advertisements and messages between the nodes.
2. The member nodes share a common group key to prevent attacks like eavesdropping.

The opportunistic route paths are extended toward the destination based on encrypted forward message with multiple keys. Multilayer commutative encryption (MLCE) [19] is a local key management technique where a node shares the secure key with the neighborhood nodes based on predefined hop count. The encrypted keys are assigned by the key certification manager. Nodes share their keys with the two-hop neighbors and encrypt the data twice with the next two-hop nodes toward the destination node. Two-level symmetric encryption between the nodes A and B with distinct keys $K1$ and $K2$ is given as follows:

$$SymE_{K2} (SymE_{K1} (A, B, msg))$$

The approved data forwarding nodes effectively trace the active and passive malicious nodes in the network. Authentic nodes are recognized based on the certified identity defined by the identity manager. Long-term secure keys are assigned to the active nodes, and additional short-term local keys are assigned to the intruded paths in the network. This protocol is applied for content-based multilayer opportunistic data forwarding to ensure a secure and trusted communication in the network.

18.5.1.4 Secure and Privacy-Preserving Opportunistic Computing Framework for Mobile-Healthcare Emergency

Opportunistic computing enhances the effective usage of available resources in delay-tolerant intermittent networks. Secure and Privacy-Preserving in Opportunistic Computing Framework (SPOC) [20] applies opportunistic-computing technique to address the emergency health care conditions, in which the personal health information (PHI) is shared between the member nodes in the network. The two-level approach in privacy maintenance includes (i) the rejection of unidentified members from accessing the valid data, and (ii) the information sharing between the authentic members with similar interests. This protocol defines the privacy preserving scalar product computation (PPSPC) scheme based on the user access control and information base. The security model works in two phases. In the first phase, privacy access control confined to authorized members limits the number of users in the network. In the second phase, the threshold parameter controls the access rights to prevent excessive traffic in the network. System initialization is executed by the trusted authority (TA) using the valid secure attributes. The symmetric access key controls are generated using random numbers for each member node. The nonmembers with emergency conditions are traced based on the random numbers from the trusted group. In the second phase, member nodes with similar interests are given access to forward or receive the data. This protocol is applied in medical health-care applications and uses the resources of neighboring nodes in emergency conditions (one smart phone can share the resources of other smart phone). The symmetric encryption by node A for a message based on personal information, current date, and secret key is given as follows:

$$SymE_K (A, \text{PersInformation}, msg, \text{CurrentDate}, K_{\text{Secret}})$$

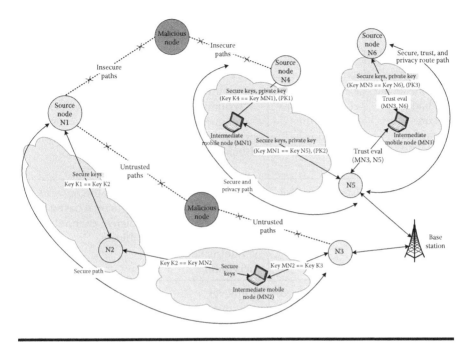

Figure 18.3   Secure route path, secure-privacy route path, and secure-trust-privacy route in OppMHWN.

This technique restricts the privacy disclosure to different users in the network. Opportunistic communication for emergency evacuation [21] considers resilience and security based on identity-based signature and content-based message verification. The intruder route paths are identified and marked as untrusted paths in the network. The intermediate mobile nodes (portable computer or mobile phone) forward the data toward the destination. Figure 18.3 illustrates the secure route path, secure-privacy route path, and secure-trust-privacy route in OppMHWNs. Table 18.3 gives the summary of security protocols in OppMHWNs.

18.5.2 Trust in OppNets

Reliability of data-forwarding nodes is the significant factor in trust-management protocols. The data forwarding is carried out by the next-hop nodes that are selected by the current forwarding node. The current active node estimates the forwarding node based on the degree of trust and packet delivery ratio. The degree of trust is a function of node reluctance factor with the malicious node, response time with the forwarding node and hop count. A proactive trust method prevents suspicious nodes in the network, and the reactive trust method identifies the compromised nodes based on periodic key exchanges.

Table 18.3 Security Protocols in OppMHWN

Security Protocols	Features	Security Attributes	Merits and Demerits
Optimistic fair exchange for secure forwarding	Reward-based cooperation to forward the data using TTP	Symmetric key encryption, asymmetric key encryption, key, TTP, and message	Rewards assigned to the sender and receiver. Complexity increases for the rewards of unsuccessful transmissions
Opportunistic encryption	Considers SNR, throughput, and bit error rate	Length of the encryption block, normalized security level, and symmetric key encryption	Improves the network performance based on predictive channel attributes
Local key management in OppNets	Content-based secure forwarding and forward look-up decisions in multihop network	Multilevel encryption: encryption at the local network and encryption at hierarchy levels toward the destination	Commutative encryption scheme applied in multilayer OppNets, but do not support advertisements in the message
SPOC	Attribute-based access control and effective sharing of resources	Secret key, session key, master key, current date, and personal information	User-centric security that achieves maximum data transfer with less privacy disclosure. Internal attacks are not resisted beyond a threshold limit

18.5.2.1 Composite Trust and Trust Management in Opportunistic Networks

Composite Trust and Trust Management in Opportunistic Networks (COTTON) [3] ensures trust and collaboration across the reliable devices in the network. Digital signatures by the source guarantee trust and reliability in the network. The forwarding nodes function in three modes: private unknown mode, public unknown mode, and trusted unknown mode. The nodes used in this protocol are as follows: overall onlooker nodes (control centers), friendly extended forwarders (helper nodes that extend services for packet forwarding), and friendly limited forwarders (helper nodes that extend limited services for packet forwarding). The friendly forwarders using registry attributes maintain the list of valid transactions performed by the trusted members, and the helper advertisement restores the services provided by the trusted members in the network. For each valid transaction the registry and advertisements are updated to create a trusted database. The maximum degree of trust is awarded to the reserve nodes with an update in the database. The OppNet node tracks the node behavior, extended services of helper nodes, and updates the trust degree in the network. This process identifies the intruders in the network. The node identification and service inventories are updated when a new forwarding node is traced in the network and updated by the current registry and overall onlooker node. The limitation of this protocol is it does not address the parsing of services for registry and advertisements in real-time XML-based web services.

18.5.2.2 Reputation-Assisted Data Forwarding for OppNets

Positive forwarding message (PFM) [22] creates a trust-based network with a unique data-forwarding scheme and monitors the status of data packets (forwarded or dropped). Reputation index is awarded to each node to increase the rate of data forwarding in the network. Reputation-assisted data-forwarding protocol for opportunistic networks (RADON) uses the *forwarding-reputation scheme* to estimate the probability of reaching the destination node. The protocol analyzes the probability of successful and unsuccessful data-forwarding rate by the authentic nodes. The competency levels estimate the degree of packet-forwarding nodes that route the data toward the destination node. The data is encrypted based on public key infrastructure (PKI) and valid node identities. The positive feedback is a function of transaction attributes and the signature assigned to the forwarding node. For large-scale real-time networks, the positive feedback message is combined to improve the packet-transmission rate. The successful transmissions by forwarding nodes are assigned the high value of *EncounterNum* and the unsuccessful transmissions by malicious nodes are assigned the low value of *EncounterNum*. The reputation module consists of direct and indirect observations that estimate the degree of reputation for the forwarding node and uncertainties in the route paths. In direct observation, if the node waits for longer duration the reputation counter is decreased. In indirection observation, the node reputation counter is a function of neighboring node reputation index and previous transactions.

The degree of trust between the nodes i and j at time t based on node belief factor, uncertain factor, and the coefficient σ is given as follows:

$$Trust^t_{i,j}(msg) = Belief^t_{i,j} + \sigma * Uncertain^t_{i,j}$$

The uncertainty factor is the rate of distrust across the nodes in the network. The reliable trust for successful data-forwarding nodes is the composite of node direct and indirect beliefs and uncertainties in the network. The nodes with higher levels of comprehensive competencies are prioritized to forward the data packets.

18.5.2.3 Social Trust in OppNets

Trust in opportunistic nodes is a function of previous transactions and node reputation. Social trust is based on node authenticity, fairness of social group, and the network connectivity factor. The degree of trust using explicit and implicit methods for social OppNets [23] is based on the distributed approach and use quality of the service features to forward the data packets. In the explicit method, the degree of trust is a function of trusted members to identify the malicious nodes and verify the hop count. Network sessions share the degree of trust with maximum number of common interest nodes in the network. In the implicit method, the degree of trust is a function of authentic nodes that forward the data packets and the number of trusted friends with common interests. The friendship tables are updated with familiar trusted nodes, and the local approximate trust weights are estimated in the network.

Trust-based message forwarding in the social OppNets [24] are cooperative in nature and operate using trust levels between relay-to-relay nodes and source-to-relay nodes. The routing paths between relay-to-relay nodes are assumed to be reliable and depend on the social behavior of the nodes. In this protocol, the source and intermediate relay nodes are assumed to be the trusted friends that have a common set of friends. Popular nodes with the high degree of trust with neighboring nodes establish the reliable routes in the network. Trust enforcement that improves the trust rate can be achieved by using level-based filters in the routing paths. The relay nodes are classified into nearest neighbor and next-to-nearest neighbor based on the distance and first-level trust filters. The nodes with common interests are grouped into the second-level filters, and the nodes with common neighbors are grouped into third-level filters. The combination of first level and second-level filters enhance the degree of trust in the network. Since the protocol does not include local history of relay nodes, the delay in message forwarding nodes is minimized. This protocol is applied for small- and medium-scale networks. Quality of message-forwarding service [25] and energy-aware attributes based on the cross-layer architecture improves message-forwarding schemes in OppNets. For Internet applications such as e-mail, social networking, and so on, the message contents are filtered to block the intruders in the network.

18.5.2.4 Trust-Based Minimum Cost Opportunistic Routing

Trusted-based Minimum COst Routing algorithm (MCOR) [26] selects the next-hop forwarding node with a maximum degree of trust and effective cost of previous trust transactions. The transaction cost is a function of network delay factor, and the trust degree is a function of interaction period and rate of successful or unsuccessful transactions. The node with maximum trust degree and optimum distance toward the destination is selected as the next-hop node. The degree of direct trust between two nodes is based on data aging and node awarding, and the degree of indirect trust is the maximum number of similar neighbors in the neighborhood. The degree of failed transactions indicates the number of malicious nodes in the network. The degree of similarity is the number of neighboring nodes that are awarded identical trust levels. MOCR uses the passive and unrestrained verification of irrelevant packets using the periodic and history-based interactions with the cooperating nodes. The malicious nodes are traced using range-based trust degree in the network. Opportunistic route cost for the trust-based minimum cost route r with probability P to forward the data packets and degree of trust in the route is given as follows:

$$\text{OpportunisticRoute}_{\text{Cost}, r} = P(\text{forward}, r) * \text{RouteCost}(r) + \text{Degree}_{\text{Trust}}(r)$$

In cooperative opportunistic multihop routing [27] for wireless mesh networks, the candidate node is selected and prioritized based on minimum expected route to enhance the throughput rate in the network, Figure 18.3 illustrates the secure route path, secure-privacy route path, and secure-trust-privacy route in OppMHWNs. Table 18.4 gives the summary of trust protocols in OppMHWNs.

18.5.3 Privacy in OppNets

The privacy protocols enhance the anonymity to protect the identity of forwarding node and message integrity based on node location and routing paths in the network. Privacy protection is the challenging task since the intermediate nodes access the neighboring nodes in active and passive states. Figure 18.4 illustrates the challenges in message integrity. The adversaries in privacy are as follows:

1. In narrow tasking, the message reports are sent based on application-specific attributes defined by the sender node. The location information of sender node is notified to the authorized nodes.
2. In selective tasking, the mobile nodes with specified tasks receive the data.
3. In task de-anonymizing, the message-forwarding nodes are traced by analyzing the tasks and received messages based on time and location attributes.
4. In report de-anonymizing, the forwarding nodes trace the intruders in the network.

The adversaries in message integrity are as follows:

1. Message tampering—the intruder modifies the message and acts as authorized legitimate node.
2. Message replaying—the intruder sends redundant data and acts as authorized legitimate node.

Anonymous opportunistic sensing (AnonySense) [28] addresses reliability and confidentiality of mobile nodes in the network. The mobile OppNet node registers

Table 18.4   Trust Protocols in OppMHWN

Trust Protocols	Features	Trust Attributes	Merits and Demerits
COTTON	Based on trusted services, history of previous transactions, and cooperation of forwarding nodes	Private and public forward nodes, degree of trust	For real-time network applications based on quality of service trust parameters
RADON	Evaluate the competency of data delivery based on positive feedback and public key infrastructure	Reliable trust with the forwarding nodes based on comprehensive belief and uncertainty	Deals with simple attacks and not applied for large-scale networks
Social trust (explicit and implicit) in OppNets	Explicit and implicit social trusts are based on node authenticity and fair member nodes	Consolidated trust is a function of explicit weight, implicit weight, trust factor, and threshold value	Used in dense social networks and modify the protocol design for sparse network with high delays
Social-based trust in mobile OppNets	Filter-based approach to establish the levels of social trust based on source–relay and relay–relay nodes	Normalized success rate, probability of successful message delivery, and normalized cost	Enhance trust in mobile networks and causes high end-to-end delays in the dense social networks
Trusted minimum cost routing	Select the potential next-hop neighbor nodes with high forwarding rate based on fairness and cost effectiveness of previous transactions	Probability of packet-forwarding, success rate, probability of unsuccessful forwarding, and trust route cost	Maximum degree of trust with minimum cost is established in the network. This protocols is not applicable for real-time networks

with registration authority (RA) to access the AnonySense software with valid IP address, and uses the task service (TS), report service (RS) to facilitate privacy in the network. This process uses the Anony task language based on an SQL template to define the message access format and message report format. RA specifies the interface and session time period. The main limitation of this protocol is it does not address the denial of service attacks.

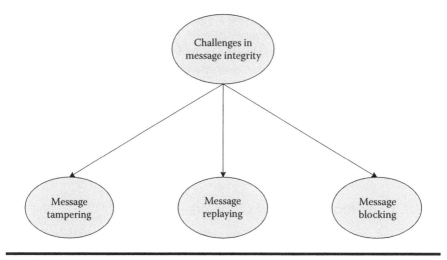

Figure 18.4   Challenges in message integrity.

18.5.3.1 Privacy-Enhanced Opportunistic Networking

Privacy Enhanced Opportunistic Networking (PEON [29] preserves the anonymity of OppNet nodes based on fixed size packet transmission. This protocol uses identity-based cryptographic method in which the intermediate nodes identify the forwarding nodes based on the node geographical attributes. Multiple private keys are used to verify the authenticity of the message. The nodes are categorized into replaceable groups and share the keys with trusted members to encrypt and decrypt the message. The intruders responding to quick succession queries from the source nodes with short time interval are identified and eliminated from the network. The intermediate nodes termed as *pawns* are categorized into groups and the nodes within the same group are assigned a unique group key. To serve the intermediate reliable forwarding nodes, the pawn groups are positioned at the routing junctions. The message is forwarded to the neighboring nodes within the same group until the packet reaches the next group. Furthermore, the encrypted key is changed, and the process is repeated until the packet reaches the destination node. The intermediate nodes are secured with varying buffer size and encrypted messages. PEON protocol addresses the attacks caused by the the intruders existing in the neighborhood of trusted nodes and the intruders that eavesdrop the message.

18.5.3.2 Enhancing Application-Level Privacy in OppNets

Enhancing application-level privacy in OppNets [30] is based the hide-and-lie principle that, prevents the malicious node from accessing the message by using time-bound message generation at the source node. The source node multicasts the message to a group of nodes and the intermediate nodes select the corresponding group for communication. The message storage cost is a function of buffer size and expiry time period of the message. The defense mechanism in this protocol is based on the message delivery factor, which is defined as the ratio of received messages and generated messages. The malicious nodes access the message periodically and build the node profile-based group membership and list of intermediate nodes to modify the route paths in the network. To resolve this problem, the intermediate forwarding nodes change their respective profiles at random intervals and reduce the rate of malicious nodes in the network. The hide-and-lie strategy range (λ) for each node is estimated based on the modified level of node profiles. The node with larger λ value prevents the malicious node from accessing the data packets.

18.5.3.3 Location Privacy in OppNets

The location privacy model [31] uses an independent dataset to establish the privacy of transmitted messages and profile information in the network. The routing schemes are classified into *privacy* and *nonprivacy* modes. The nodes are categorized based on the degree of sharing dataset, location, and identity of intermediate forwarding nodes. The modes used in the location privacy model are as follows: (i) nonpreferential mode: the nodes exchange their messages globally, (ii) responsive mode: the nodes exchange their messages based on the social networking attributes, and (iii) dual mode: the nodes exchange their messages based on the current state and network traffic conditions. Based on the degree of privacy, the respective mode is used. In this protocol, the mobile nodes trace the intruders in the local neighborhood.

18.5.3.4 Privacy-Preserving Interest-Casting in OppNets

Opportunistic data forwarding enables the sender node to reveal its private information to the forwarding nodes. In privacy-preserving interest-casting in OppNets [32], the source node and the intermediate nodes participate in data forwarding without disclosing their respective private information. Nodes selected with the common interest profiles establish the privacy preserving route paths in the network. Interest casting enables the group nodes to share the common interest profiles and route the data toward the destination. The interest of individual nodes and neighboring nodes are exchanged with the forwarding nodes. This protocol uses two schemes to preserve privacy. In the first scheme, the *direct delivery* method forwards the message to the neighboring node based on the maximum privacy precision and coverage ratio. In the second scheme, the *restricted-forwarding* method with two-hops transmits the message to the common interest of two-hop nodes. The main limitation is this protocol does not consider the effect of message modification by the malicious nodes, Figure 18.3 illustrates the secure route path, secure-privacy route path, and secure-trust-privacy route in OppMHWNs. Table 18.5 gives the summary of privacy protocols in OppMHWNs.

Table 18.5   Privacy Protocols in OppMHWN

Privacy Protocols	Features	Privacy Attributes	Merits and Demerits
AnonySense	Energy-aware privacy-preserving scheme for secure data forwarding is based on the task type, mobile nodes, and digital signature	RA, TS, RS, and finder node to locate the intruder nodes in the network	Receives data with less control overhead. The compromised nodes can monitor the privacy of the network
PEON	Group-based message encryption for privacy uses public and private key pairs with respect to the private group and session duration	Encrypted public key and private key based on request and member identity	Suitable for small-scale networks. The cost and control overhead are more for large-scale networks
Hide-and-lie: enhancing application-level privacy in OppNets	Time-bounded message diffusion with identical probability for common interest profiles	Probability of common interest and probability of success	Message delivery ratio is enhanced for real-time applications. Increases the number of active nodes in data dissemination processes
The impact of location privacy in OppNets	Local privacy based on node density and size of intermediate nodes	Friend nodes, group node set, and public–private nodes	Privacy is achieved with respect to bounded delays in the network
Privacy-preserving interest-casting in OppNets	Message is broadcasted to the interested members based on user dataset and relevant destination set	Maximum possible privacy precision and coverage ratio	Increases the coverage and privacy factor of the data-forwarding node in the network

18.6 Conclusions

OppNets is self-organized infrastructure-based ad hoc network comprising of distributed controller nodes, heterogeneous intermediate nodes, and mobile agents. Since, the routing paths are not known in advance, the forwarding nodes establish the reliable route paths based on opportunistic resources in the network. This chapter explains the architecture, metrics, and protocols for security, trust, and privacy in opportunistic MHWNs. Intrusion-detection methods identify the malicious nodes and avoid faulty messages in the network. Security protocols based on optimistic fair-exchange, opportunistic encryption, and local key management are explained in this chapter. Trust management protocols based on composite and social trust, reputation-assisted data forwarding and trust-based minimum cost routing are elucidated in this chapter. Privacy protocols based on message content, location, group size, and interest-casting methods based on common interests are described in this chapter. Comparative analysis of security, trust, and privacy protocols based on characteristic features, design attributes, advantages, and limitations are highlighted in this chapter.

18.7 Future Research Directions

Future research directions can be the design of application specific adaptable protocols based on the attributes of security, trust and privacy. Authentic intermediate nodes can be used to compress the data arriving from multiple nodes and transmit it to the forwarding node. Additional secure keys can be used to maintain the message integrity in communication channel. Key distribution and key-management techniques need to be distinctly developed for security, trust, and privacy protocols in OppNets. Protocol design for the mobile nodes to forward the data based on resource sharing, scheduling, and distributed-computing techniques need to be developed for opportunistic wireless networks. Prioritized queue-based multihop techniques need to be explored for the design of fault-tolerant OppNets. Further research enhancements can be to resolve the network congestion in resource scarce OppNets.

Abbreviations

AnonySense	Anonymous opportunistic Sensing
CFB	Cipher Feed Back mode
COTTON	COmposite Trust and Trust management in Opportunistic Networks
CTR	CounTeR mode
FFD	Fast-Forward Dissemination
FEC	Forward Error Correction Codes
MFR	Makeup-For-Reliability
MCOR	Minimum COst Routing algorithm
MHWN	Multi-Hop Wireless Network
MLCE	Multi-Layer Commutative Encryption
NDI	Node-to-Node Direct Interaction
NII	Node-to-Node Indirect Interaction
NSI	Node-to-Node Selfish Interaction

OppCast	Opportunistic broadCast protocol
OppMHWN	Opportunistic Multi-Hop Wireless Network
OppNets	Opportunistic Networks
OFB	Output Feed Back mode
PRR	Packet Reception Ratio
PHI	Personal Health Information
PFM	Positive Forwarding Message
PEON	Privacy Enhanced Opportunistic Networking
PPSPC	Privacy Preserving Scalar Product Computation
PKI	Public Key Infrastructure
RA	Registration Authority
RS	Report Service
RADON	Reputation-Assisted Data forwarding protocol for Opportunistic Networks
SMART	Secure Multilayer Credit-Based Incentive Scheme for Delay-Tolerant Networks
SPOC	Secure and Privacy-Preserving in Opportunistic Computing Framework
SNR	Signal-to-Noise Ratio
TS	Task Service
TA	Trusted Authority
TTP	Trusted Third Party

Author Biography

M. Bala Krishna earned a bachelor of engineering degree (BE) in computer engineering from the Delhi Institute of Technology (presently Netaji Subhash Institute of Technology), University of Delhi, Delhi, India and a master of technology degree (MTech) in information technology from the University School of Information Technology (presently University School of Information and Communication Technology), GGS Indraprastha University, Delhi, India. He earned a doctor of philosophy (PhD) in computer engineering from JMI Central University, Delhi, India. He earlier worked as a senior research associate and project associate at the Indian Institute of Technology, Delhi, India in the areas of digital systems and embedded systems. He also worked in projects related to communication networks. He is presently an assistant professor at the University School of Information and Communication Technology, GGS Indraprastha University, Delhi, India. His teaching and research areas include computer networks, wireless networking and communications, mobile and ubiquitous computing, and embedded system design. He has published book chapters, and articles in international journals and conferences. His current research areas include wireless ad hoc and sensor networks, green networking and communications, cognitive networks, and smart grid communications.

References

1. Mieso K. D, *Mobile Opportunistic Networks Architectures, Protocols and Applications*, Auerbach Publications, Taylor & Francis Group, Boca Raton, FL, USA, 2011.
2. Leszek L, Zille Huma K, Vijay B and Ajay G, Chapter 5. The concept of opportunistic networks and their research challenges in privacy and security, In *Mobile and Wireless Network Security and Privacy*, S. K. Makki, P. Reiher, K. Makki, N. Pissinou, and S. Makki (Eds.), Springer Science + Business Media, LLC, 233 Spring Street, New York, NY 10013, USA, 2007, pp. 85–117.
3. Elvira B T, Isaac W, Leszek L and Mieso K D, Trust management in opportunistic networks: A semantic web approach, In *Proceedings of IEEE World Congress on Privacy, Security and Trust and the Management of e-Business (PST)*, Saint John, New Brunswick, Canada, 25–27 August 2009, pp. 235–238.
4. Pelusi L, Passarella A and Conti M, Opportunistic networking: Data forwarding in disconnected mobile ad hoc networks, *IEEE Communications Magazine,* November 2006, 44(11), pp. 134–141.
5. Abdolbast G, Stuart M A and Roger M W, RFP: Repository based forwarding protocol for opportunistic networks, In *Proceedings of IEEE Third International Conference on Next Generation Mobile Applications, Services and Technologies (NGMAST)*, Cardiff, Wales, UK, 15–18 September 2009, pp. 329–334.
6. Vijay E and Mark C, Forwarding in opportunistic networks with resource constraints, In *Proceedings of ACM Third Internation Workshop on Challenged Network (CHANTS)*, San Francisco, California, USA, 15 September 2008, pp. 41–48.
7. Davide C, Osvaldo S and Michele Z, Throughput and energy efficiency of opportunistic routing with type-I HARQ in linear multihop networks, In *Proceedings of IEEE International Conference on Global Telecommunications Conference (GLOBECOM)*, Miami, FL, USA, 6–10 December 2010, pp. 1–6.
8. Jianwei N, Jinkai G, Qingsong C, Norman S and Shaohui G, Predict and spread: An efficient routing algorithm for opportunistic networking, In *Proceedings of IEEE International Conference on Wireless Communications and Networking Conference (WCNC)*, Cancun, Quintana Roo, Mexico, 28–31 March 2011, pp. 498–503.
9. Yanhua L, Debin Z, Hao W, Zheng Z, Yuan-An L and Yan Q, LOR: Localized opportunistic routing in large-scale wireless network, In *Proceedings of IEEE International Conference on Global Telecommunications Conference (GLOBECOM)*, Miami, FL, USA, 6–10 December 2010, pp. 1–5.
10. Chung-Ming H, Kun-chan L and Chang-Zhou T, A survey of opportunistic networks, In *Proceedings of IEEE Twenty Second International Conference on Advanced Information Networking and Applications Workshop (AINAW)*, Okinawa, Japan, 25–28 March 2008, pp. 1672–1677.
11. Qiang Z, Jing Y and Minghui W, A detection method for uncooperative nodes in opportunistic networks, In *Proceedings of IEEE Second International Conference on Network Infrastructure and Digital Content (IC-NIDC)*, Beijing, China, 24–26 September 2010, pp. 835–838.
12. Haojin Z, Xiaodong L, Rongxing L, Yanfei Fan and Xuemin Sherman S, SMART: A secure multilayer credit-based incentive scheme for delay-tolerant networks, *IEEE Transactions On Vehicular Technology*, October 2009, 58(8), pp. 4628–4639.

13. Leszek L, Ajay G, Zill-E-Huma K and Zijiang Y, Opportunistic resource utilization networks—A new paradigm for specialized ad hoc networks, *Elsevier Journal of Computers and Electrical Engineering*, March 2010, 36(2), pp. 328–340.

14. Poersch A M, Macedo D F and Nogueira J M S, Resource location for opportunistic networks, In *Proceedings of IEEE Fifth International Conference on New Technologies, Mobility and Security (NTMS)*, Istanbul, Turkey, 7–10 May 2012, pp. 1–5.

15. Ming L, Kai Z and Wenjing L, Opportunistic broadcast of event-driven warning messages in vehicular ad hoc networks with lossy links, *Elsevier Journal of Computer Networks*, 14 July 2011, 55(10), pp. 2443–2464.

16. Onen M, Shikfa A and Molva R, Optimistic fair exchange for secure forwarding, In *Proceedings of IEEE Fourth Annual International Conference on Mobile and Ubiquitous Systems: Networking & Services (MobiQuitous)*, Philadelphia, PA, USA, 6–10 August 2007, pp. 1–5.

17. Haleem M. A, Mathur C. N, Chandramouli R and Subbalakshmi K. P, Opportunistic encryption: A trade-Off between security and throughput in wireless networks, *IEEE Transactions on Dependable and Secure Computing*, October–December 2007, 4(4), pp. 313–324.

18. Shikfa A, Onen M and Molva R, Privacy in content-based opportunistic networks, In *Proceedings of IEEE International Conference on Advanced Information Networking and Applications Workshops (WAINA)*, Bradford, UK, 26–29 May 2009, pp. 832–837.

19. Abdullatif S, Melek O and Refik M, Local key management in opportunistic networks, *Inderscience International Journal of Communication Networks and Distributed Systems*, July 2012, 9(1/2), pp. 97–116.

20. Rongxing L, Xiaodong L and Xuemin (Sherman) S, SPOC: A secure and privacy-preserving opportunistic computing framework for mobile-healthcare emergency, *IEEE Transactions On Parallel And Distributed Systems,* March 2013, 24(3), pp. 614–624.

21. Gokce G and Erol G, Resilience and security of opportunistic communications for emergency evacuation, In *Proceedings of ACM Seventh Internatioanl Workshop on Performance Monitoring and Measurement of Heterogeneous Wireless and Wired Networks (PM2HW2N)*, Paphos, Cyprus Island, 21–25 October 2012, pp. 115–124.

22. Na L and Sajal Kumar D, RADON: Reputation-assisted data forwarding in opportunistic networks, In *Proceedings of ACM Second International Workshop on Mobile Opportunistic Networking (MobiOpp)*, Pisa, Italy, 22–23 February 2010, pp. 8–14.

23. Trifunovic S, Legendre F and Anastasiades C, Social trust in opportunistic networks, In *Proceedings of IEEE Conference on Computer Communications Workshops (INFOCOM)*, San Diego, CA, USA, 15–19 March 2010, pp. 1–6.

24. Mtibaa A and Harras K. A, Social-based trust in mobile opportunistic networks, In *Proceedings of IEEE Twenth International Conference on Computer Communications and Networks (ICCCN)*, Maui, HI, USA, 31 July 2011–4 August 2011, pp. 1–6.

25. Wei Y, Weifa L and Wenhua D, Energy-aware real-time opportunistic routing for wireless ad hoc networks, In *Proceedings of IEEE International Conference on Global Telecommunications (GLOBECOM)*, Miami, FL, USA, 6–10 December 2010, pp. 1–6.

26. Wang B, Huang C and Y. W. Li L, Trust-based minimum cost opportunistic routing for ad hoc networks, *Elsevier Journal of Systems and Software,* December 2011, 84(12), pp. 2107–2122.

27. Yu-Shan L, Wei-Ho C, Hongke Z, and Sy-Yen K, Throughput improvement of multi-hop wireless mesh networks with cooperative opportunistic routing, In *Proceedings of IEEE Wireless Communications and Networking Conference (WCNC)*, Paris, France, 1–4 April 2012, pp. 3035–3039.

28. Minho S, Cory C, Dan P, Apu K, David K and Nikos T, AnonySense: A system for anonymous opportunistic sensing, *Elsevier Journal of Pervasive and Mobile Computing*, February 2011, 7(1), pp. 16–30.

29. Zhengyi L, Gauri V and Matthew W, PEON: Privacy-enhanced opportunistic networks with applications in assistive environments, In *Proceedings of ACM Second International Conference on Pervasive Technologies Related to Assistive Environments (PETRA)*, Corfu, Greece, 9–13 June 2009, pp. 76:1–76:8.

30. László D and Tamás H, Hide-and-Lie: enhancing application-level privacy in opportunistic networks, In *Proceedings of ACM Second International Workshop on Mobile Opportunistic Networking (MobiOpp)*, Pisa, Italy, 22–23 February 2010, pp. 135–142.

31. Parris I and Henderson T, The impact of location privacy on opportunistic networks, In *Proceedings of IEEE International Symposium on a World of Wireless, Mobile and Multimedia Networks (WoWMoM)*, Lucca, Italy, 20–23 June 2011, pp. 1–6.

32. Costantino G, Martinelli F and Santi P, Privacy-preserving interest-casting in opportunistic networks, In *Proceedings of IEEE International Wireless Communications and Networking Conference (WCNC)*, Shanghai, China, 1–4 April 2012, pp. 2839–2834.

Index

Printed and bound by CPI Group (UK) Ltd, Croydon, CR0 4YY

18/10/2024

01776257-0018